THE LEGO® MINDSTORMS® ROBOT INVENTOR ACTIVITY BOOK

A Beginner's Guide to Building and Programming LEGO Robots

Daniele **Benedettelli**

no starch press

San Francisco

Printed in the United States of America

First printing

24 23 22 21   1 2 3 4 5 6 7 8 9

ISBN-13: 978-1-7185-0181-2 (print)
ISBN-13: 978-1-7185-0182-9 (ebook)

Publisher: William Pollock
Production Manager and Editor: Rachel Monaghan
Developmental Editor: Nathan Heidelberger
Cover Design: Monica Kamsvaag and Maureen Forys
Interior Design: Octopod Studios
Technical Reviewer: Xander Soldaat
Cover Photograph: Matteo Neri
Copyeditor: Paula L. Fleming
Compositor: Jeff Lytle, Happenstance Type-O-Rama
Proofreader: Sharon Wilkey

For information on book distributors or translations, please contact No Starch Press, Inc. directly:
No Starch Press, Inc.
245 8th Street, San Francisco, CA 94103
phone: 1-415-863-9900; info@nostarch.com
www.nostarch.com

Library of Congress Control Number: 2021940449

[V]

To Maria Sole and Libero

about the author

Daniele Benedettelli is an Italian robotics engineer known worldwide for his original LEGO MINDSTORMS robots, including the first NXT Rubik's Cube solver, various humanoid robots, and working car factory models. From 2006 to 2013, he helped the LEGO Group test and develop software for the LEGO MINDSTORMS product line. He works as a freelance LEGO designer, holds robotics lectures and workshops around the world, and teaches high school robotics.

about the technical reviewer

Xander Soldaat is a former Mindstorms Community Partner for LEGO MINDSTORMS. He was an IT infrastructure architect and engineer for 18 years before becoming a full-time software developer. He has recently come full circle and has gone back to his Linux roots at Red Hat. In his spare time, he likes to tinker with robots, 3D printing, and home-built retro-computers.

brief contents

contents in detail

acknowledgments

I was just collecting my energy and ideas, waiting for this new LEGO MINDSTORMS set to come out, because a new LEGO MINDSTORMS set means a new book to write! There are many people I need to thank for making this possible.

First, thanks to my family: to my wife, **Lucia**, for bearing with me; to my daughter, **Maria Sole**, who beta tested all the models of the book with ever growing interest and helpful feedback; to my super-young son, **Libero**, who still can't understand which LEGO toys are his and which are mine (he struggles to stick to his DUPLO bricks with all the LEGO robotics stuff at hand!).

Thanks to all the **grandparents**, who took care of the kids, giving me some daytime hours to work on this book (the rest were night hours, alas). A big thanks to my brother **Alessandro**, talented guitarist and music teacher, for his excellent hints on how to make guitar scales and techniques accessible to everyone, even on a LEGO guitar with no strings.

Thanks to the No Starch Press team, especially to **Nathan Heidelberger**, a precise, creative and trustworthy editor, and to **Bill Pollock** for his steady and resolute lead.

A huge thanks to **Xander Soldaat**—man, I realize you haven't missed a chance to tease me with your wacky *Genuine Italian™* dishes since 2006 by now!—who is as precious, serious a technical reviewer as he is a witty friend when hanging out, unfortunately only remotely in recent years. Thanks to **Maureen**, a dear friend who has never stopped encouraging me throughout this project (sorry for my occasional whining!).

I wish to thank the dear **Amelia** for her invaluable early input about the general mood of the book and the fun factor it needed. Thanks to all the kids in my casual focus group that self-generated over summer vacation, when the book was just a few scribbled pages. Thanks to **Adele** for her feedback on the project list.

Thanks to photographer and friend **Matteo Neri** (*https://matteoneriphoto.com/*) for the gorgeous photo on the book's cover.

A big thanks to all the LDraw community members who developed the bits and tools to create high-quality building instructions. Special thanks to master builder and book author **Philippe Hurbain** (Philo), a master in modeling 3D LEGO elements; to **Roland Melkert** for his awesome LDCad software; and to **Trevor Sandy** for his work on developing LPub3D.

1

getting started

Let's go already!
—Bender, *Futurama*

All the models in this book can be built with a single LEGO set: the LEGO MINDSTORMS Robot Inventor set #51515. With this LEGO set and a device that can run the LEGO MINDSTORMS App, you're ready to start creating any robot you can imagine. From a remote-control car that transforms into a walking humanoid robot to a guitar that you can really play, the sky's the limit on what you can invent!

what's in the box?

The LEGO MINDSTORMS Robot Inventor set comes with almost 1,000 LEGO Technic elements, including large Technic frames, wheels, and connectors that make it super easy to quickly build real working robots that can do all sorts of cool things. Besides plain LEGO Technic elements, the set includes some electronic devices, shown in Figure 1-1: four Medium Motors with built-in rotation sensors, a Color Sensor, a Distance Sensor, and the Hub (short for the LEGO Technic Large Hub), which will be the brain of all the robots you'll create.

the hub

The Hub is a computer, a *programmable* LEGO brick for making your robots work. You can use the Hub to run *programs*, instructions you'll write for the robots you'll build. The programs tell the Hub how to control the motors and sensors, which you can plug into the Hub's six ports (three on each side). The Hub is smart enough to know what kind of motor or sensor is connected to each port, without needing to be told.

The Hub has a 5×5 *matrix display*, made of five rows and columns of light-emitting diodes (LEDs), and a built-in speaker. It has a six-axis *inertial measurement unit* (IMU) consisting of a three-axis *accelerometer* and three-axis *gyroscope*. An IMU is the same kind of device that tells a smartphone or tablet which way is up, and you can similarly use the Hub's IMU to detect how the Hub is rotating or oriented in space.

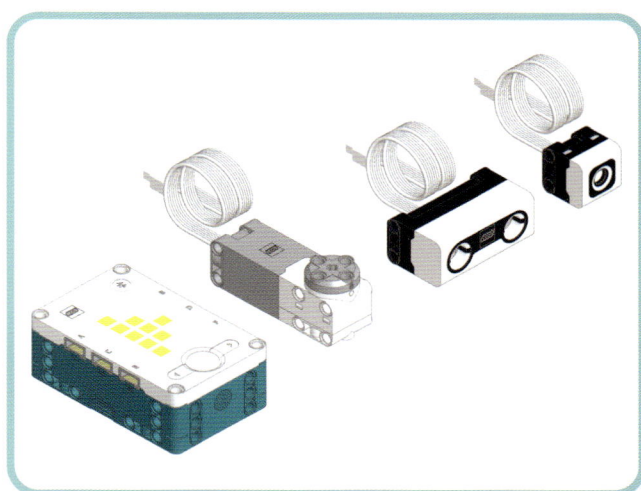

Figure 1-1: The electronic devices included in the LEGO MINDSTORMS set: (from left) the Hub, one of the four Medium Motors, the Distance Sensor, and the Color Sensor

The Hub also has three buttons for menu navigation and one button to enable Bluetooth pairing. You can connect the Hub to other devices for programming and remote control using USB or Bluetooth. Last but not least, the Hub has a battery that can be recharged by attaching the Hub to any USB power supply with the included USB-to-microUSB cable. That's really handy and eco-friendly!

the motors and sensors

The four Medium Motors are a pleasure to work with, thanks to their compact shape and many mounting holes. They're *servomotors*, meaning you can turn them to specific positions and control their speed. Each motor has a zero-position, marked on the motor body with a circle. The motor's built-in rotation sensor measures the position of the motor shaft (the part of the motor that spins) relative to that zero-position.

This handy feature means you don't need to bring the motor to an initial known position.

The Distance Sensor measures the distance of an object placed in front of it up to 2 meters away. It can measure in centimeters or inches. It measures distances just like bats do. The sensor emits a burst of *ultrasonic sound waves* (sounds that are too high for us humans to hear) from its transmitter "eye" and measures the time it takes for the sound waves to bounce off the object and come back to the other receiver "eye." The sensor calculates the distance using this time measurement and the speed of sound in air. The distance measurements are quite precise for flat, hard objects, but the sensor is blind to clothes or other soft objects that absorb sound waves. The sensor also features four programmable lights around the eyes. Cool—you can make your robots wink!

The Color Sensor can shine a light from its LED onto an object placed about 2 centimeters (a little less than 1 inch) in front of it and measure the percentage of light reflected by that object. This kind of *grayscale measurement* is useful, for example, to detect the edge of a black line on a white surface. The sensor can also detect the color of an object, returning a number that corresponds to one of the LEGO colors (white, blue, black, green, yellow, red, teal, violet, or no color).

the app

The LEGO MINDSTORMS App is where you'll create the programs for your robots. The app can run on many devices, including Windows 10 computers, macOS computers, and Android and iOS smart devices (both smartphones and tablets). Before buying the Robot Inventor set, you should have checked if your device satisfies the minimum requirements to run the app. You can find information about device compatibility and minimum requirements at the official LEGO support page: *https://www.lego.com/ en-us/service/device-guide/mindstorms-robot-inventor/*.

Even if you're already been using the app to work with the five robots that come with the Robot Inventor set, you could have missed some detail or feature of the app, or you may not have created any programs in the app yourself. That's why I'll describe the programming blocks and the features of the app as we need to use them in the following chapters.

using this book

This book takes a project-based approach. In each chapter, you'll create one complete robot. I chose the most captivating, fun, and interactive models I could think of, from a Transformer, to a pinball machine, to an electric guitar and beyond. Each project builds on knowledge gained from the previous ones, so I recommend going through the chapters in order.

Part of each chapter will contain step-by-step instructions for building the chapter's robot. As you follow the instructions, you'll learn many cool building techniques that you can apply to your own LEGO creations. The other part of each chapter will explain how to tell the robot what to do using the Scratch programming language. We'll talk through each program in detail so you can see how it works and how you can create your own programs. However, if you're in a hurry, or you just can't wait to start playing with the robots in the book, you can download all the programs for the robots (their *source code*) at *https:// www.nostarch.com/lego-mindstorms-robot-inventor-activity -book/*. Don't forget to bookmark my website *https://robotics .benedettelli.com/* for updates and extra materials about this book and LEGO MINDSTORMS in general.

conclusion

The LEGO MINDSTORMS Robot Inventor set includes everything you need to make awesome, original robots. It features many plastic LEGO Technic elements, but the true heart—or rather, the brain—of the set is the Hub, a smart programmable brick that you can use to control the set's motors and sensor.

In the next chapter, you'll learn how to program a baseball-playing robot from scratch (pun intended).

2

baseball batter

"Take me out to the ball game . . ."
—J. Norworth and A. Von Tilzer

The simple contraption you're about to build is a baseball-playing robot (Figure 2-1). This project will introduce you to the three stages of any robotic system: input, processing, and output.

Input is data from the world outside the robot. In this case, the input comes through the LEGO sensors, which will detect the ball when it comes close. *Processing* happens when we take that input and do something with it. Processing happens inside the Hub, which will run a program that we'll write. The program will tell the robot to wait for the ball, then swing the bat.

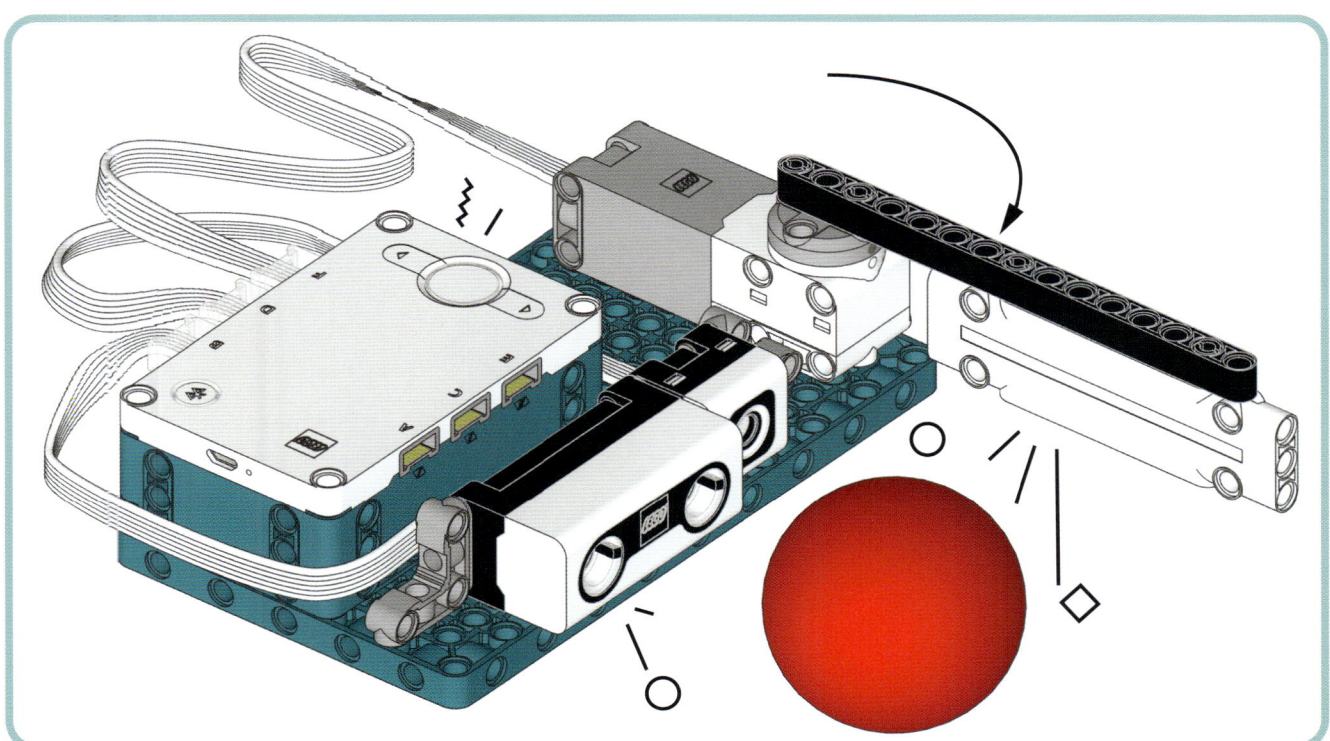

Figure 2-1: The Baseball Batter is a simple contraption that will help you take the first steps in robot programming.

Output is what the robot does as a result of the processing. Our output will be an action: the motor's swinging the bat to hit the ball. We can make the robot even more fun by adding other outputs, like playing a sound effect and showing a nice animation on the Hub's display.

building the baseball batter

In this section, you'll find step-by-step instructions for building the Baseball Batter. For each step, the first box shows the parts you'll need for that step.

Throughout the book, sometimes you'll see a number inside a circle (for axles or panels) or inside a square (for beams) next to a part. This number tells you the part's length in LEGO units, or *modules*. You can measure a beam's length by counting ts holes. You can measure axles by putting them next to a beam and counting the holes on the beam. For example, Figure 2-2 shows a 5-module (5M) axle next to a 9M beam.

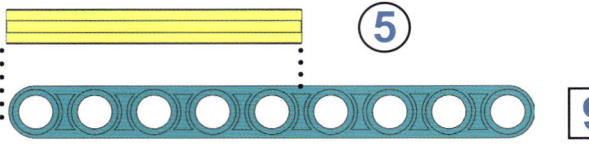

Figure 2-2: You can measure a beam's length by counting its holes. To find the length of an axle, place it next to a beam and count the holes on the beam.

The white dots help you put the black Technic pins in the right holes. The teal Technic baseplate is not shown in the parts list to save space.

Attach the motor to port F of the Hub.

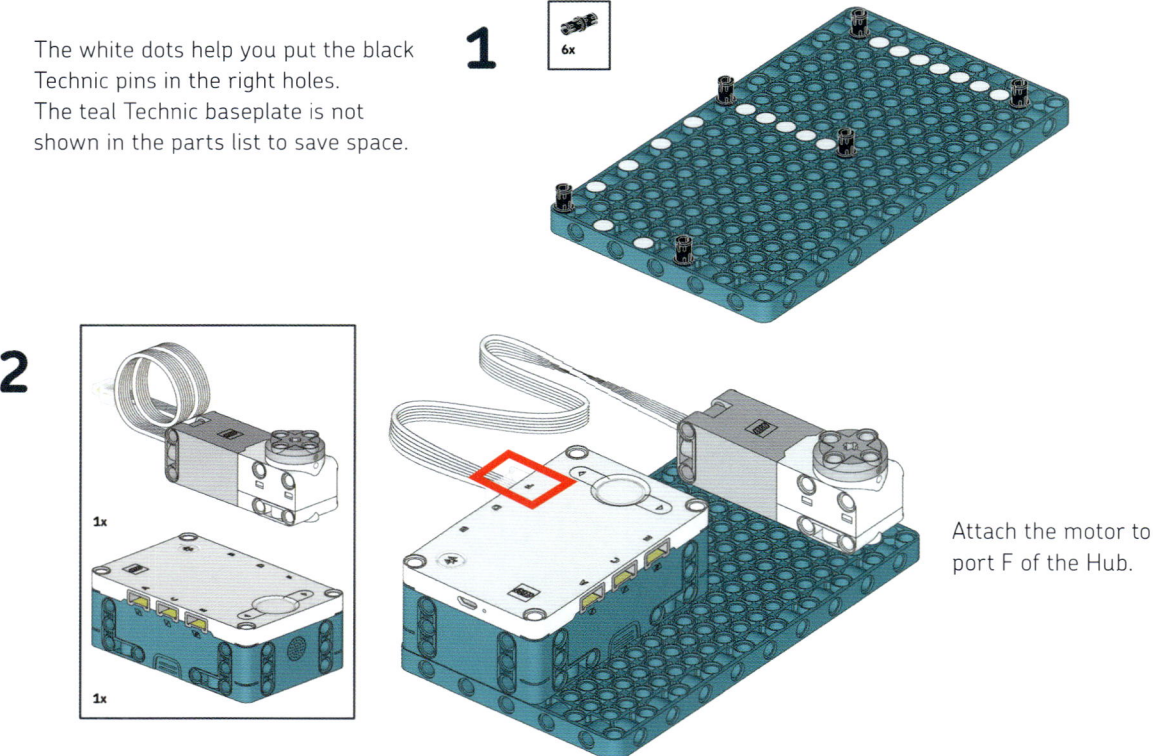

3

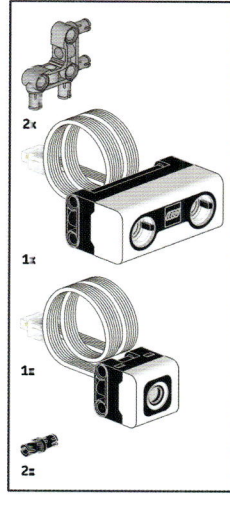

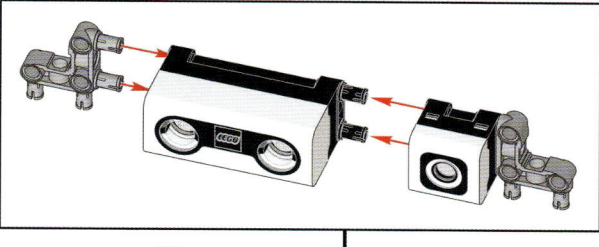

When you see a submodel in its own box like this, always assemble it *before* attaching it to the main model.

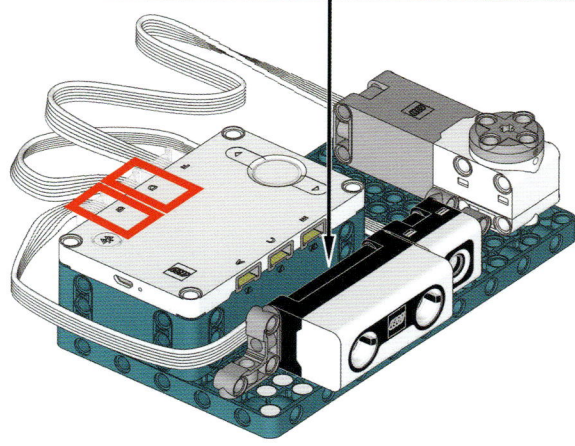

Attach the Distance Sensor to port B and the Color Sensor to port D.

4

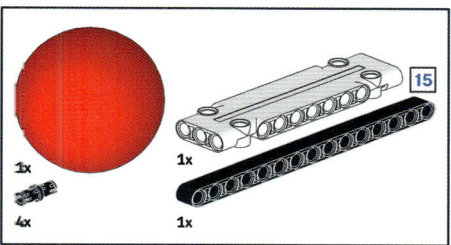

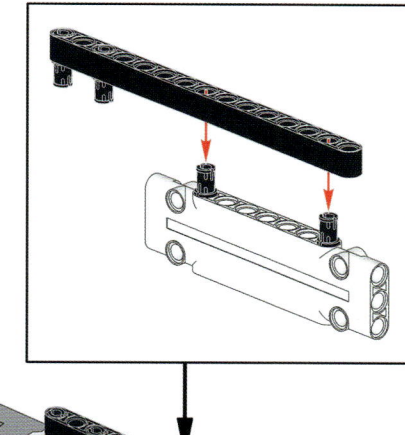

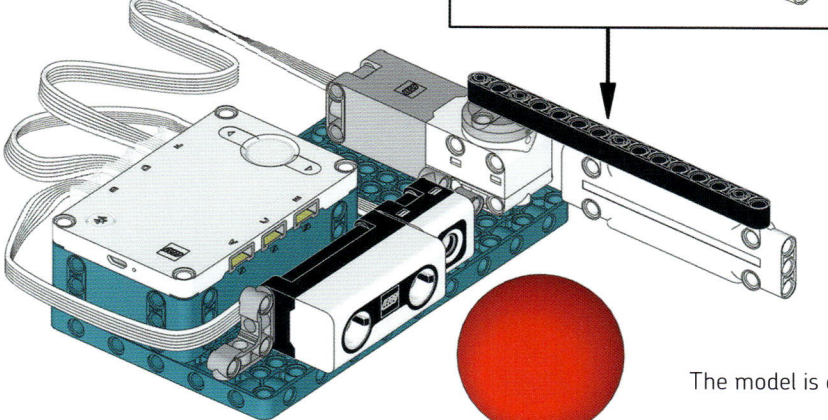

The model is complete!

programming the batter

Now that you've built the robot, it's time to tell it what to do. In other words, you need to program it. You might know programming by another name: *coding*. When you program, you're writing code with instructions that a machine like a robot can understand. I'll guide you step-by-step through the process of programming the Baseball Batter so you can see how to write code in the MINDSTORMS App. Remember, you can also download all the code used in this book at *https://nostarch.com/lego-mindstorms-robot-inventor-activity-book/*.

NOTE *If you're an expert MINDSTORMS App user, you can just take a look at the complete program in Figure 2-5 before going straight to "Understanding the Program" on page 12.*

getting around the app

Launch the LEGO MINDSTORMS App by double-clicking or tapping its icon, depending on the device you're using. After a few moments, the lobby screen should show up. On a computer, it should look like Figure 2-3. On smaller screens, you'll see only one robot at a time. Either way, you should see four buttons at the bottom of the screen, and one at the top right of the screen. The Home button brings you to the lobby, Community shows a list of inventions made by LEGO fans, Projects shows you a list of the projects you made, and Code brings you straight to the programming environment. The Settings button, the gear shape at the top-right corner of the screen, shows the settings and help.

Figure 2-3: The LEGO MINDSTORMS App lobby. From here, you can start the activities for the five robots included in the box, manage your projects and settings, or start a new program. The app's look may vary depending on the device it's running on.

Click the **Code** button at the bottom right of the screen to open the programming area. It should look like Figure 2-4.

the word blocks

The LEGO MINDSTORMS App lets you program your robots using Scratch 3.0, a block-based visual programming language. On the left of your screen, you'll see all the *word blocks*, the building blocks of your programs. Each word block represents one instruction for your robot to carry out. The blocks are grouped into *palettes* based on their function. For example, one palette has blocks just for controlling the motors, another has blocks for the sensors, and so on.

You can drag blocks from the palette to the programming area on the right. By combining blocks into stacks, you can build programs that tell your robot to do all sorts of things. Each stack of blocks reads from top to bottom.

connecting to the hub

To control your robot, you must connect your computer, phone, or tablet to the Hub. You can use a USB cable or Bluetooth. When the Hub is connected to your computer via USB, the app finds the Hub automatically. If you want to connect via Bluetooth, click the **Open Hub Connection** button 🔲 and follow the onscreen instructions.

NOTE If you have trouble connecting to the Hub through Bluetooth, see the official LEGO support page: *https://education.lego.com/en-us/product-resources/spike-prime/troubleshooting/bluetooth-connectivity/*. This page refers to SPIKE Prime, a different LEGO robotics kit. The SPIKE Prime Hub is exactly the same as the Robot Inventor Hub, except it's yellow.

Once the Hub is connected, the dot on the Hub icon at the top right of the app turns green, and real-time sensor readings appear beside that icon. These readings tell you exactly what each sensor is picking up at any given moment. Try moving your hand back and forth in front of the Distance Sensor and you should see the numbers change on the screen.

creating a program from scratch

When you open the programming area or create a new project, you should already see one block on the programming canvas: `when program starts`. This is an example of a *hat block*. Like a hat, a hat block goes on top of other blocks. Every block stack must start with a hat block, which tells the robot when to run that stack. For now, you can find all the hat blocks in the yellow Events palette. (Later, when we create a remote controller, more hat blocks will appear in the Remote Control palette.)

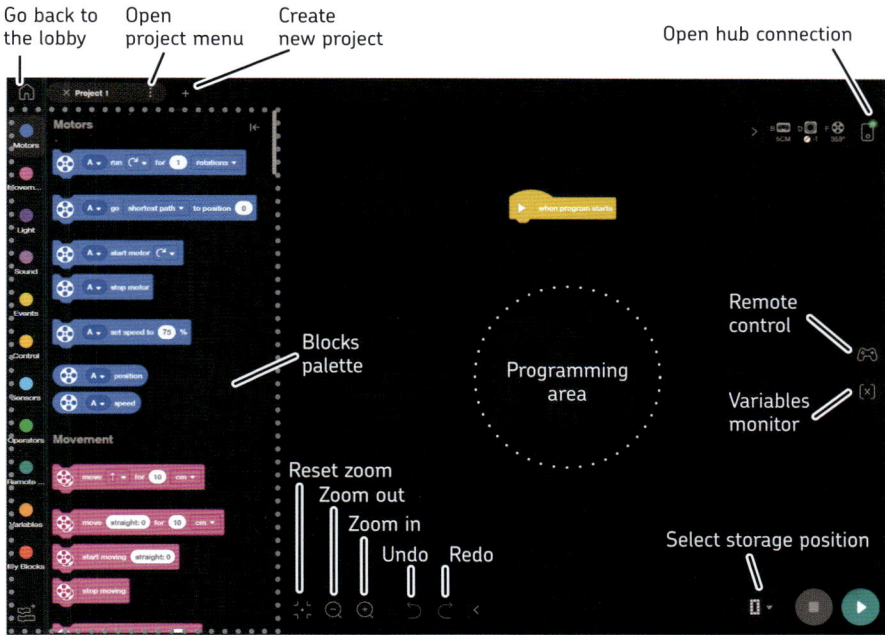

Figure 2-4: The programming area shown on a laptop screen. The LEGO MINDSTORMS App lets you create programs by stacking blocks, just like you stack LEGO parts.

Any blocks attached to when program starts will automatically run as soon as the program begins. You can have more than one when program starts block in your program. In that case, each of those stacks will start at the same time.

Let's create the program for the Baseball Batter step-by-step. All we want the robot to do is swing the bat when it sees the ball coming. Figure 2-5 shows the complete program. We'll talk through building it one block at a time.

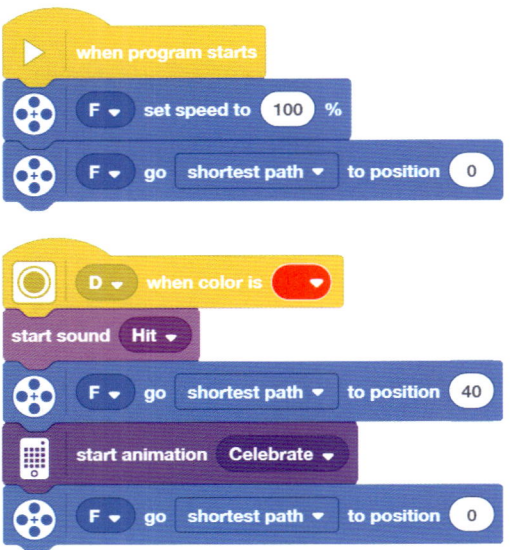

Figure 2-5: The complete program for the Baseball Batter uses the color sensor to detect the ball.

1. Open the **Motors** palette, drag the **set motor speed** block to the programming canvas, and attach it under the yellow hat block. Notice how the block snaps into place even if you just drop it near the yellow block.

    This block doesn't actually turn any motors on. It just tells them how fast to go once they're turned on. You can choose any speed from –100% to 100%. Negative numbers make the motors spin backward. Without this block, the motors will try to run at 75% speed by default.

If only one motor is attached to the Hub and the Hub is connected to the app, the set motor speed block should already show the port the motor is attached to. If you don't already see port F selected on the block, choose t using the port selector drop-down menu, as shown in Figure 2-6. Then set the speed to **100%**. Now the motor should run at full speed when the program is run.

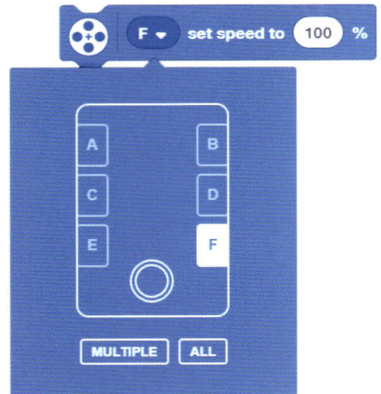

Figure 2-6: Each motor block has a port input menu where you can specify which port the motor you want to control is attached to. You can select a single port or multiple ports.

2. From the same palette, drag and drop the **motor go to position** block below the previous blue block and set the port to **F**. This block tells the motor to spin until the mark on the shaft reaches a certain angle relative to the round mark on the motor body. An angle of 0 aligns with the mark. Angles then increase clockwise, up to 359.

    To reach the specified angle, you can tell the motor to spin clockwise, counterclockwise, or whichever direction makes the shortest path, as shown in Figure 2-7. In our case, choose **shortest path** and set the position to **0**. We do this because it's possible for the motor shaft to move away from the zero-position while the program isn't running. Setting the position to 0 at the start of the program puts the LEGO Technic beam into the best position for hitting the ball.

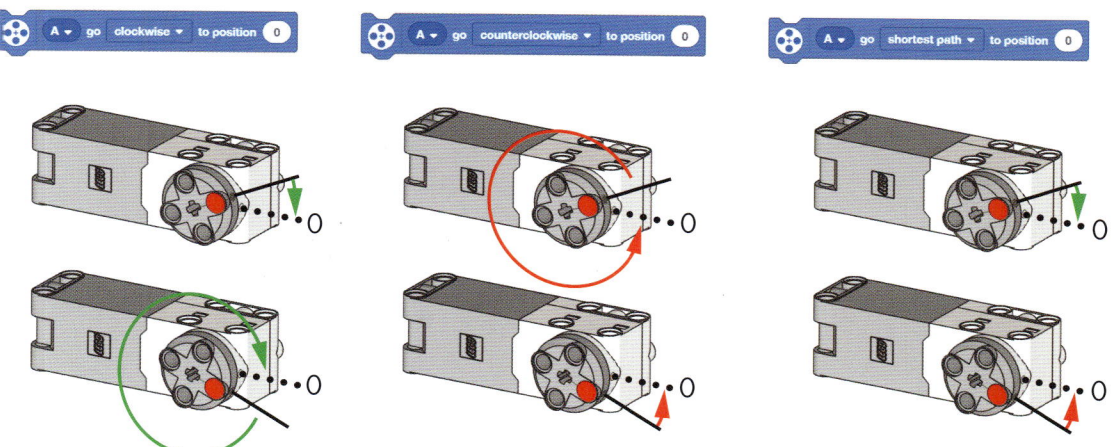

Figure 2-7: The `motor go to position` block spins the motor until it reaches a certain position. You can make it spin clockwise or counterclockwise, or whichever way is shorter.

3. Now let's make the robot sense the ball. From the **Events** palette, drag and drop the `when color is` block. Set the port letter to **D** and the color to **red** (Figure 2-8).

    The block `when color is` is another hat block. It will run the stack of blocks beneath it when the Color Sensor detects a certain color. As you can see in Figure 2-8, the sensor can look for several colors besides red. When the sensor sees the right color, it runs the block stack only once. If the sensor still sees the same color when the stack is finished, it won't run the code again.

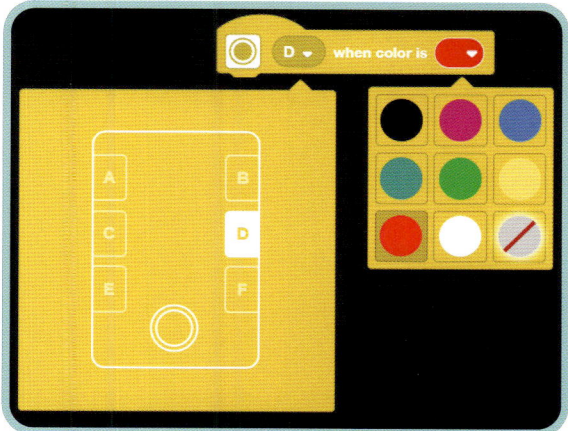

Figure 2-8: Setting the `when color is` hat block to look for red with the Color Sensor attached to port D

4. From the **Sound** palette, drag and drop a `start sound` block under the `when color is` hat block. Then, following the numbered steps in Figure 2-9, choose the sound **Hit** from the **Tricky** tab of the Sound Library.

    This block starts playing a sound of your choice and lets the program continue running as the sound plays. The app has a library of sounds, including Hit, already stored in the Hub. These will play from the Hub speaker. You can also select sounds from your own device or even record new sounds, but these will play from your device's speakers, not from the Hub.

5. Add another `motor go to position` block below the sound block. Set the port to **F**, choose **shortest path**, and set the position to **40**. This block will make the motor rotate to an angle 40 degrees clockwise from the zero-position, moving the bat to hit the ball.

6. Add a `start animation` block from the **Lights** palette and follow the steps in Figure 2-10 to select the **Celebrate** animation from the Animation Library. This block starts an animation on the Hub's display. The program will continue to run while the animation plays.

    The `start animation` block also lets you design your own animations with the Animation Editor. Each animation is a sequence of patterns on the Hub's 5×5 grid of lights. You can choose which lights to light up and how bright they should be.

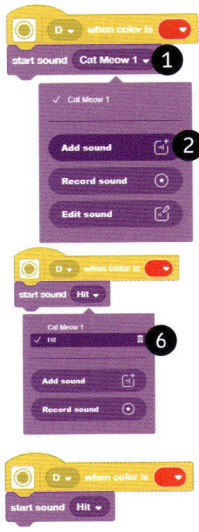

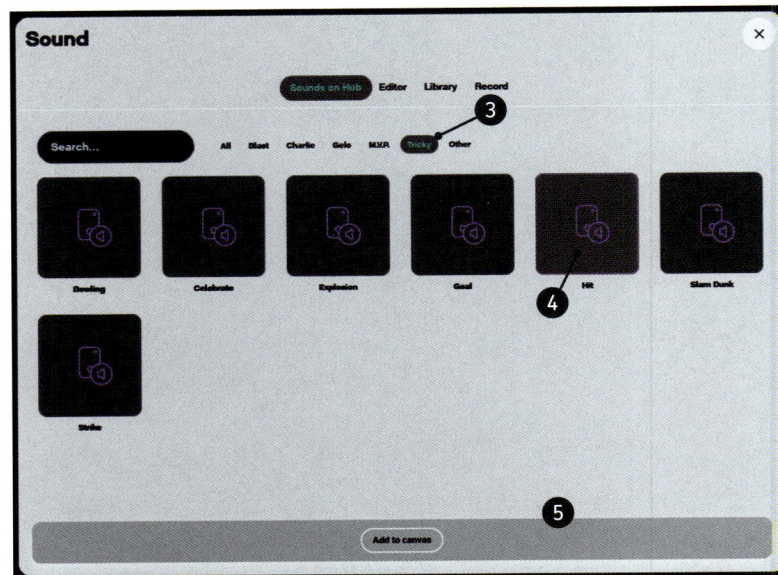

Figure 2-9: The start sound block plays sounds stored on the Hub or on your device.

7. Finally, add another **motor go to position** block below the animation block. Set the port to **F** and the position to **0**. Then choose **shortest path**. This block will bring the motor back to the zero-position, ready to hit the next ball you throw.

saving and running the program

Now you're ready to save the program and try it out! Here's how:

1. In case your Hub went to sleep while you created the program, turn it on again by pressing the round button and reconnect it to the app.

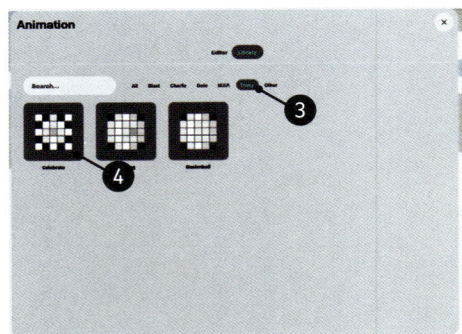

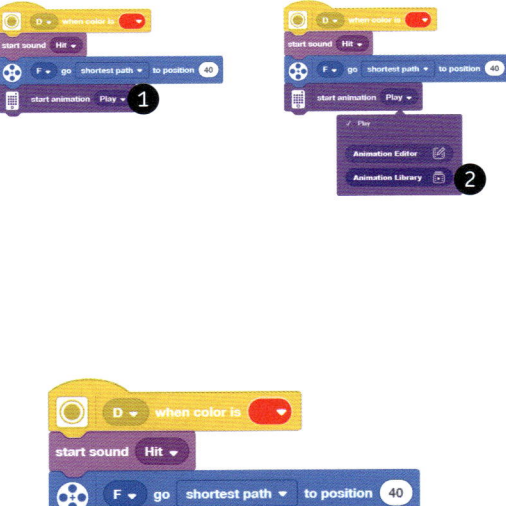

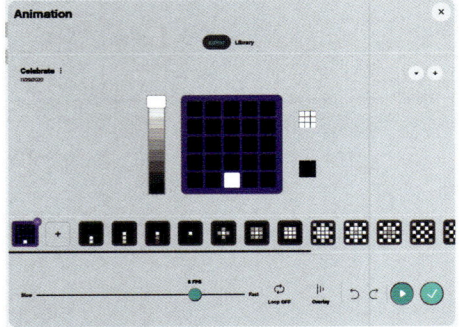

Figure 2-10: The start animation block plays animations on the Hub's display.

2. Using the **Project** menu, which you can access by clicking the three dots next to the project name, rename the project *baseball_batter* (Figure 2-11).

3. To run the program, click the **Select Storage Position** button in the bottom-right corner of the programming area, beside the gray Stop button. Select **Download** mode, choose program slot number **0**, and click the **Play** button, as shown in Figure 2-12. The app will download and start the program on the Hub.

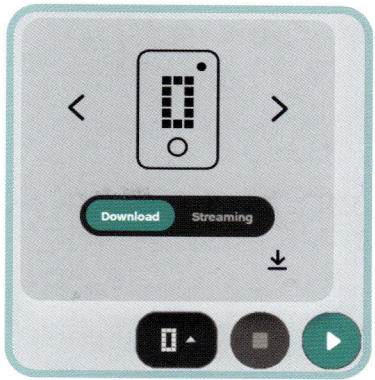

Figure 2-12: Select Download mode by clicking the Select Storage Position button before running your program.

Figure 2-11: Select Rename Project from the Project menu.

The program will now be visible in the Hub Program view, as shown in Figure 2-13. You can get there by clicking the Hub Monitor button at the top right of the screen (see Figure 2-4).

4. Throw the ball at the robot and see how it reacts. Was it a strike, a ball, or a home run? How far from the Color Sensor can the ball be and still be detected?

5. To stop the program, press the round button on the Hub or press the **Stop** button 🔴 in the app.

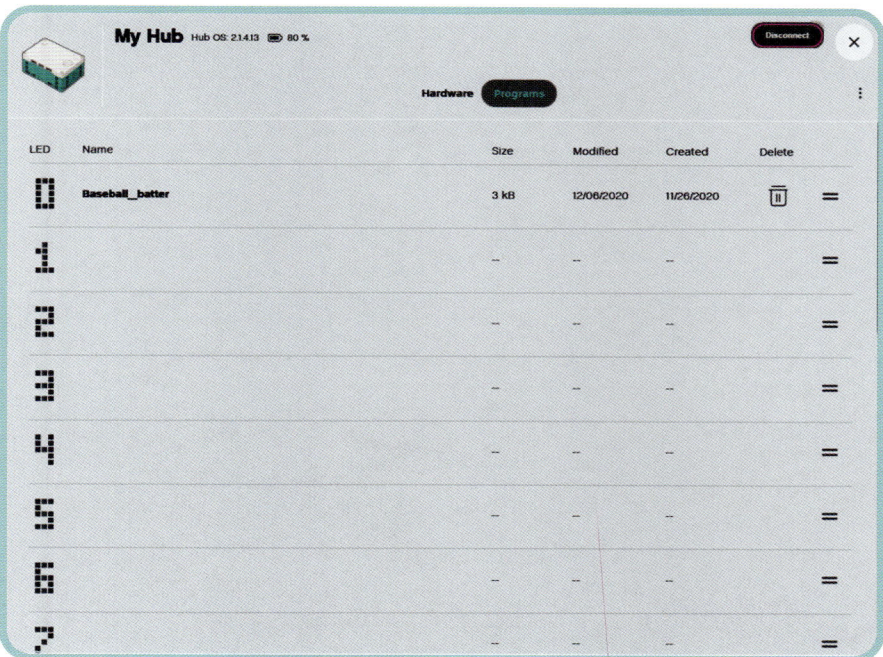

Figure 2-13: The Hub Programs menu, accessible from the Hub Monitor button, shows the programs stored on your Hub and lets you reorder and delete them.

understanding the program

The cool thing about word blocks is that they let you read your programs as you would read ordinary sentences. Here's how you might read the stack attached to the `when program starts` hat block. Each line represents one of the word blocks:

```
When the program starts,
set motor F's speed to 100% and then
tell motor F to run the shortest path to position 0.
```

Here's how to read the other stack in our program:

```
When the Color Sensor attached to port D reads RED,
start the Hit sound,
tell motor F to run the shortest path to position 40,
start the Celebrate animation, and
tell motor F to run the shortest path to position 0.
```

The first stack runs only once at the beginning of the program, while the other stack is triggered every time the Color Sensor detects red. In other words, detecting the color red is the *event* that triggers the sequence of actions.

NOTE When a program is written with ordinary words, it's called *pseudocode*. Pseudocode is a really important programming tool. In fact, before you create an actual program for a robot, it's important to plan out what you want your program to do using pseudocode. Once you have a plan, you're ready to start writing your actual code.

EXERCISE 2-1

Try using the Distance Sensor instead of the Color Sensor to detect the ball. Which block should you change? What should you change it to? Does the robot work better or worse than before?

ANSWER KEY: Replace the `when color is` hat block with another block from the **Events** palette: `when distance is`. Set the port to B, the threshold to 6, and the units to cm.

what you've learned

We've covered a lot in this chapter! By building and programming the Baseball Batter, you learned the key steps in any robotic system: reading sensor inputs, processing them, and producing outputs. You learned how to use event blocks, how motor blocks can rotate a motor to a precise angle, how to play sounds from the Hub, and how to show animations on the Hub display.

In the next chapter, you'll build and program the Gobbler, a robot that can help you cope with troubling thoughts and answer the important questions of life.

3

the gobbler

The legend is that if you're given to lying, you put your hand in there, it'll be bitten off.
—Joe (Gregory Peck) in *Roman Holiday* (1953)

Once upon a time, there was a sorcerer who molded a creature out of clay. To bring it to life, the sorcerer wrote a spell on a piece of paper and put it into the creature's mouth. It gobbled up the paper, then fell still again. All the creature could do was eat strips of paper. The sorcerer, initially frustrated, started feeling more and more at peace with each paper strip the creature ate. Finally, the sorcerer took another piece of paper and wrote a question for the creature to swallow: "Will you ever work?" The Gobbler replied: MAYBE.

In this chapter, you'll create the Gobbler (Figure 3-1), a desktop pal that swallows and crumples strips of paper you put in its mouth. Write down troubling thoughts or things that make you feel sad. Then let the Gobbler do its job: gobble them, crease them, crumple them, tear them, and shred them. You might feel a bit better afterward.

The Gobbler can also answer important questions if you write them down on strips of paper and let it gobble them. Make sure to ask only questions that can be answered YES, NO, or MAYBE. Those are the only words the Gobbler can use to share its wisdom with us mere mortals.

While building the Gobbler, you'll learn how to use the bracing technique to make sturdy assemblies, how to use gears, and how to transform rotation into an up-and-down motion. As you program the Gobbler, you'll learn how to control the flow of your program by repeating blocks or by choosing what to do depending on certain conditions. You'll also learn how to detect a blocked motor, how to write text on the Hub display, and how to make a robot behave randomly.

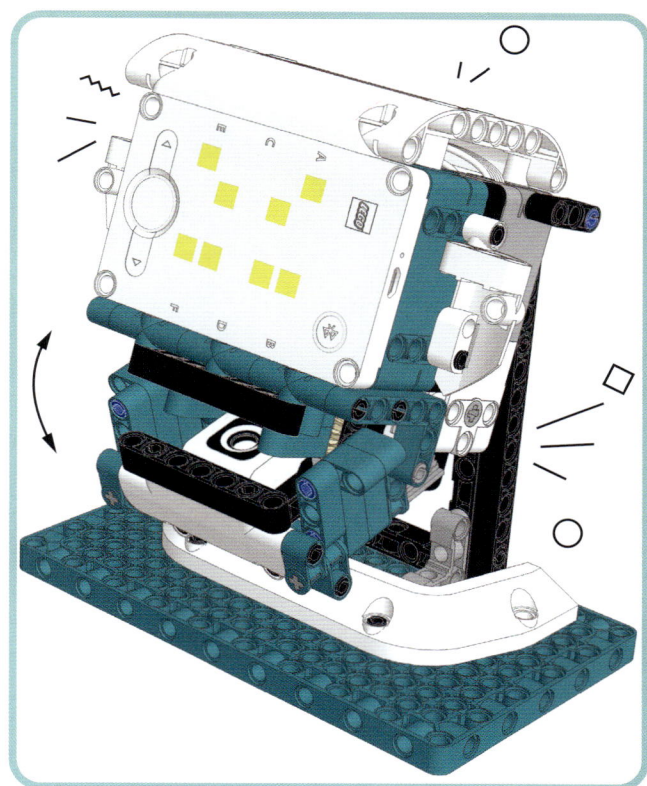

Figure 3-1: The Gobbler is a robot head that swallows and crumples pieces of paper.

building the gobbler

In this section, you'll build the Gobbler. Along the way, you'll find some notes describing important design choices and interesting techniques.

1

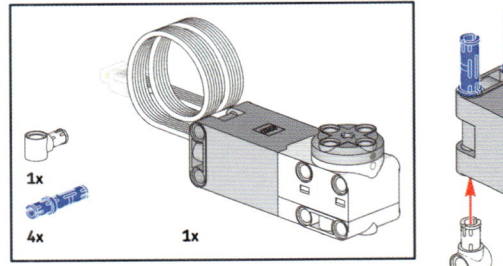

2

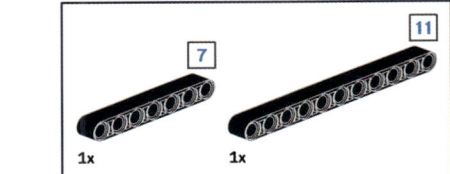

3

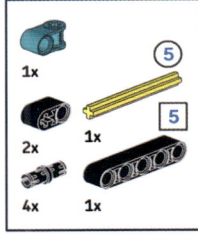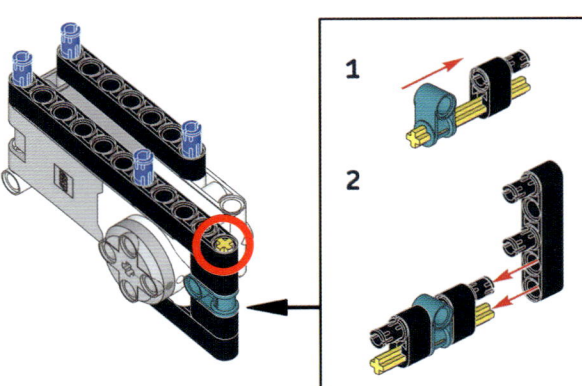

4

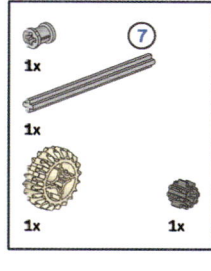

1x
1x
1x   1x

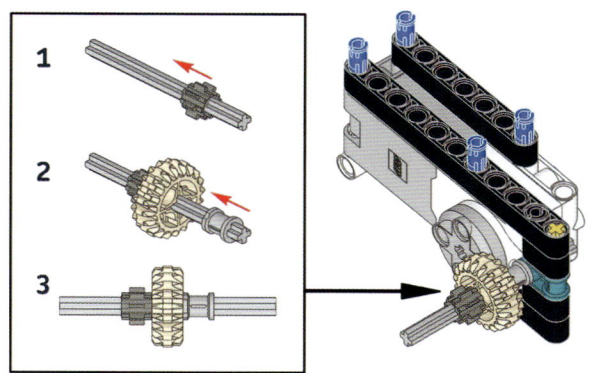

1

2

3

5

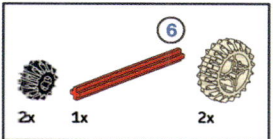

2x   1x   2x

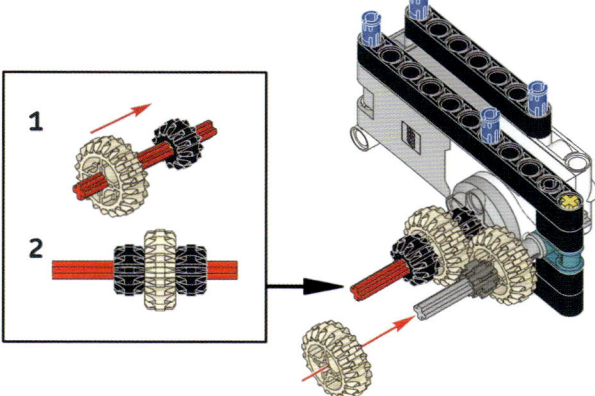

1

2

6

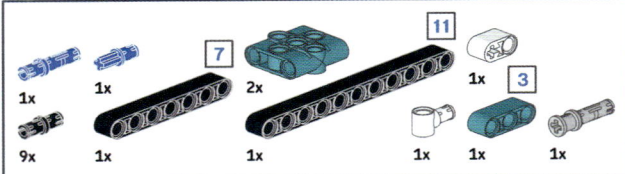

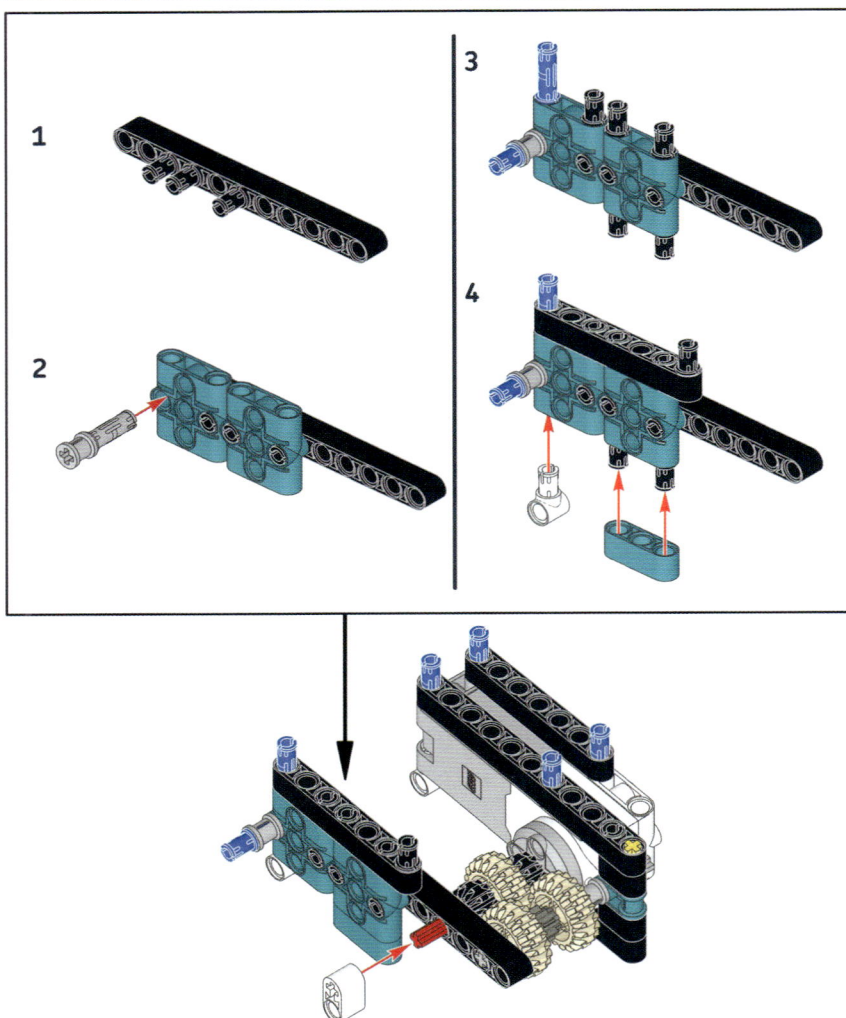

7

1x

The Hub is not shown in the parts list for space reasons.

8

2x 2x

9

2x 2x 2x 1x

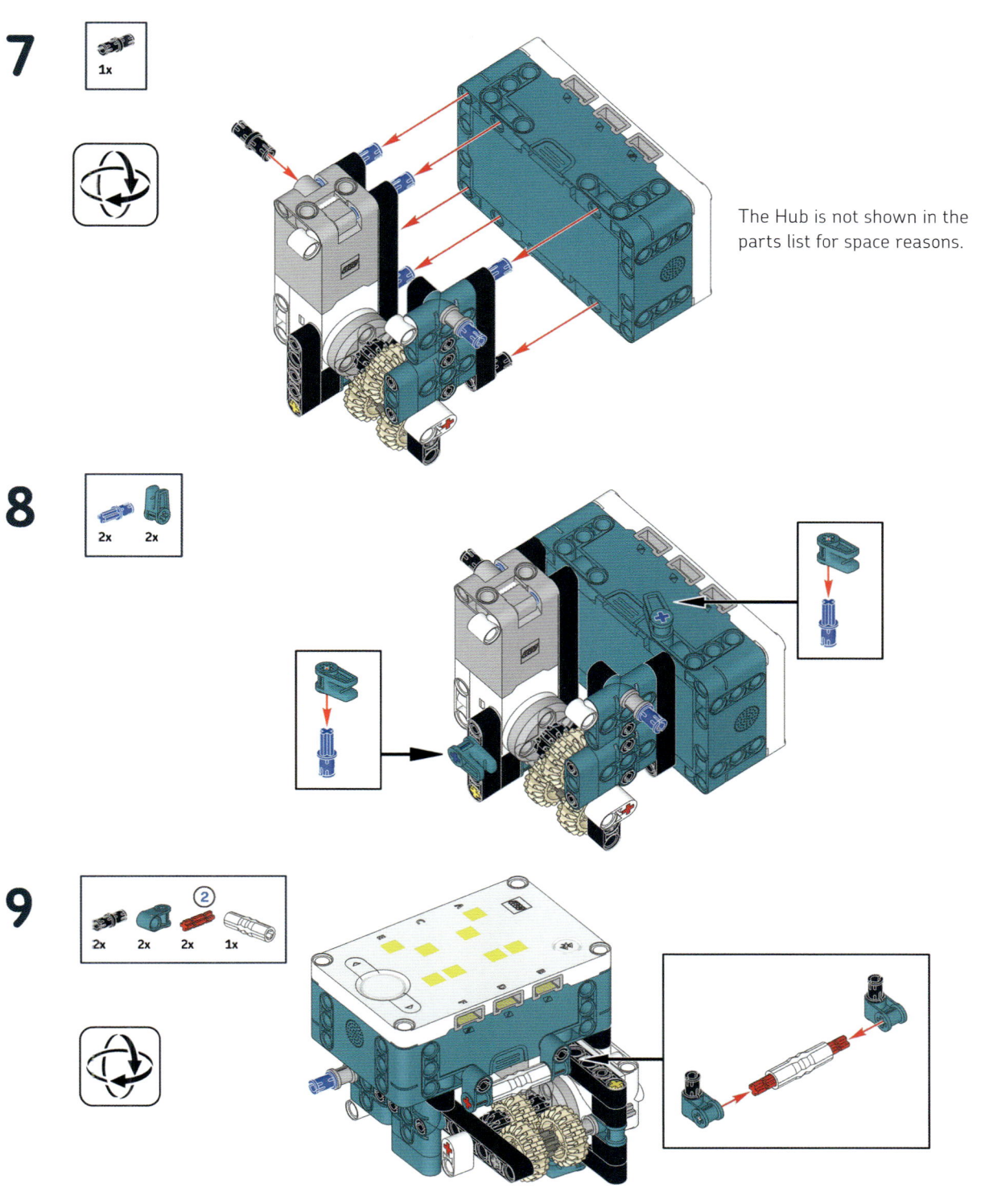

10

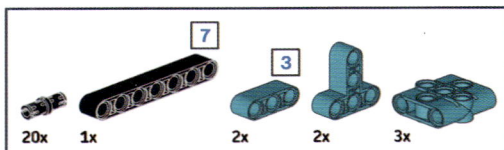

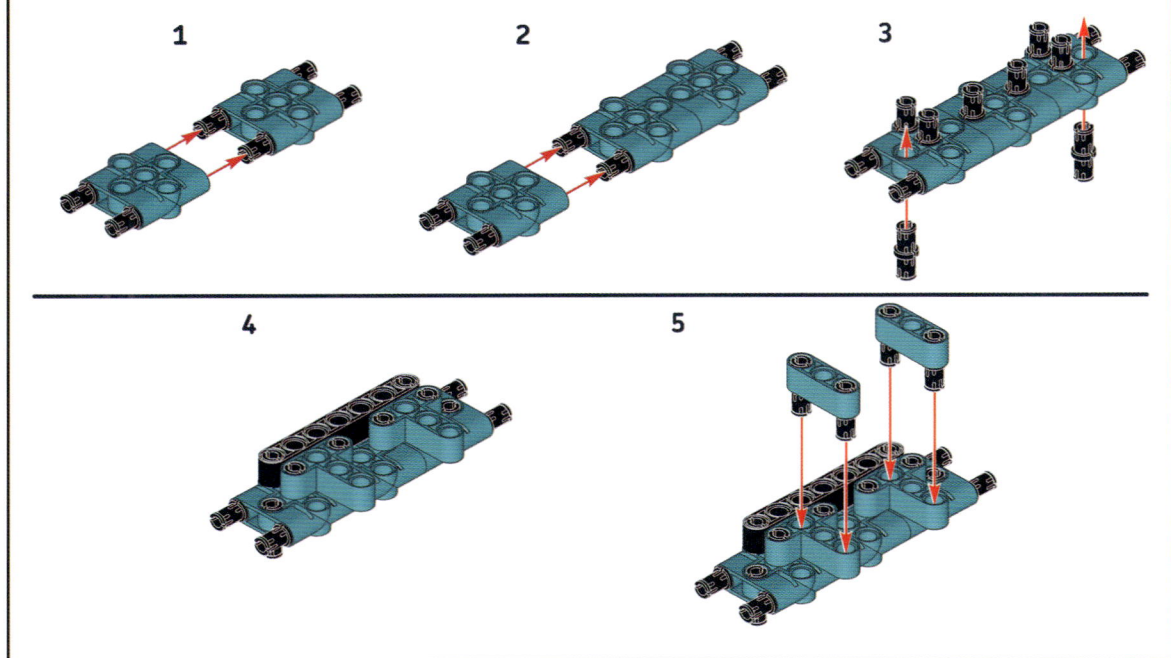

1

2

3

4

5

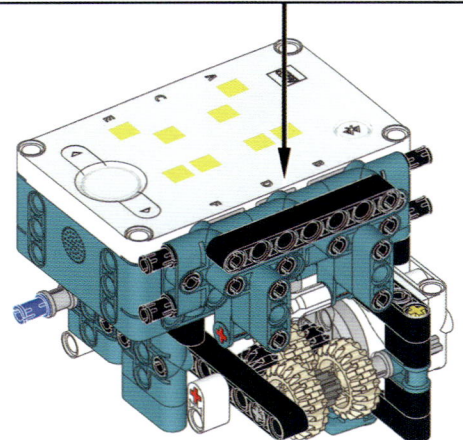

jaw subassembly

11

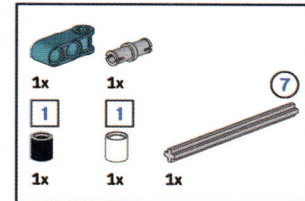

1x   1x
[1] [1]
1x   1x   ⑦
1x

Use the black and white 1M beams to correctly position the teal beam on the axle, and then remove them.

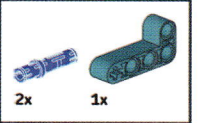

12

2x   1x

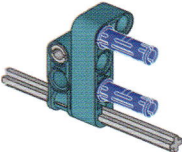

13

[3]
1x   1x

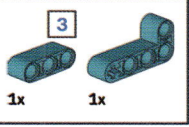

14

2x

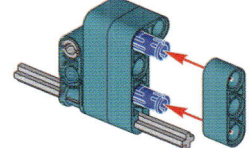

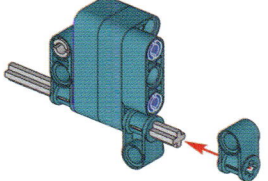

15

2x

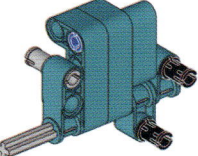

16

[7]
1x

6x   1x

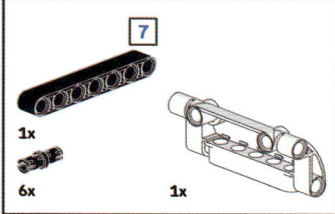

17

2x   2x   1x

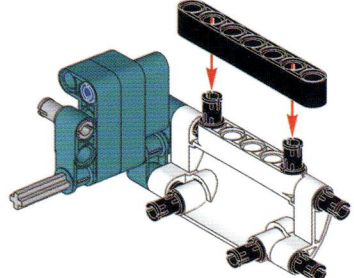

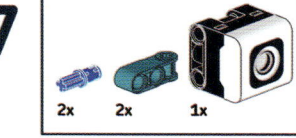

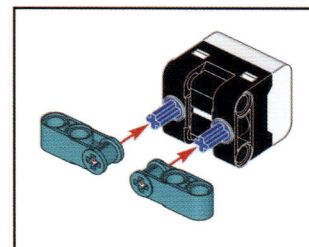

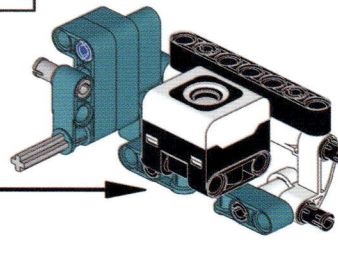

18

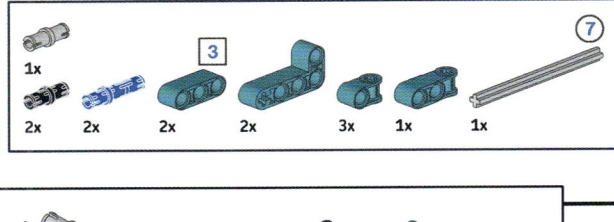

1x 2x 2x [3] 2x 2x 3x 1x 1x ⑦

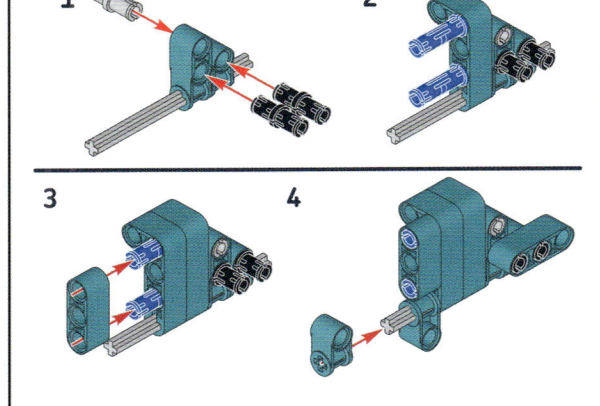

1 **2** **3** **4**

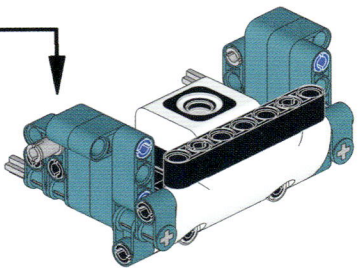

19

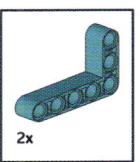

2x

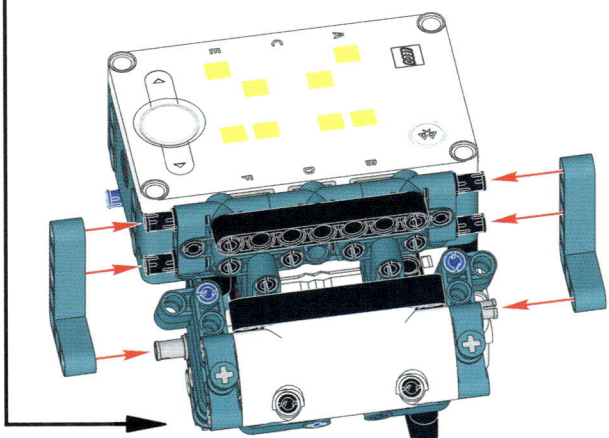

20

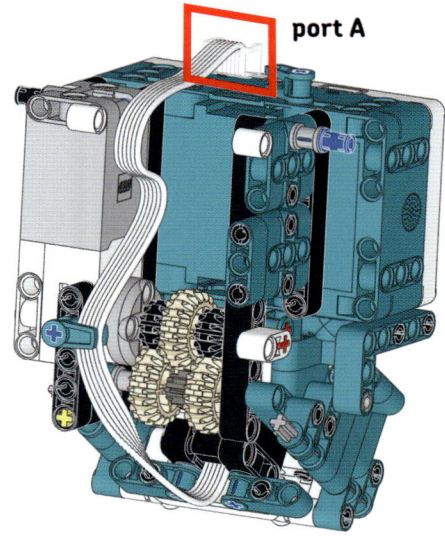

port A

21

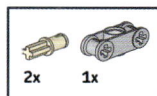

2x 1x

E

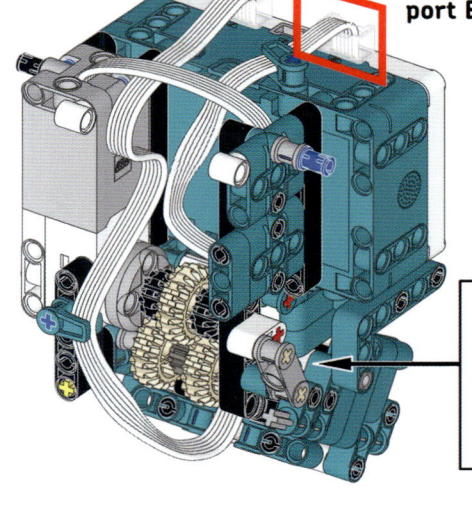

port E

The gears will crumple and ruin paper strips. The white 2M beam works as a *crank*, transforming the motor's circular motion into a seesaw motion that opens and closes the Gobbler's jaw.

22

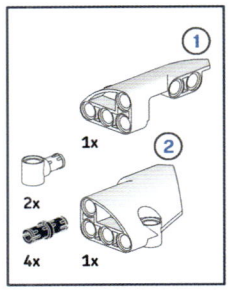

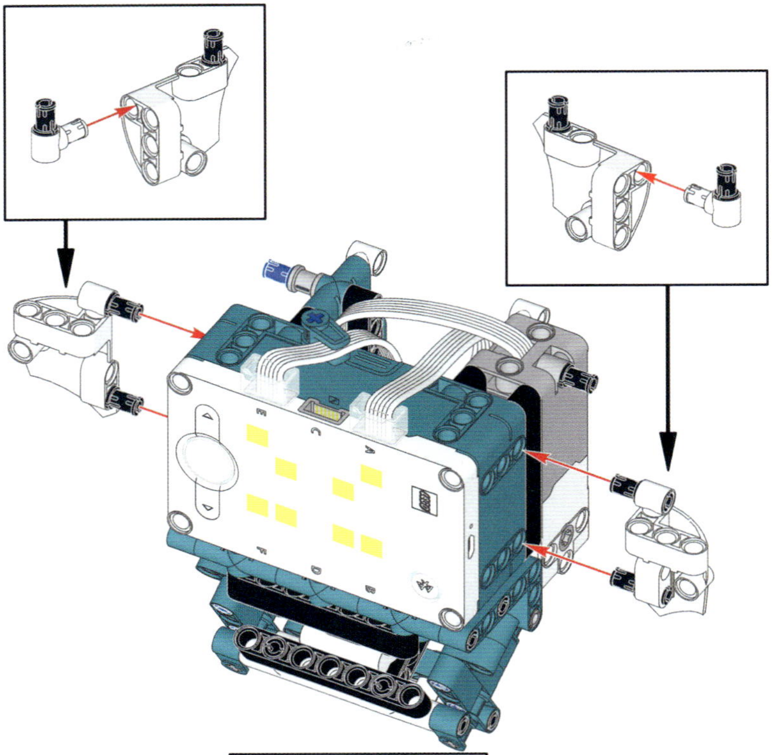

1x
2x
4x 1x

①
②

23

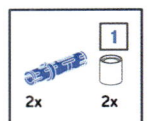

2x 2x

①

x2

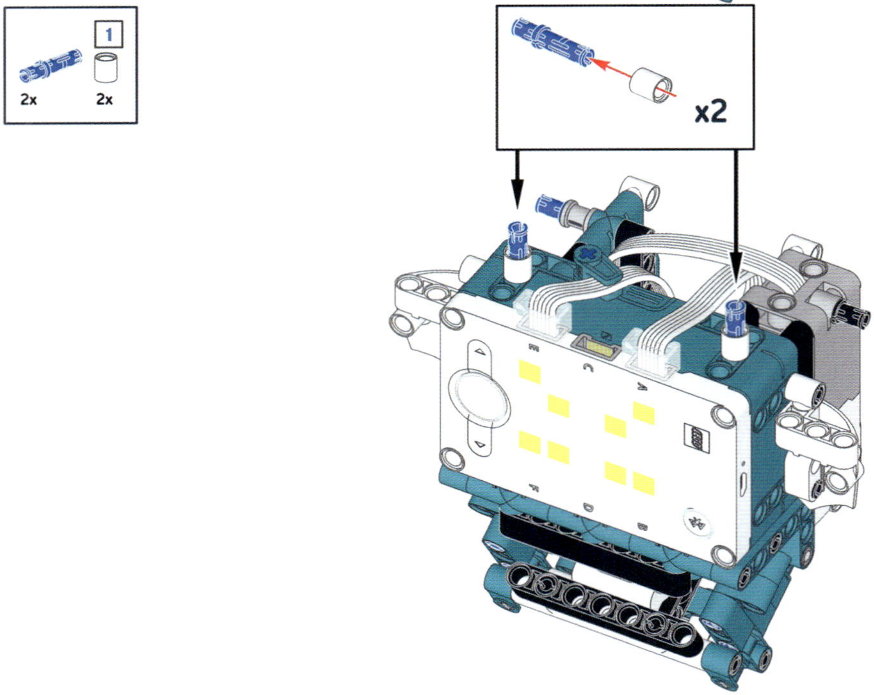

neck subassembly

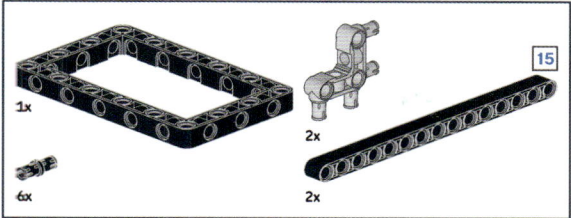

24

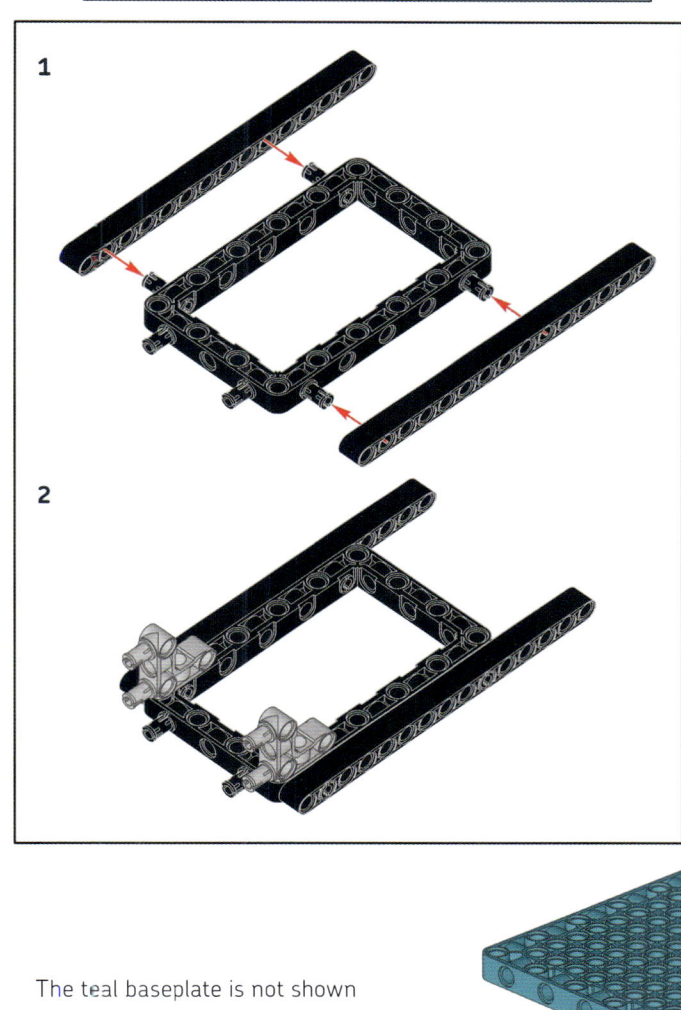

The teal baseplate is not shown in the parts list for space reasons. Use the white dots as guides for inserting the neck blocks onto the baseplate correctly.

25

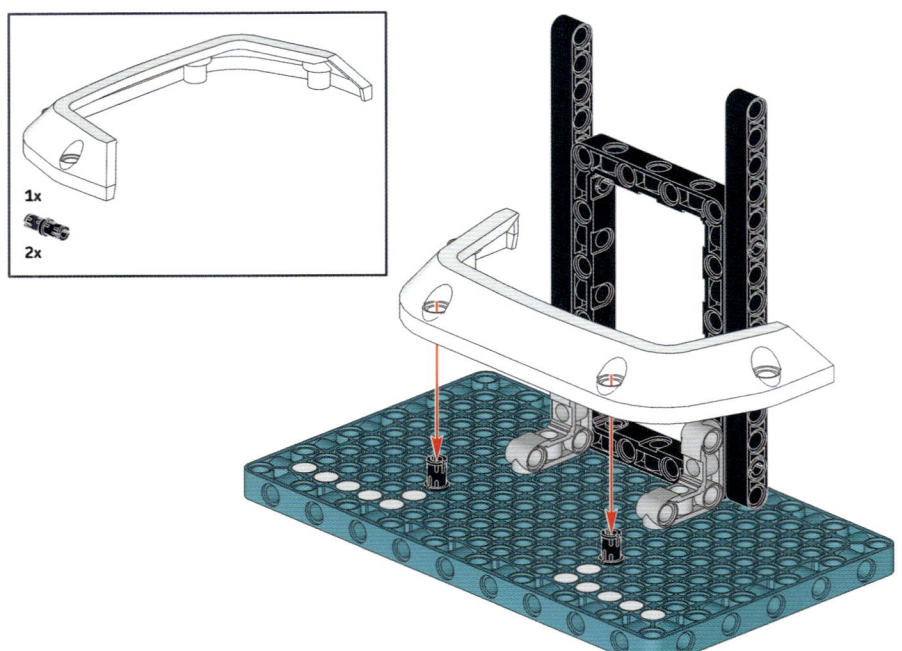

26

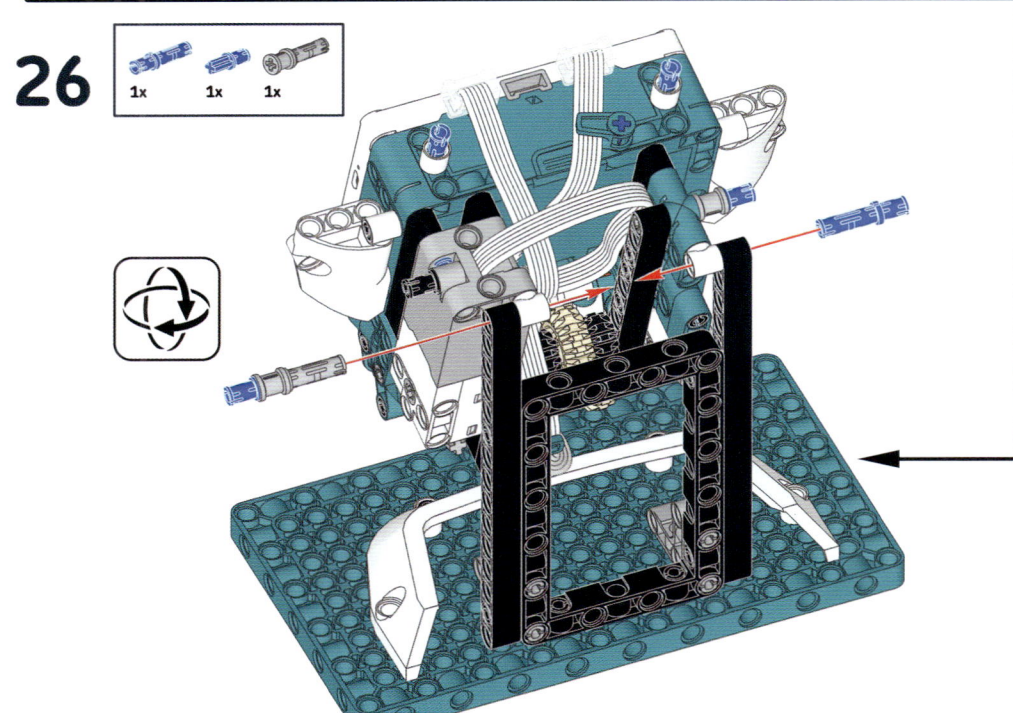

27

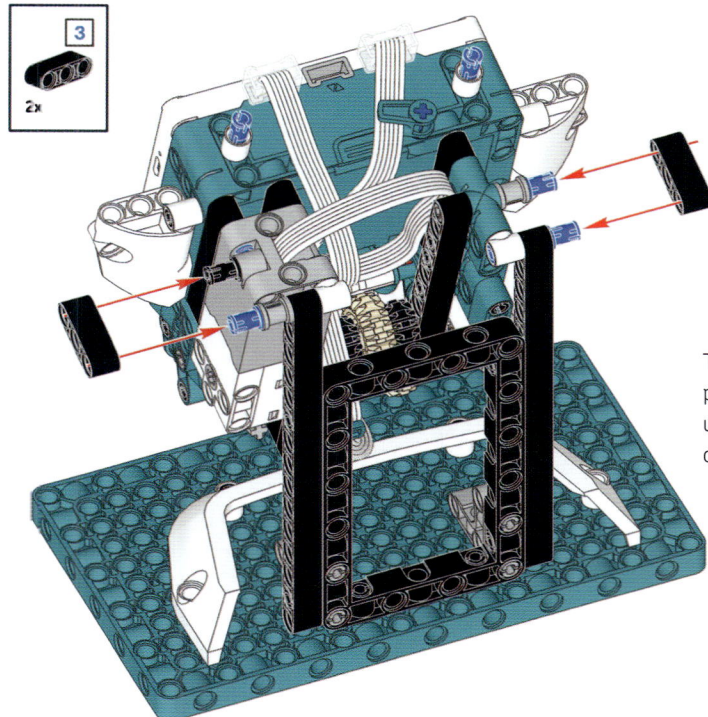

The 3M black beams join the neck with the head, preventing the assembly from coming apart under the head's own weight. This technique is called *bracing*.

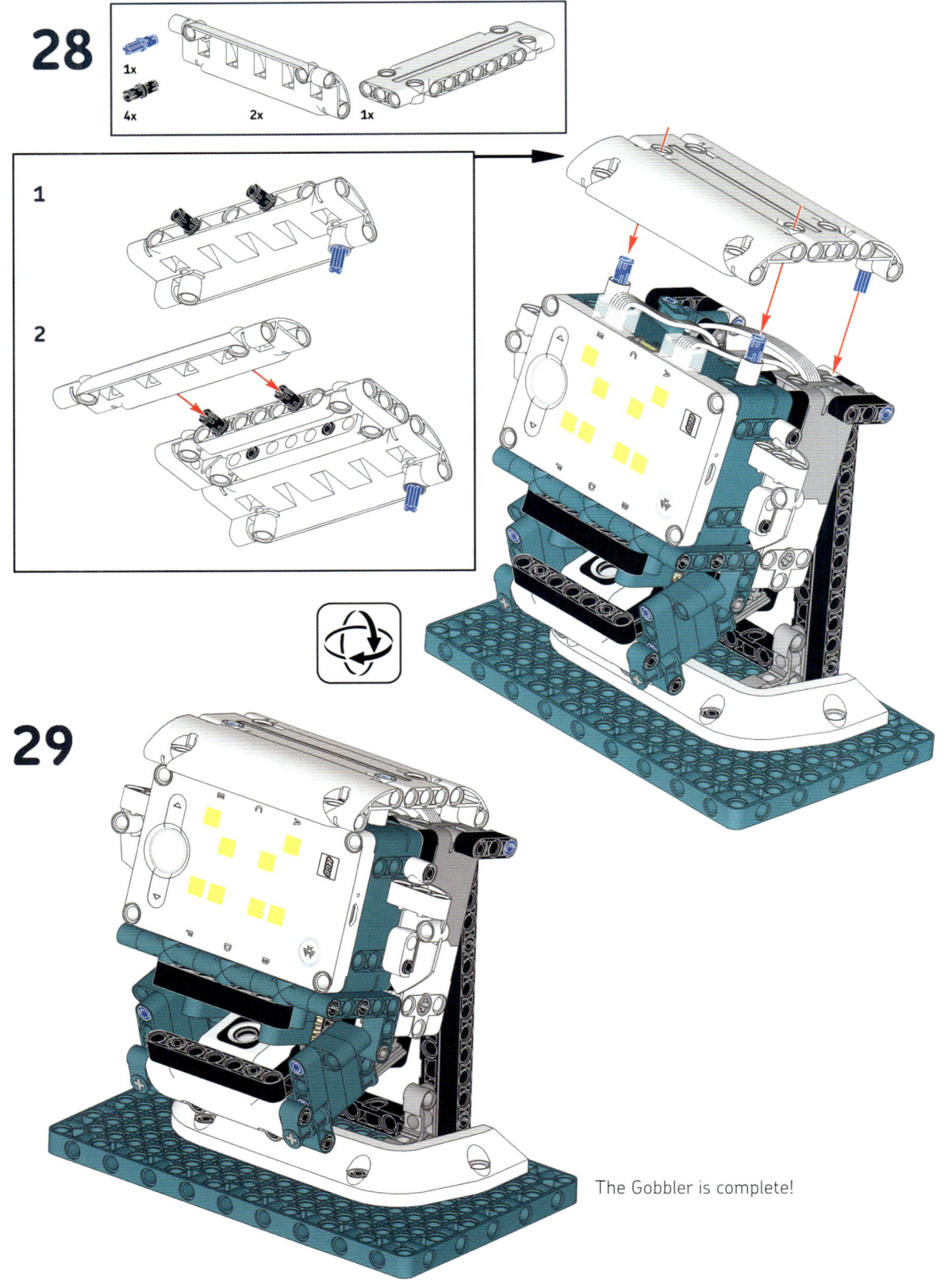

The Gobbler is complete!

programming the gobbler

We'll create three programs for the Gobbler. The first, our base program, allows the Gobbler to detect a knock on its head, wake up, open its mouth, and wait for a white piece of paper to be inserted. Once it detects the paper, the Gobbler will munch it until the paper has gone completely through its digestive system (the gears!).

The second program improves on the first. It lets the robot detect paper of colors besides white. It also checks for paper getting stuck between the gears.

The last program will give the Gobbler the oracular ability to answer even the deepest existential questions with cryptic, random answers: YES, NO, or MAYBE.

creating the base program

Let's create the base program for the Gobbler. Start a new project by clicking the **Code** button in the MINDSTORMS App lobby. Name the project *Gobbler_base*. Then take your time reproducing the code shown in Figure 3-2.

Figure 3-2: The finished base program for the Gobbler

You've already met some of the word blocks in this program. Others are new, and we'll go over them soon.

The blocks are color coded, to help you find them in the various palettes. For example, the `wait until` block is in the orange Control palette. Similarly, the `is hub (tapped)?` block is in the cyan Sensors palette, although it's listed as `is hub (shaken)?` in the palette—you'll have to choose **tapped** from the block's drop-down menu. You'll find all three animations needed for the program in the Charlie tab of the Animation Library.

Once you've finished the program, select a program slot from the **Select Storage Position** button. Then download and execute the program by pressing **Play**. The Gobbler should close its mouth and start looking around. Prepare some narrow strips of white paper on which you can write sad thoughts like "I got a bad grade," "There's no more ice cream," or "Jack Skellington has been blown to smithereens." The strips should be at most 24 millimeters (0.9 inches) wide. Knock on the Gobbler's head. Its eyes should turn angry and its mouth should open. Slide a piece of paper in and let your robot gobble away. Don't you feel better now?

understanding the base program

Our base program works because of *control structures*, pieces of code that guide the flow of a program. A control structure can make a program stop or pause, repeat a sequence of actions, or choose between two possible actions, to name a few examples. Specifically, our program uses two kinds of control structures: loops and conditional statements. You'll find all the Scratch control structures in the Control palette of the MINDSTORMS App.

loops

Often there are parts of a program that you want to repeat more than once. For example, we don't just want the Gobbler to eat a single strip of paper and turn off. We want to be able to feed it strips of paper over and over again. The easiest way to make this happen is to use a *loop*. A loop tells a robot that a sequence of blocks should be repeated.

Our program uses a `forever loop` block (❶ in Figure 3-2), which is an example of a *C block*. Any blocks inserted inside the "C" of the `forever loop` block will keep repeating over and over again as long as the program keeps running. Our forever loop contains the instructions for gobbling one piece of paper. Thanks to the loop, the Gobbler will keep crumpling up as many unpleasant thoughts as you can feed it.

NOTE You can stop a forever loop by shutting down your program. To do this, press the Hub's center button or press the **Stop** button in the app. You can also halt a forever loop without ending a program by using the `stop all` or `stop other stacks` blocks.

conditional statements

Our forever loop is a simple control structure: it just keeps repeating as long as the program is running. Most other control structures are a little more interesting: they guide the flow of a program based on a certain *condition*, such as a reading from a sensor or a certain amount of time going by. A piece of code that makes a decision based on conditions like these is called a *conditional statement*.

For example, our program uses a `wait until` Control block near the beginning of the forever loop (❷ in Figure 3-2). This block pauses the program until a specified condition becomes true. It's like a loop that does nothing but wait for the condition to occur. The block has a hexagonal (six-sided) slot where you can place another word block, which provides the *logic data*, or the condition that must be met for the program to resume. All the blocks that can provide logic data have a hexagonal shape, to match the slot where they'll be inserted.

In this case, we've inserted the `is hub (tapped)?` sensor block into the `wait until` block. It checks for a change in the Hub's built-in accelerometer readings due to a knock or a tap on the Hub. When a tap is detected, the `wait until` block's condition is met, and the program can continue. Thanks to this control structure, the Gobbler opens its mouth to eat a piece of paper only when you tap it on the head. Otherwise, it keeps its mouth closed.

Logic data can be only true or false. For example, either the Hub has been tapped or it hasn't. Either the Color Sensor detects a certain color or it doesn't. The `wait until` blocks in our program (we use three of them) make the program wait for a condition to be true to proceed. Later in this chapter, we'll use another kind of conditional statement, an `if then else` block, to execute one sequence of blocks if a condition is true and another sequence if the condition is false.

NOTE The `is hub (tapped)?` block can be configured to detect a different motion, such as when the Hub is shaken or falls.

reading through the program

Now that we understand what loops and conditional statements are, we're ready to read through the program in plain language to see how it works. The program starts like this:

```
When the program starts,
set motor E's speed to 50%,
set the display orientation to right, and
play the Wake Up animation once before continuing the
program
```

We've already used the `when program starts` and `set motor speed` blocks in Chapter 2. The `set orientation to` block controls which side of the Hub display is considered up. We need

this block because the Hub is lying on its right side, not sitting upright. Setting this block to **right** ensures that animations from the Animation Library will appear as if the Hub is straight, even though it's actually sideways.

Next, we play the Wake Up animation to make the Gobbler seem like it's coming to life by using the `play animation until done` block. Unlike the `start animation` block that we used in the previous chapter, this block won't let the program continue until the animation has finished playing.

The rest of our program is inside the forever loop:

```
Repeat the following forever in a loop:
    Tell motor E to run the shortest path to position 0,
    start the Looking animation,
    wait until the Hub is tapped,
    start the Angry animation,
    tell motor E to run the shortest path to position 90,
    wait until Color Sensor A detects WHITE,
    start spinning motor E clockwise,
    wait 2 seconds,
    wait until Color Sensor A detects TEAL,
    wait 1 second, and
    stop motor E.
Go back to the beginning of the loop.
```

First we tell the motor to move to position 0, which closes the Gobbler's mouth. Next we play the Looking animation. Thanks to the `start animation` block, the program can move on to the next block in the stack while the animation keeps playing.

Next comes the `wait until` block that we've already discussed. Once the Hub's accelerometer detects a tap, the program proceeds, playing the Angry animation and moving the motor to position 90. The movement of the motor opens the Gobbler's mouth so you can insert a strip of paper.

Next we have another `wait until` block. This time, it gets its logic data from the `is color?` block, which checks the current reading of the Color Sensor. The program waits for the sensor to detect white, indicating there's a white piece of paper inside the Gobbler's mouth. This condition lets the program continue with the execution of the `start motor` block, which spins the motor in one direction—in our case, clockwise—until the motor is told otherwise. The motor turns the gears, driving the paper inside the mouth while the jaw moves up and down as if the Gobbler is chewing. The motor runs at the last speed we set (50% in this program).

The `wait for seconds` block pauses the flow of the program for a certain amount of time, specified in seconds. This block doesn't actually stop any actions that are already in progress. It only stops the program from moving on to the next block. Therefore, the motor keeps spinning while we wait for 2 seconds, giving the gears some time to start crumpling the paper.

The next `wait until` block allows the motor to keep running until the Color Sensor sees the teal color of the Gobbler's empty mouth. This means the paper has fully passed through the mouth. After waiting one more second to let the paper be ejected, we use the `stop motor` block to turn the motor off. Then the loop begins all over again, with the `motor go to position` block closing the Gobbler's mouth.

improving the base program

The base program we've created for the Gobbler *should* work well. However, robots often encounter unexpected problems that affect their performance. The best robotics programmers try to plan for these problems and write code that gets around them. Doing this makes a robot more *robust*, or better able to cope with real-world disturbances.

Let's update our base program to make the Gobbler more robust. We'll help it get around two challenges. First, since not all paper is white, we'll teach the Gobbler to detect paper strips of any color. Second, we'll tell the Gobbler what to do when paper gets stuck in its gears. When this happens, the motor can *stall*, or stop spinning even though it's still turned on. Stalling makes a motor use more power than it should and can lead to damage, so it's important to turn off a motor when it stalls.

To improve the Gobbler, start with the base program we've already made. Click the three dots next to the project name to expand the **Project** menu, select **Save As**,

and type **Gobbler_enhanced** as the new name for the program. Then update the block stack as shown in Figure 3-3. The blocks to be changed are marked with arrows.

As you can see, two of the `wait until` blocks have been updated. They now check for more-complex logic conditions, which we'll discuss soon. Figure 3-4 shows how to build up the two `wait until` blocks.

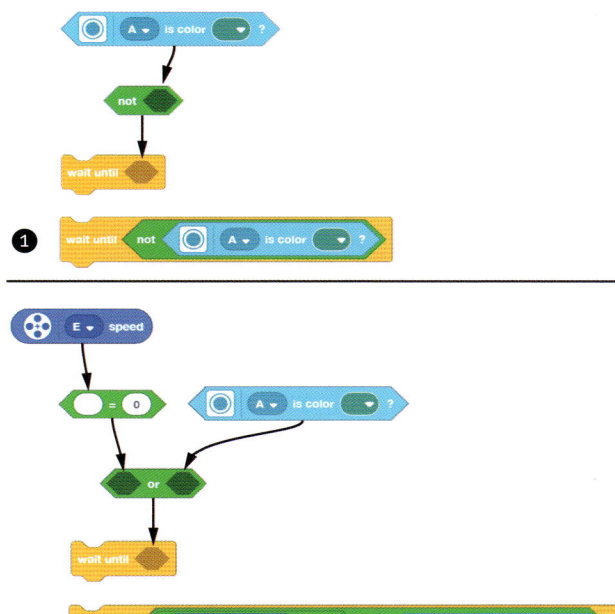

Figure 3-4: How to build the composite logic conditions for the enhanced Gobbler program

Once you've updated the program, test it out. Try putting a colored strip of paper in the Gobbler's mouth. Does it work as expected? What happens if you block the motor while the Gobbler is chewing?

understanding the enhanced program

Our enhanced program uses *operators*, pieces of code that perform a task and report, or *return*, the result. The result can then be used in the program. An operator's task can be as simple as adding or subtracting two numbers, for example, or it might involve comparing two values or checking whether multiple conditions are true. There are two kinds of operators in our enhanced program: *comparison operators* and *logic operators*. You can find all the MINDSTORMS operators in the green Operators palette.

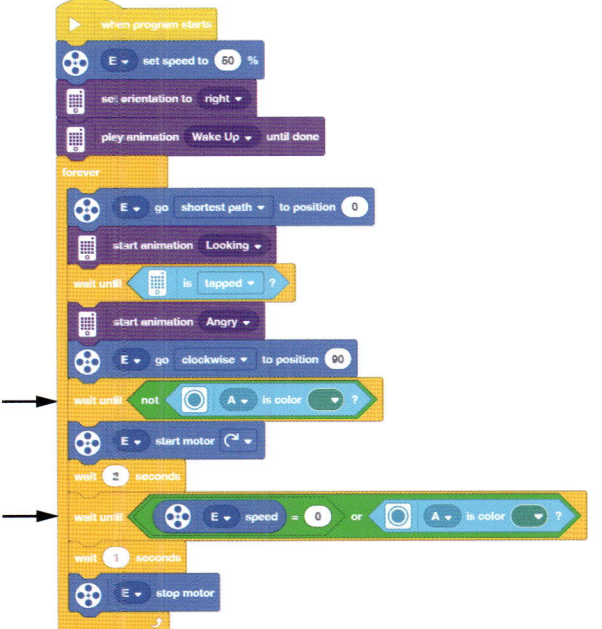

Figure 3-3: The enhanced Gobbler program uses logic operators to combine logic conditions based on sensor readings.

comparison operators

A comparison operator takes two values as inputs, compares them, and returns the result of the comparison as a logic value (true or false). The three comparison operator blocks are shown in Figure 3-5. The `less than` block (left) returns `true` if the first value is less than the second value. Otherwise, it returns `false`. The `equal` block (center) returns `true` if the first value is equal to the second value. The `greater than` block (right) returns `true` if the first value is greater than the second value.

Figure 3-5: The comparison operators check whether the first input is less than, equal to, or greater than the second input.

The examples in Figure 3-6 show how the comparison operators work. The result is reported inside a balloon that should appear if you click the operator while the app is in Streaming mode. As you can see, `2 = 1` returns `false` because the two inputs are not equal to each other, while `1 = 1` returns `true`.

NOTE In Streaming mode, the program isn't executed on the Hub, but rather on the device that runs the app. The commands to move the motors or read the sensors are then sent to the Hub. You can switch to Streaming mode by clicking the Select Storage Position button shown in Figure 2-4 in Chapter 2.

In addition to numbers, you can compare letters or words. When a piece of data is in the form of text rather than numbers, programmers call that data a *string*. The string that comes first in alphabetical order is considered the smaller one. For example, A < B is true because A comes before B in the alphabet. Comparison operators ignore whether letters are uppercase or lowercase. That's why `SPAM = spam` returns `true`.

The hexagonal shape of Scratch comparison blocks is a clue that they return a logic value. By placing a comparison block inside another block with a corresponding hexagonal input slot (for example, the `wait until` block), you can feed the result of the comparison into the other block. Then the other block will make a decision based on whether the comparison is true or false.

logic operators

Logic operators take in one or more logic conditions, evaluate them, and return `true` or `false`. The three logic operator blocks are and, or, and not, as shown in Figure 3-7. These blocks let you combine and transform logic conditions to create advanced behaviors for your robots.

Figure 3-7: The three logic operator blocks

Each logic operator can be represented by a *truth table*. This kind of table has a row for every possible combination of inputs and shows what the resulting output will be. For example, Table 3-1 shows the truth table for an and block. As you can see, the and block takes in two logic conditions, input A and input B. It outputs `true` only when *both* input A *and* input B are true. If either input is false, or if both inputs are false, the and block returns `false`. Using this operator, you can have your robot do something only if two conditions are satisfied.

Table 3-1: Truth Table of the AND Operator

Input A	Input B	Output = A and B
False	False	False
False	True	False
True	False	False
True	True	True

The or logic operator block takes in two logic conditions and returns `true` when *at least one* input is true (input A *or* input B, or both), as shown in Table 3-2. The block returns `false` only when both inputs are false. You can use this operator to check whether one or another condition is true, and perform some action in either case.

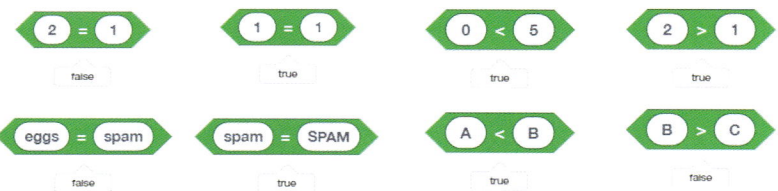

Figure 3-6: When the app is in Streaming mode, you can see the result of the comparison operator blocks by clicking them.

Table 3-2: Truth Table of the OR Operator

Input A	Input B	Output = A or B
False	False	False
False	True	True
True	False	True
True	True	True

The not logic operator block inverts a logic condition, returning the opposite of the input. As you can see in Table 3-3, the output is true if the input is false, and vice versa. If you want to perform some action when a certain logic condition is *not* true, insert that condition into a not operator block.

Table 3-3: Truth Table of the NOT Operator

Input	Output = not (Input)
True	False
False	True

NOTE As with comparison operators, you can see the result of a logic operator by clicking it while the app is in Streaming mode.

reading through the program

Now that you know how comparison and logic operators work, we can read the new logic conditions inside the wait until blocks in our enhanced program. Before we updated the program, the first wait until block could be read as follows:

```
Wait until Color Sensor A detects color WHITE
```

Remember, the point of this block was to wait for a white piece of paper. However, we want the Gobbler to detect other colors of paper as well. Put another way, we want the Gobbler to start gobbling when it sees *anything but* the teal color of the roof of its mouth. To accomplish this, we take an is color? block, set it to **teal**, and negate the result by inserting it into a not logic operator. The new condition becomes the following:

```
Wait until Color Sensor A DOES NOT detect color TEAL
```

This way, the Gobbler can eat paper of any color (as long as the paper isn't the exact color of the roof of the Gobbler's mouth, that is).

The other wait until block used to read as follows:

```
Wait until Color Sensor A detects color TEAL
```

This block sensed when the paper had passed through the Gobbler's mouth so the program could prepare to turn off the motor. However, we also want to turn off the motor if the paper gets stuck and the motor stalls. That is, we want the wait until block to be triggered by *one or the other* of two conditions—the perfect job for an or operator!

We insert the original is color? block on one side of the or operator to test for the color teal. To test for a stalled motor, we need a second operator: the equal comparison block. We place the motor speed block, which reports the current speed of the motor, on one side of the comparison, and set the other side of the comparison to 0. Then we place the equal block into the other slot of the or operator. If the motor stops running due to jammed paper, its speed will be equal to 0, and the wait until block will know to move on.

Here's how the entire logic condition reads:

```
Wait until motor E's speed is equal to 0 OR Color Sensor
A detects color TEAL
```

We've tested for one logic condition using the equal comparison operator and combined it with another logic condition using the or logic operator. If either condition is met, the program can proceed.

making the gobbler answer questions

Let's add another part to the program so the Gobbler will reply to questions with random YES/NO/MAYBE answers. Figure 3-8 shows the complete program, with the new section enlarged.

To make the new program, start with the *Gobbler_enhanced* program file, select **Save As** from the **Project** menu, and rename it *Gobbler_answers*. Then delete the stop motor block at the end of the forever loop.

NOTE You can delete a block by dragging it back into the palette area or by right-clicking it and selecting Delete Block.

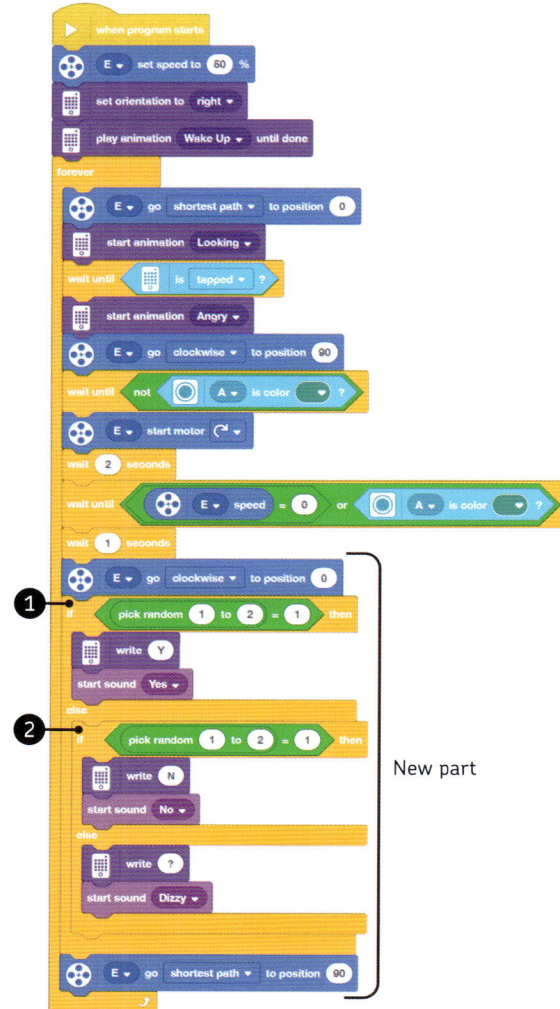

Figure 3-8: With this program, the Gobbler can respond to questions with random answers.

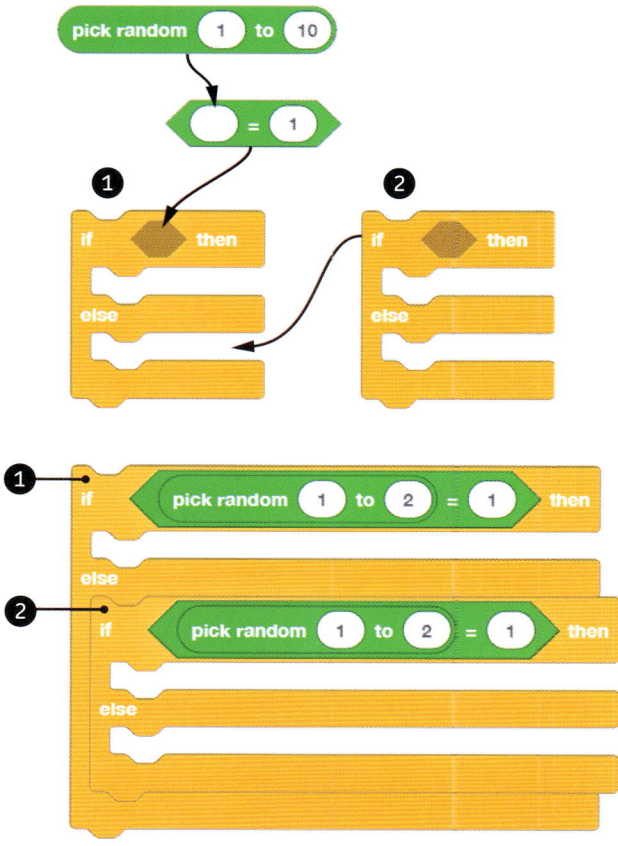

Figure 3-9: Here's how to assemble the part of the program that lets the Gobbler answer questions. You can put an *if then else* block inside another to create a three-way choice.

Begin adding the new blocks, after the `wait 1 seconds` block. Figure 3-9 shows how to start building up the new part of the program. The sounds played by the `start sound` blocks (Yes, No, and Dizzy) are all in Charlie's sound library.

Try asking the Gobbler a few questions to see how the program works. Are you able to predict what answer the Gobbler will give each time, or do its responses seem random?

condition-based choices

The new part of our program uses `if then else` blocks to execute different block stacks depending on some logic conditions. The `if then else` control structure is another kind of conditional statement: *if* the specified logic condition is true, *then* the robot will perform one action, or *else* it will perform a different action.

As you can see in Figure 3-9, our program nests one `if then else` block ❷ inside the `else` branch of another `if then else` block ❶. This splits the program flow into three possible cases, one for each of the three possible answers (YES/NO/MAYBE) the Gobbler can give.

NOTE In the Control palette, you'll also find the `if then` block. It works just like the `if then else` block, but it doesn't have a second space for specifying what actions should be executed when the condition is false. If the condition is false, the program skips over the `if then` block and moves on to the next block in the stack.

generating random numbers

Robots become more lifelike if their behavior is a bit unpredictable. The easiest way to introduce unpredictability is to make a robot choose at random from a set of actions. Programs make random choices using a *random number generator*. Much like rolling dice or picking a name out of a hat, a random number generator outputs a random value picked from a range of numbers. There's an equal chance of the random number generator drawing any of the numbers in the range you give it.

The MINDSTORMS random number generator is the `pick random number` block `pick random 1 to 10`. You'll find it in the Operators palette. The block outputs a random number from within the range you specify. Both the upper and lower limits are included in the range. In our program, we set a range of 1 to 2 to make a random either/or choice, similar to flipping a coin.

understanding the new part of the program

Let's read through the new part of our program to see how the Gobbler can randomly answer YES, NO, or MAYBE:

```
[previous blocks]
    Tell motor E to run clockwise to position 0.
❶ If the random number picked in the range [1-2]
    is equal to 1, then:
        write 'Y' on the Hub display and
        start the Yes sound,
    else:
❷       if the random number picked in the range [1-2]
        is equal to 1, then:
            write 'N' on the Hub display and
            start the No sound,
        else:
            write '?' on the Hub display and
            start the Dizzy sound.
        End if.
    End if.
    Tell motor E to run the shortest path to position 90.
Go back to the beginning of the loop.
```

First, the motor moves to position 0, closing the Gobbler's mouth. Next we enter the nested `if then else` control structure. Two random draws between 1 and 2 determine which answer the Gobbler will give. Each random draw is fed into an equal comparison block to test whether the number chosen is 1. If the first random number is 1, this satisfies the condition of the first `if then else` block (❶ in Figures 3-8 and 3-9). The Gobbler will answer "Yes," and the rest of the blocks won't be executed because they're included in the `else` space. If the first random number is 2, we move on to the `else` space, which contains the second `if then else` block (❷ in Figures 3-8 and 3-9). We pick a second random number. If it's equal to 1, the Gobbler will answer "No." Otherwise, it will answer "Maybe."

The Gobbler communicates its chosen answer using the `write on 5×5 matrix` block, which lights up the specified text on the Hub display. Longer texts will scroll across the display, but a single character (such as our Y, N, or ?) will remain still. We also play a sound for each answer. Both the `write on 5×5 matrix` and `start sound` blocks let the program stack continue when they execute, so the text will appear on the display as soon as the sound starts playing, while the last `motor go to position` block opens the Gobbler's mouth as if it's speaking. Finally, after this new part of the program ends, the loop will repeat from the beginning.

EXERCISE 3-1

Can you change the *Gobbler_answers* program so that the Gobbler will give an answer only if you put a red piece of paper into its mouth? Hint: use an `if then else` block after you detect the paper insertion.

what you've learned

While building the Gobbler, you learned some interesting building techniques, such as bracing, using gears, and transforming rotation into alternating motion with a crank. While programming the Gobbler, you learned how to detect knocks, sense when the motor has stalled, and show text on the Hub display. You also learned how to control the flow of the program, using loops to repeat a sequence of blocks and conditionals to choose what to do depending on whether a statement is true or false. You saw how to use logic and comparison operators and how to make your robot behave unpredictably with a random number generator.

In the next chapter, you'll build a fast remote-control car that can transform into a walking humanoid robot!

SARKIAP-1 the transformer

It's been an honor serving with you all. Autobots, roll out!
—Optimus Prime in *Transformers* (2007)

Transformers are robots from another planet that can hide on Earth by transforming into vehicles. They've become a cult for '80s kids like me. Ever since I played with my first Transformer toys, I dreamed of making a car that could transform into a humanoid robot out of LEGO bricks. In the '80s, LEGO released an official set featuring a Technic truck that could transform into a robot (set #8852), but the result was disappointing.

Finally, thanks to the fantastic assortment of pieces in the LEGO MINDSTORMS Robot Inventor set, I was able to realize my childhood dream: a remote-control car that can transform into a humanoid robot that can walk. And I mean really walk, not just roll around on the same wheels it uses as a car.

While making this robot, you'll learn how a biped robot can walk and turn without falling down, and you'll see how to render the complex shape of a car's trunk. As a programmer, you'll learn how to create custom blocks and how to write code that responds to a remote control. Are you ready to make the Transformer shown in Figure 4-1?

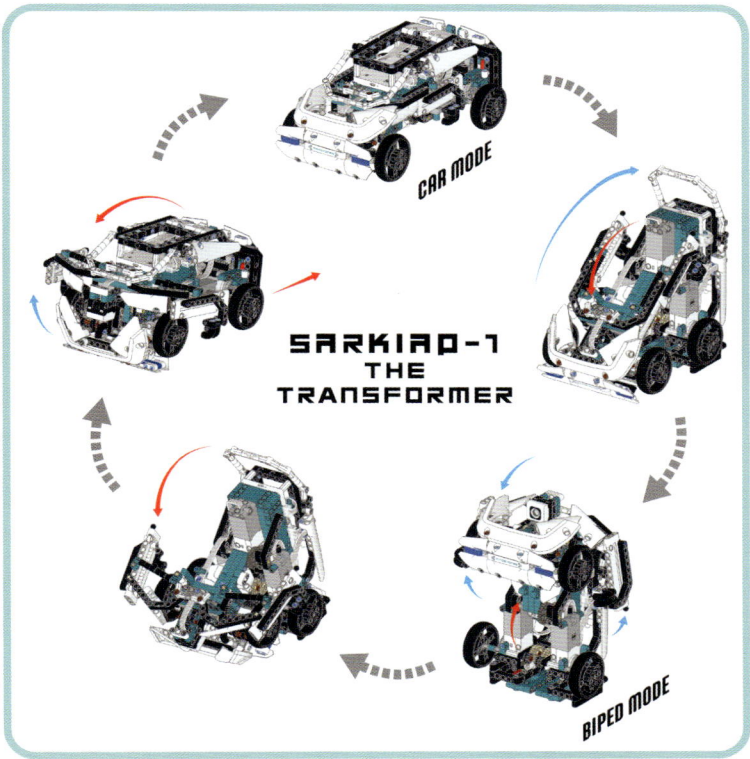

Figure 4-1: SARKIAP-1 is a remote-control robot that can change from a car to a biped and back.

building the walking core robot

In this section, you'll start building SARKIAP-1 the Transformer by assembling its legs, feet, and torso. During the assembly process, you'll find some notes describing important design choices and interesting techniques. Once you've built the core of the walking robot, you'll test it out by programming it to walk around and avoid obstacles. After that, you'll finish building SARKIAP-1, adding the parts needed for it to turn into a car.

right leg subassembly

1

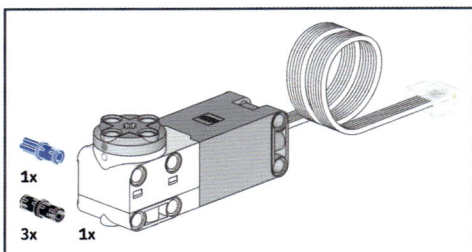

2

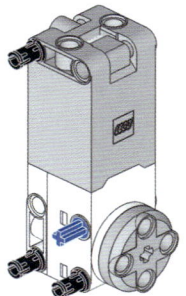

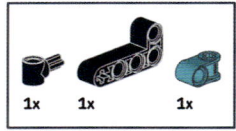

3

4

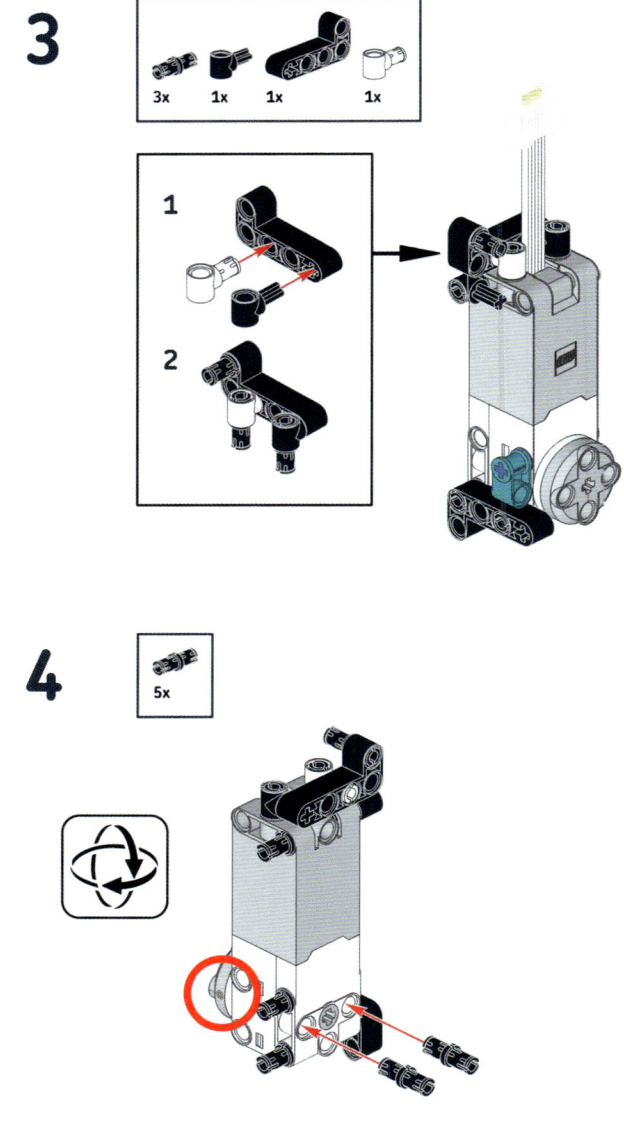

5

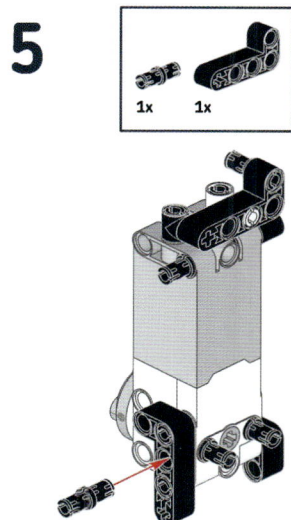

6

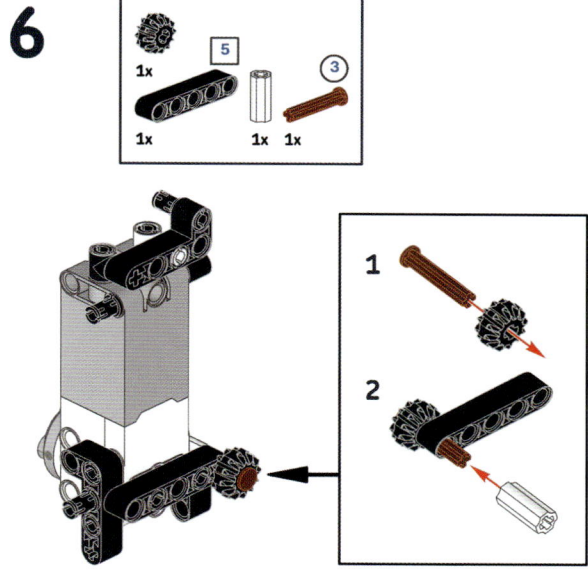

7

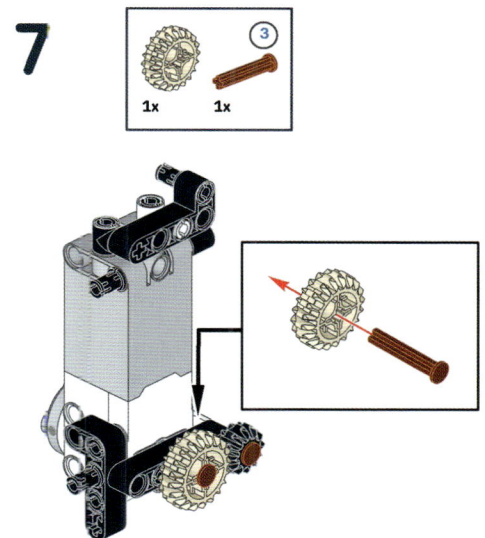

8

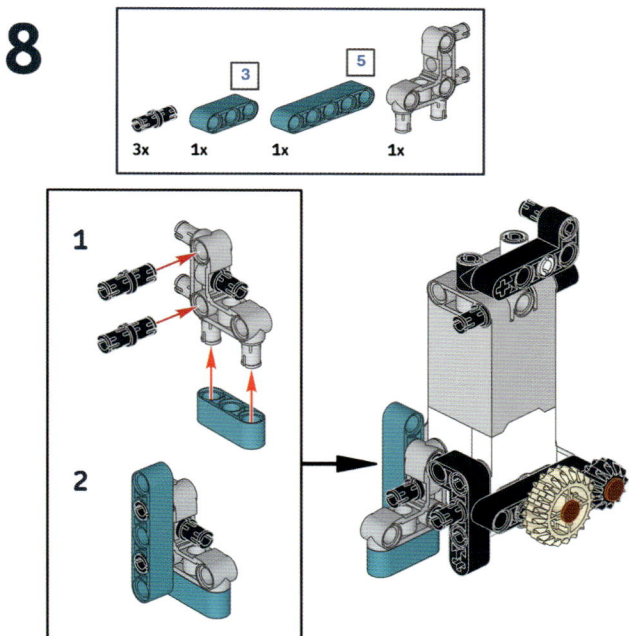

The *driving gear* (the one on the motor shaft) has 20 teeth, and the *driven gear* (on the axle the wheel will be mounted on later) has 12 teeth. The gear ratio is 12:20 = 3:5. That means that the wheels make five turns for every three turns of the motor shaft. This gearing increases the speed of the wheel with respect to the motor speed and reverses the direction of rotation.

left leg subassembly

9

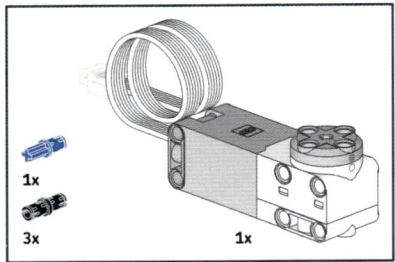

1x

3x   1x

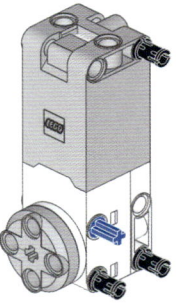

10

1x   1x   1x

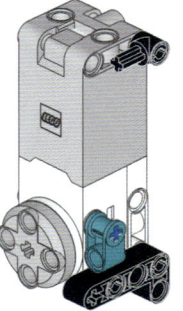

11

3x   1x   1x   1x

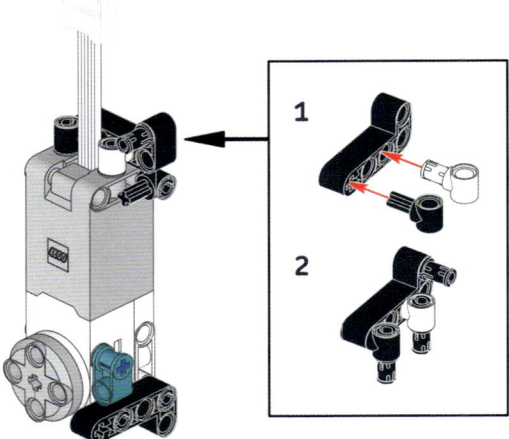

1

2

12

5x

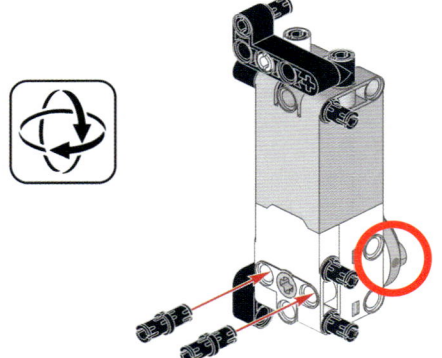

13

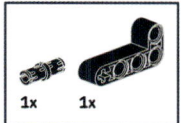

1x 1x

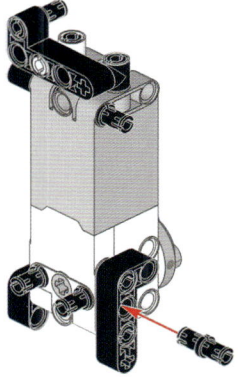

14

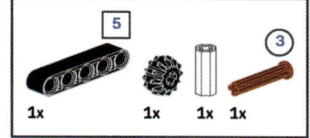

5

3

1x 1x 1x 1x

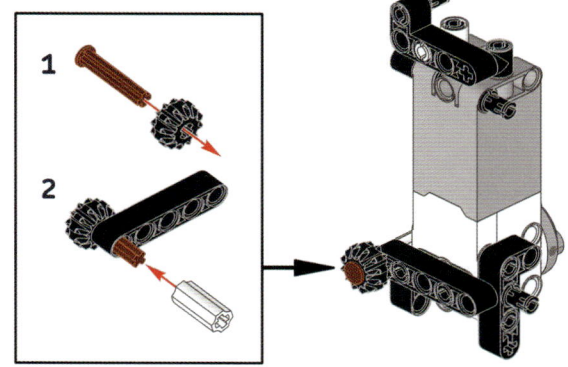

1

2

15

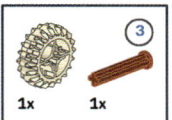

3

1x 1x

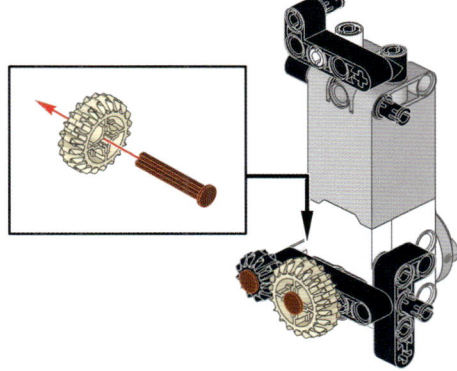

16

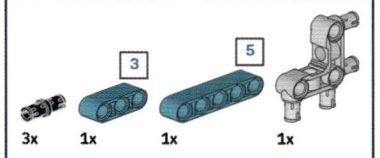

3   5

3x 1x 1x 1x

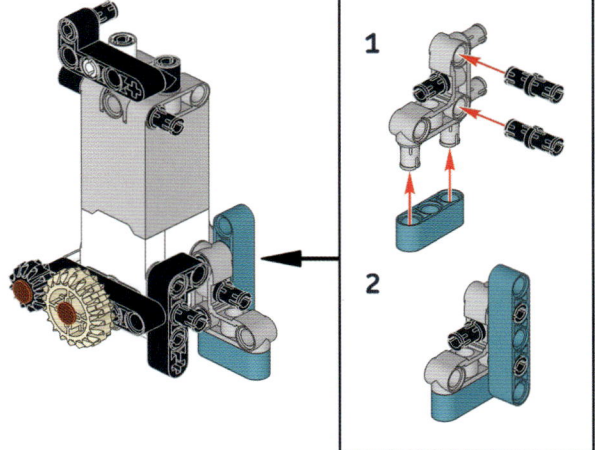

1

2

17

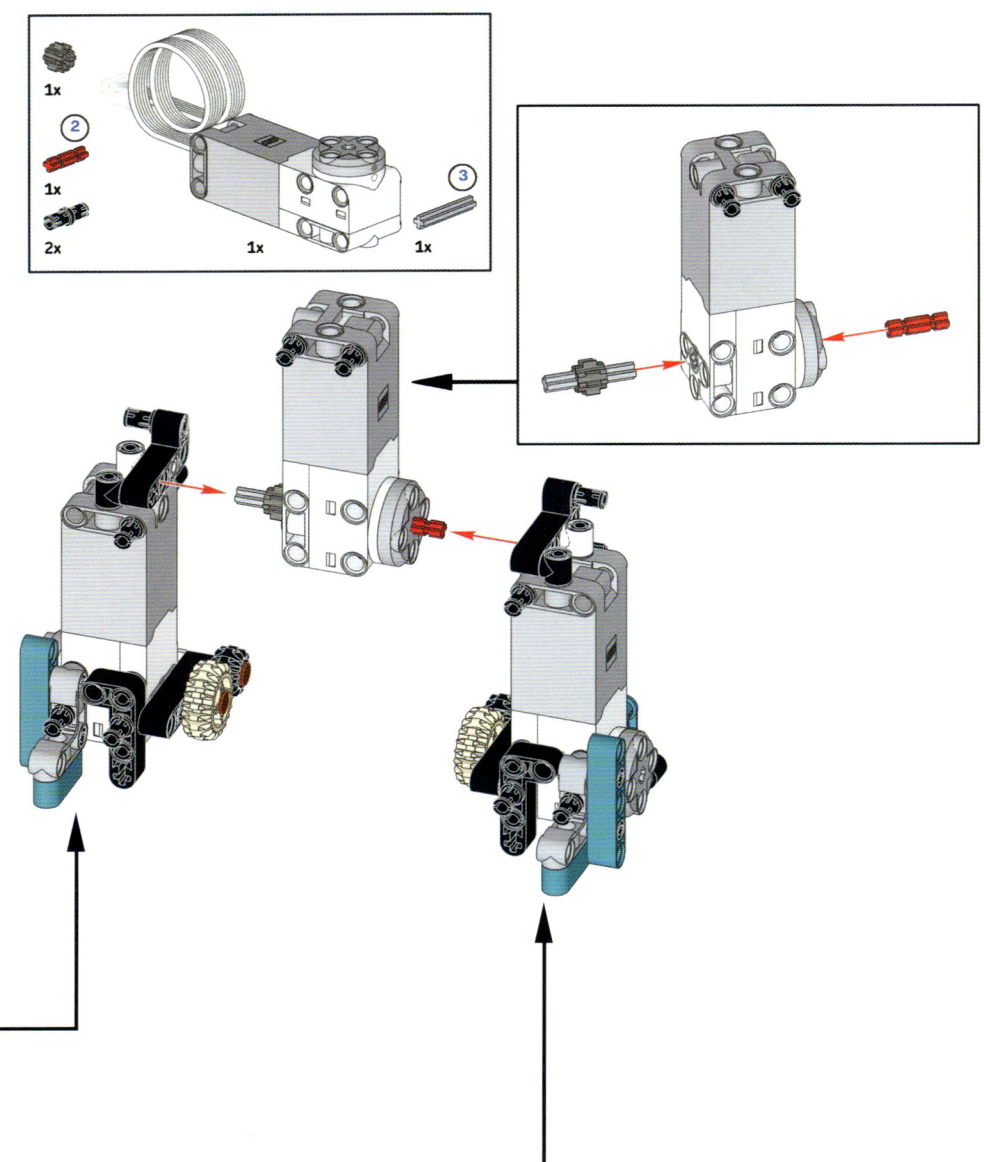

18

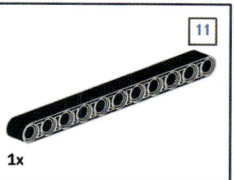

1x

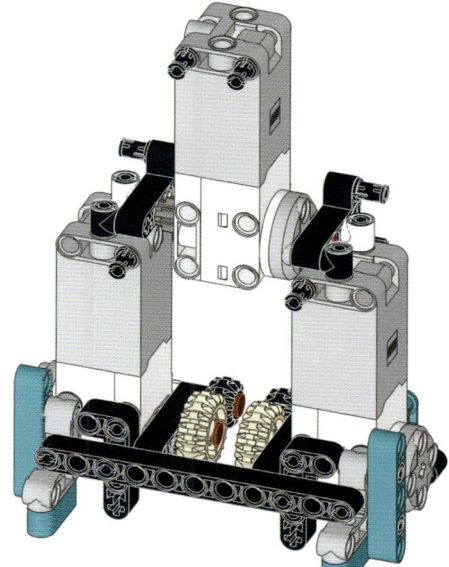

19

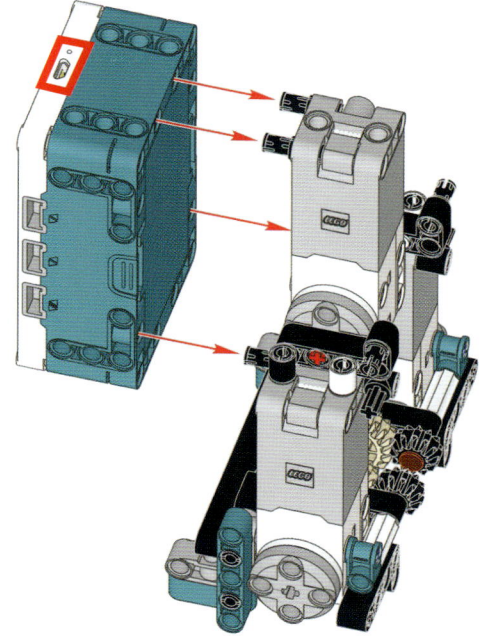

20

1x

E

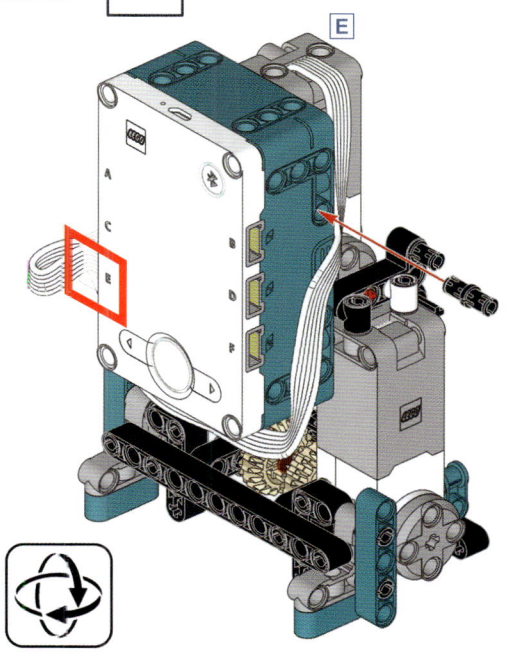

21

1x 1x

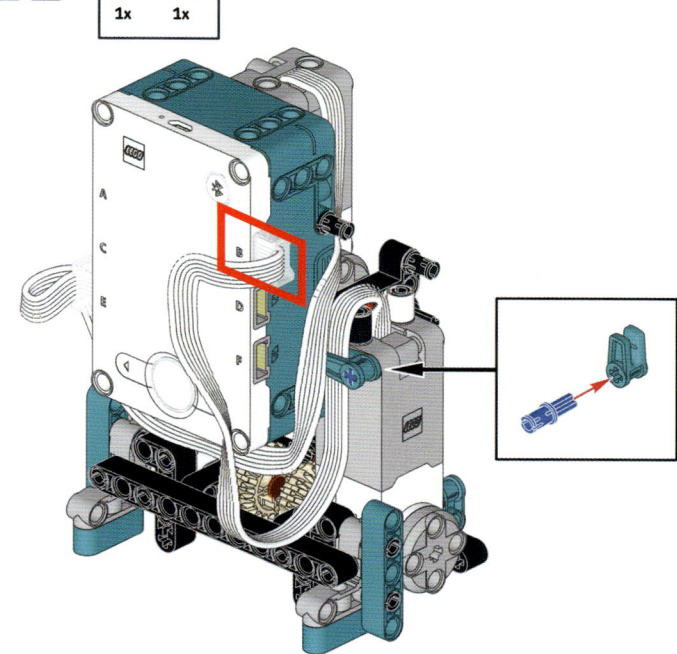

right foot subassembly

22

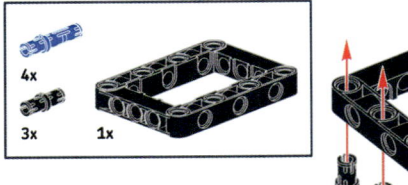 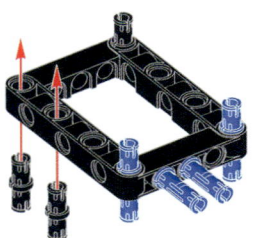

23

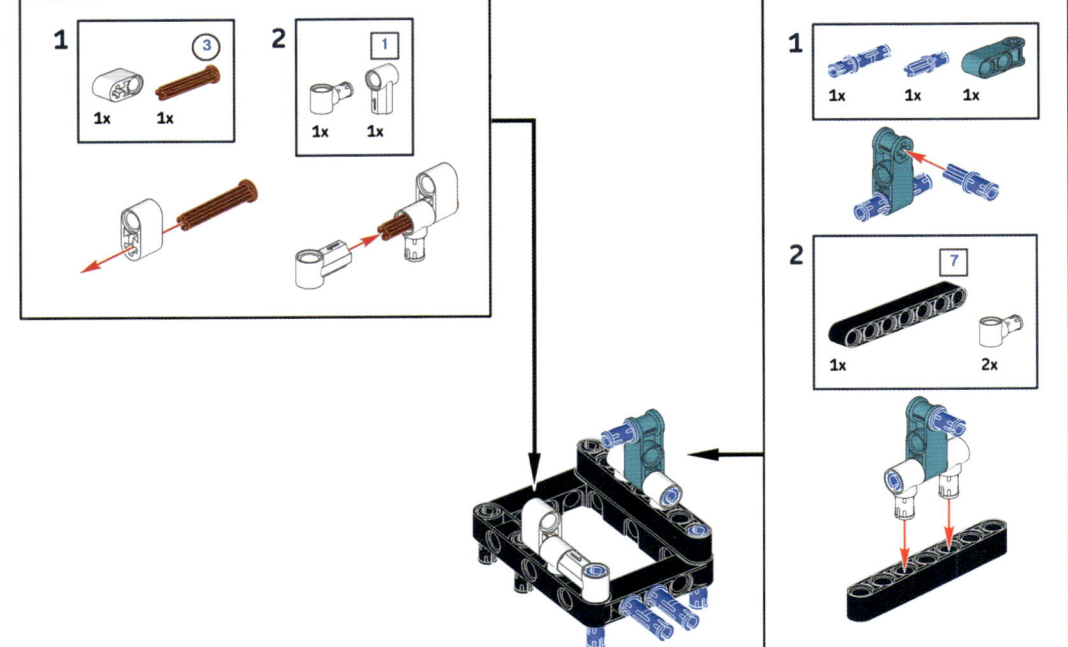

24

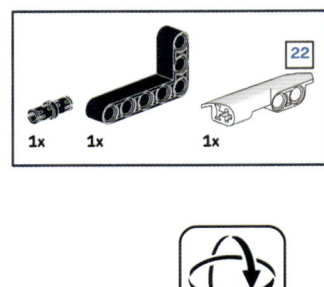  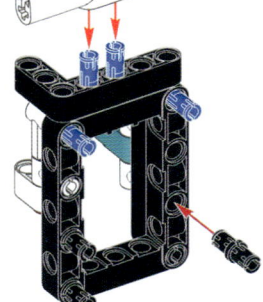

25

1x

1

2x

4x 1x

9

2

2x 1x

3

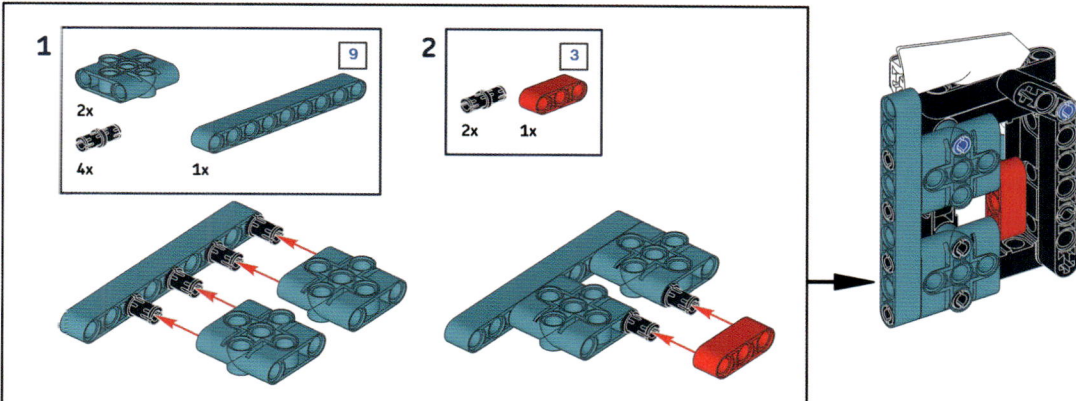

26

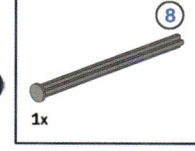

1x

8

Insert the blue pin of the ankle joiner in the marked hole of the shaft. The motor shaft works as a crank. Its rotation is transformed into a back-and-forth motion of the foot. In one turn of the motor, first the ankle is bent so that the weight of the robot shifts onto the foot, lifting the other foot from the ground. Then the foot slides backward, moving the whole robot forward. Finally, the ankle is bent to put the other foot on the ground again. The same happens for the other foot (step 31), but offset by half a turn.

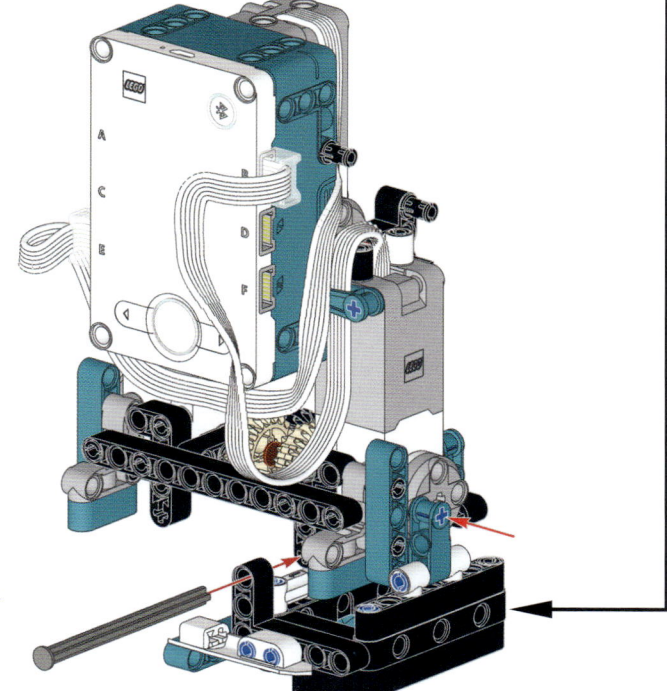

left foot subassembly

27

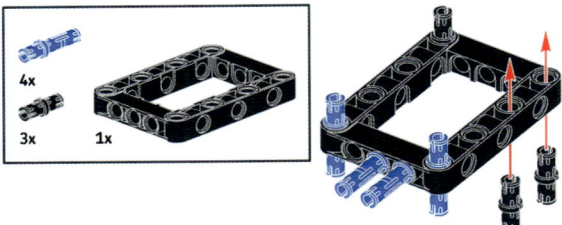

28

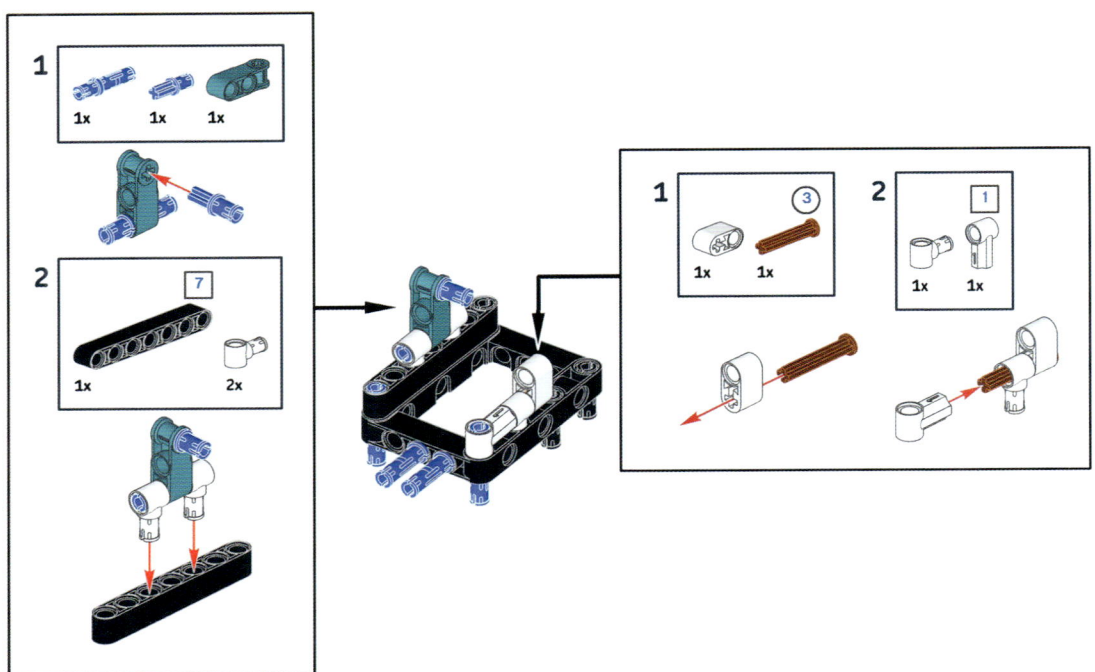

29

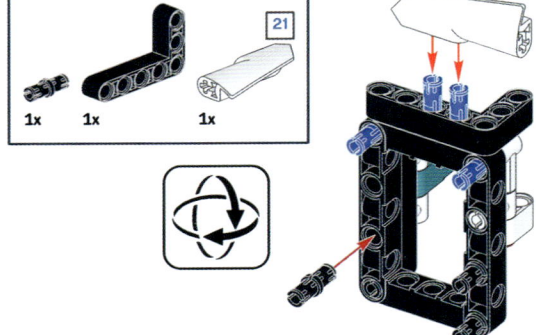

30

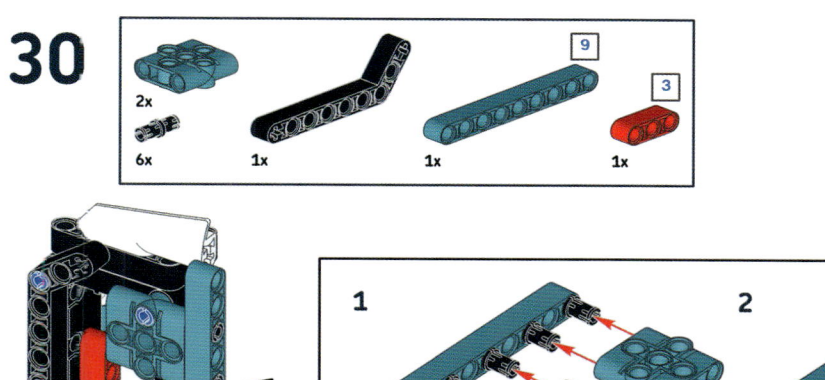

31

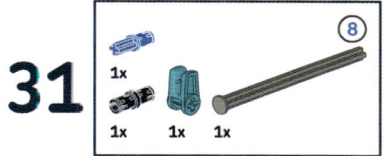

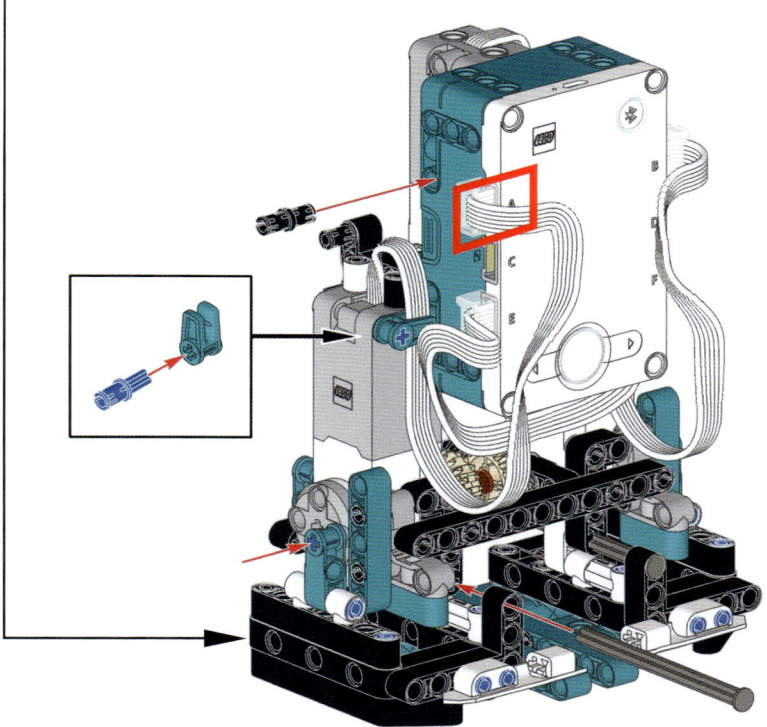

Insert the blue pin of the ankle joiner
in the marked hole of the shaft.

32

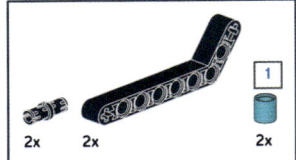

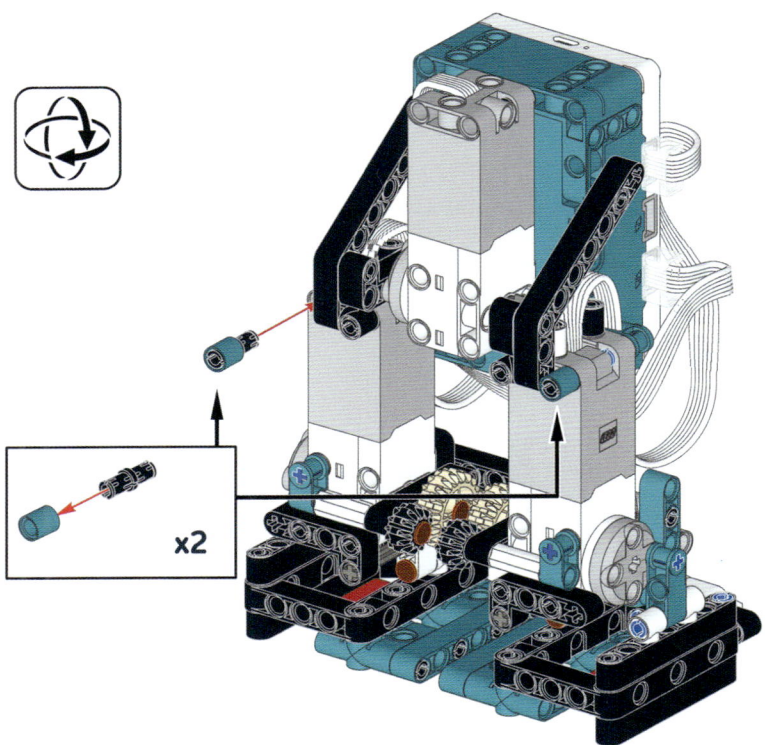

x2

33

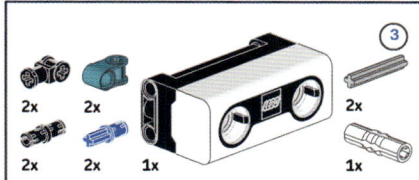

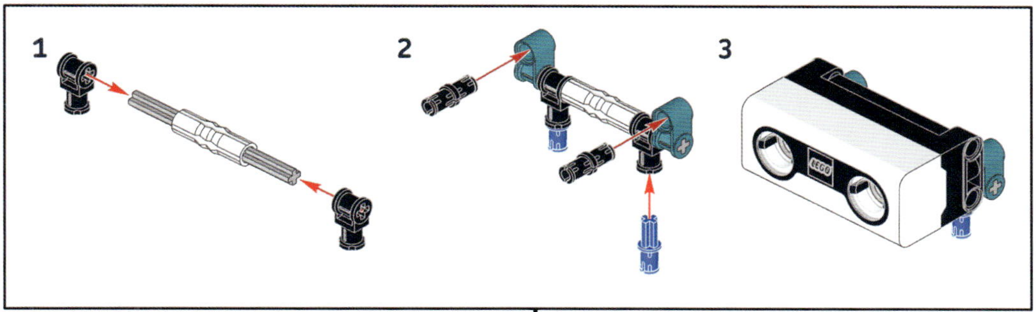

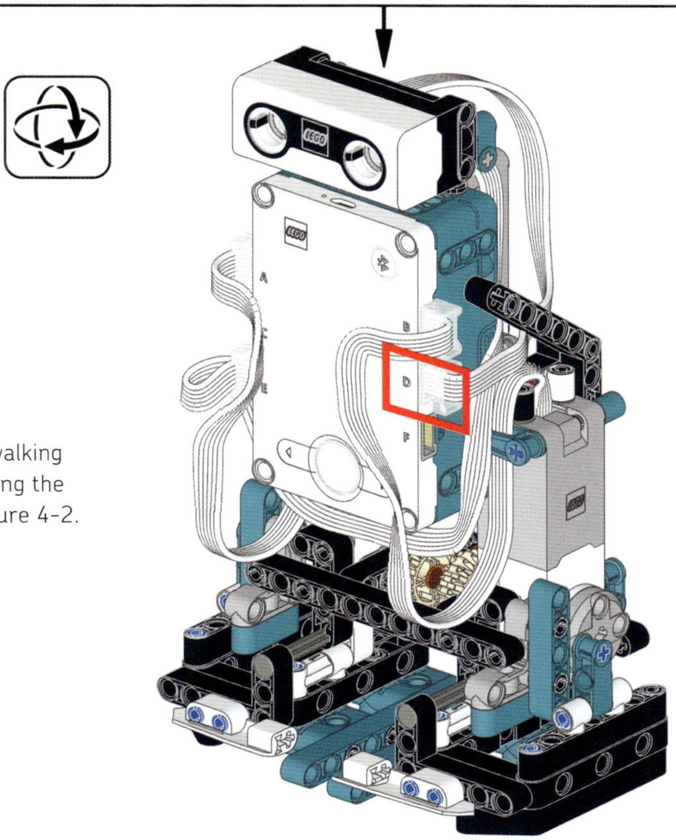

1

2

3

You can test out this walking humanoid robot by using the program shown in Figure 4-2.

programming the walking core robot

At this point, the robot has everything it needs physically to walk and avoid obstacles, though if you want, you can customize it to make it look more interesting—for example, by adding arms. The program to make the robot walk is shown in Figure 4-2. The robot will walk straight by shifting its weight from one foot to the other and stepping forward smoothly. When it sees an obstacle with the Distance Sensor, the robot will turn to the left by three steps (each step turns the robot about 18 degrees), and then it will begin walking straight again.

Create a new program and rename it *walking_robot*. Then reproduce the walking core program shown in Figure 4-2. Here are a few tips to get you started:

* Most of the magenta blocks are in the Movement palette. Blocks in this palette let you coordinate the motions of two motors. Usually they're used to operate the wheels of a car, but here we'll use the blocks to make our robot walk.

* The magenta set movement acceleration block in stack A is in the More Movement extension palette. See the Adding an Extension Palette box to learn how to access this block.

* The violet light up distance sensor blocks in stack A are in the Light palette. These blocks control the lights around the "eyes" of the Distance Sensor.

* The red walk forward and turn left by 3 steps blocks in stack A are custom blocks. We'll discuss how these blocks work and how to create them in the next section.

* The orange repeat loop blocks in stacks C and E are in the Control palette. A repeat loop block repeats the block stack inside it a certain number of times.

ADDING AN EXTENSION PALETTE

To use the set movement acceleration block, you need to add an extension palette. Use Figure 4-3 as a reference. It shows the available extensions and how to add the More Movement extension.

1. Click the **Show Block Extensions** button at the bottom-left of the programming environment, beneath the palette names. A box should pop up with the different extension palettes.

2. Click the **Add** button beneath More Movement.

3. The button should be replaced by the text *Extension is added*.

4. Click the X at the top right of the Extensions box to close it.

5. The More Movement palette should now appear as the last category in the block palettes.

Now you can drag the set movement acceleration block into your program. Later, you'll need other programming blocks from this extension palette.

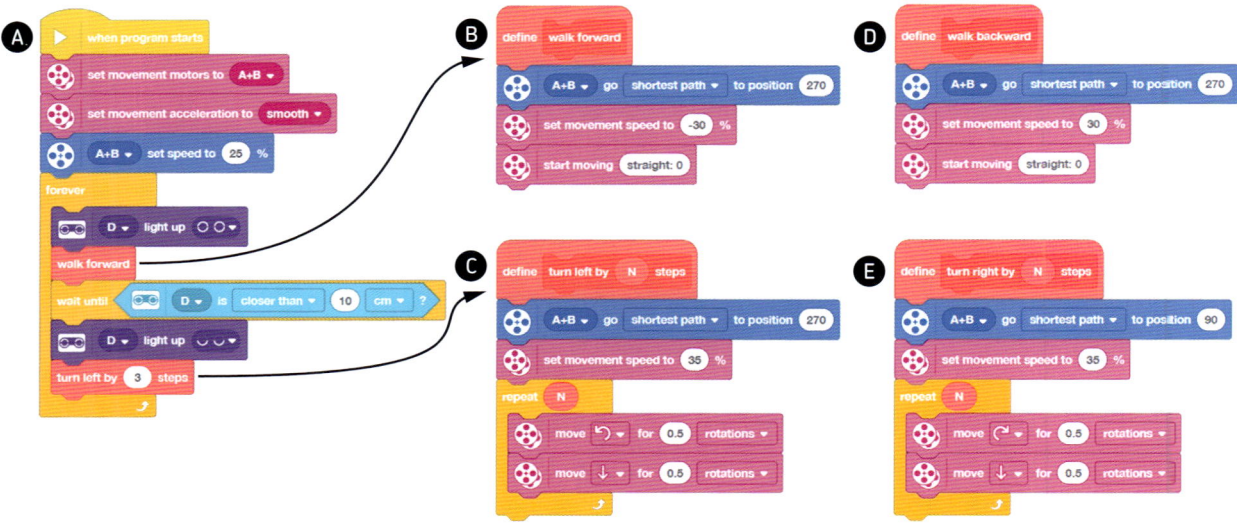

Figure 4-2: This program allows the walking core of SARKIAP-1 the Transformer to walk, detect obstacles, and turn.

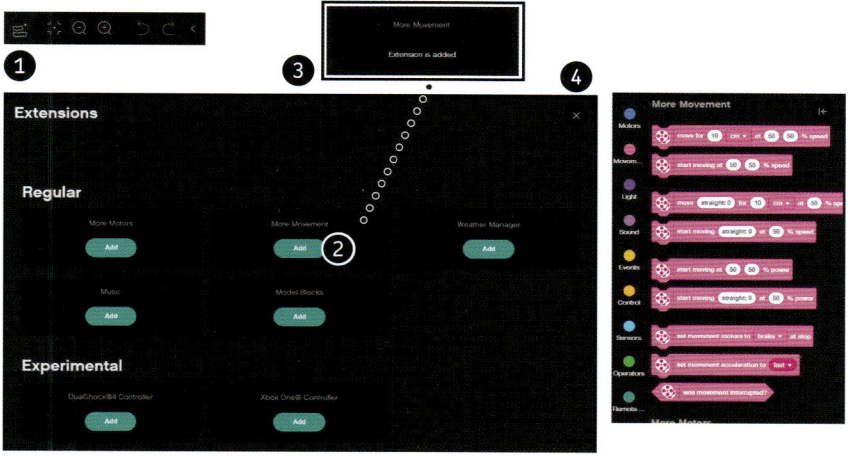

Figure 4-3: Follow this procedure to add the More Movement extension palette so you can add blocks from this palette to your program.

custom blocks

You can use a *custom block* to combine multiple Scratch blocks that define a sequence of commands into a single block. This way, you can reuse the same sequence of blocks repeatedly in your program just by placing the one custom block, instead of having to place the entire stack of many blocks each time you need it.

A custom block has two parts: the custom block itself and the custom block *definition*. The definition is a separate block stack containing all the blocks you want to group together into the custom block. In our walking core program in Figure 4-2, for example, `walk forward` in stack A is a custom block, and stack B is the definition for that custom block. Notice that stack B begins with a hat block called `define walk forward`. Every time you use the `walk forward` block in your program, the blocks under `define walk forward` in stack B will run.

Custom blocks are a great way to organize your code, keeping it clean and readable. When custom blocks have clear names, such as `walk forward`, anyone looking at your program can immediately see what the code is supposed to do. By contrast, if stack A of your program included the three blocks from stack B rather than the `walk forward` block, people (including your future-self) might be confused about what those blocks are for.

Let's find out how to create custom blocks!

creating a custom block

We'll go step-by-step through the process of creating the `walk forward` custom block. Use Figure 4-4 as a reference.

1. In the block palettes, select the **My Blocks** category, which should currently be empty, and click the **Make a Block** button. The Make a Block box should appear.

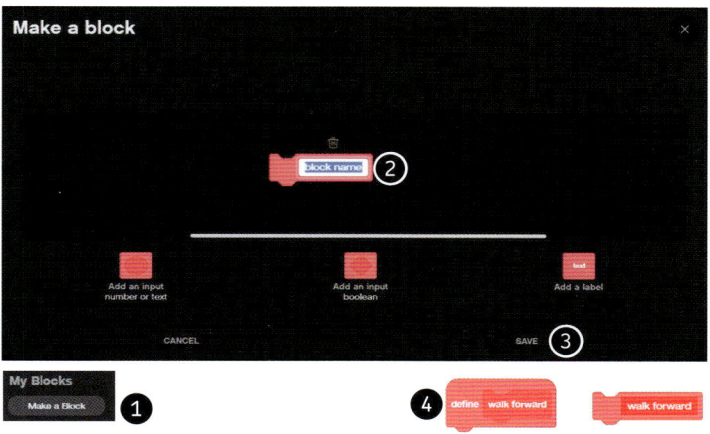

Figure 4-4: How to create a custom block

2. The block label should be highlighted. Type in the new name for the block, **walk forward**.

3. Click the **Save** button.

4. You should now see the `define walk forward` hat block in the programming area. You can drag it where you want and attach blocks to build up the custom block definition. Refer to Figure 4-2 to see what blocks you should add.

5. The `walk forward` custom block should now be available in the My Blocks palette. You can drag it into stack A of your program.

creating a custom block with inputs

The `walk forward` custom block will execute the same way every time it's called. However, you can also create custom blocks with *inputs*. An input is a piece of information that gets passed to the custom block definition. By specifying different inputs, you can change some aspect of how the custom block is executed. For example, our walking robot program uses a `turn left by N steps` custom block. In this case, N is an input that lets you specify how many steps the robot should pivot to the left. You can type any positive number into the block to replace the N, and that's the number of steps to the left the robot will turn.

We'll go through the process of creating the `turn left by N steps` block so you can see how to create a custom block with an input. Use Figure 4-5 as a reference.

1. Click the **Make a Block** button in the My Blocks palette. The Make a Block box should appear.

2. The block name should be highlighted.

3. Type **turn left by** into the text label.

4. Click the **Add an Input Number or Text** button to add an input. Notice the round sides of the input label, indicating that the input will have text or numeric data, not logic data (true/false).

5. The new input label should be highlighted. Notice the trash bin icon above it. It allows you to remove the selected item from the custom block if you change your mind or make a mistake.

6. Type in **N** as the input name.

7. Click the **Add a Label** button to continue adding text to the block.

8. The label text should be highlighted.

9. Type in **steps** and press **ENTER**.

10. The block definition is complete. Notice how labels can be used to form sentences that make sense. Having a block named `turn left by N` or `turn left by steps N` would have sounded weird.

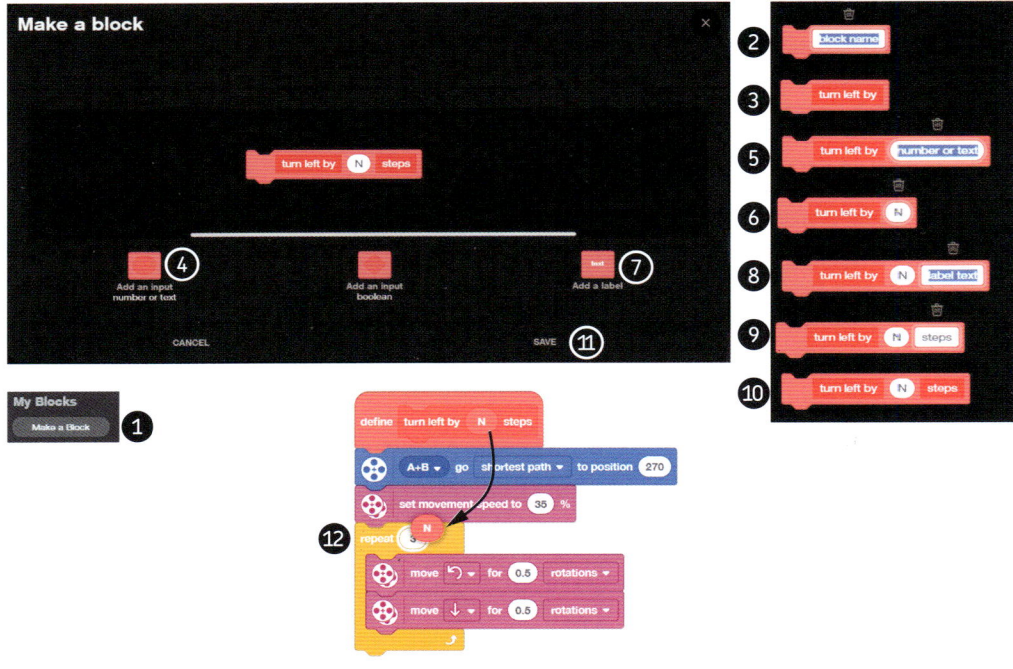

Figure 4-5: How to create a custom block with an input

11. Click the **Save** button to create the custom block.

12. Start building the custom block definition for `turn left by N steps`, shown as stack C in Figure 4-2. When you get to the `repeat loop` block, drag the `N` block from the `define` hat block into the round slot on the `repeat loop` block.

The `N` in the custom block definition will take on whatever input value you specify when you place the `turn left by N steps` block in your program. For example, when we place the block in stack A of our program, we specify the input value 3. That way, whenever the `turn left by 3 steps` block executes, `N` in the block definition will take on the value 3, and the `repeat loop` in the definition will repeat three times.

If we wanted, we could use the `turn left by N steps` block a second time somewhere else in our program and specify a different input value—for example, 5. In that case, the `repeat loop` would execute five times. That's the beauty of custom blocks with inputs: you can reuse them as many times as you want with different inputs.

Now that you understand how to create custom blocks, try creating two more: `walk backward` and `turn right by N steps`. The definitions for these blocks are shown as stacks D and E in Figure 4-2. Although we don't actually use the blocks in our walking core program, you'll need them to complete Exercise 4-1 on page 52. We'll also use them in our final Transformer program.

NOTE You can edit a custom block at any time by right-clicking (long-pressing on touchscreen devices) the block definition, the custom block in the programming area, or the custom block in the palette. The Make a Block box should appear, where you can make your modifications. Notice that you can also add a logic input by clicking the **Add an Input Boolean** button in the middle (*Boolean* is another name for a true/false logic value). In that case, the input will have pointy sides instead of round ones.

understanding the walking core program

Let's talk through the walking core program to see how it works. The program starts by executing block stack A. First the `set movement motors` block defines which two motors the Movement blocks in the program will control: in this case, motors A and B. As mentioned, the magenta Movement blocks allow two motors to run in sync with each other. They're essential for making our robot walk properly when it's in humanoid form or drive precisely when in car form.

Next, we use the `set movement acceleration` block to control how the motors will handle changes of speed. We choose `smooth` so the motors' speed will change slowly. If the motors started and stopped abruptly, the robot would tend to lose its balance. Then we use a `set motor speed` block to set the motor speed to 25%. This setting will apply only to the blue Motor blocks in the other block stacks. For all the magenta Movement blocks, the motor speed is set by a `set movement speed` block.

After all the motor setup, the program starts a `forever loop`. In the loop, the `light up distance sensor` block turns the Distance Sensor lights on. Then the custom `walk forward` block starts the robot walking forward. As we've discussed, the custom block is defined in stack B. First motors A and B move to position 270 to move one foot fully forward and the other foot fully backward, so that they are at opposite ends of the walking cycle. Then we use the `set movement speed` block in combination with the `start moving` block to make the robot move forward. We set the speed to -30% because the walking core robot is facing backward compared to the final version of the Transformer.

NOTE The `start moving` block uses two synchronized motors to move a robot forward, with the possibility of steering forever along a curve. As you can see in Figure 4-6, the block's drop-down menu allows you to set a steering angle. Positive numbers steer to the right, and negative numbers steer to the left, while 0 makes the robot go straight ahead. The higher the steering value, the sharper the curve. The values 100 and -100 will make the robot spin in place.

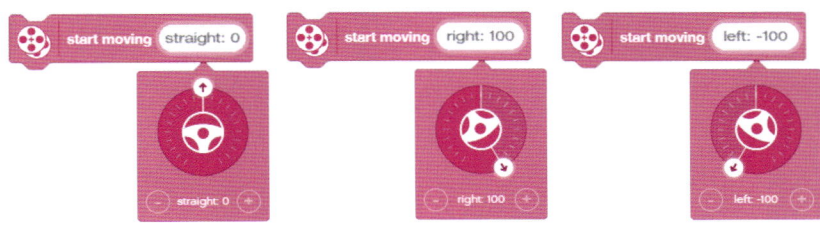

Figure 4-6: Setting the steering angle of the start moving *block*

The blocks in stack B are executed in an instant, without stopping the program flow of stack A. Since the next block in stack A is a `wait until` block, the robot continues walking forward until the `wait until` condition is met. We set that condition with the `is distance?` block, which tests the readings from the Distance Sensor. When the sensor detects an object closer than 10 cm, the program moves on to the next block, which turns off the top lights of the Distance Sensor. Then, the custom `turn left by N steps` block executes stack C.

In stack C, first the `motor go to position` block displaces the motors by half a turn, so one foot is ahead and the other behind. In the loop, the `move left for 0.5 rotations` block moves the feet in opposite directions to make the robot spin in place about 18 degrees to the left. Then the `move backward for 0.5 rotations` block makes the robot take one step forward. Thanks to the input from our custom block, the loop repeats three times.

Once stack C finishes, the `forever loop` starts over from the beginning. The robot will continue walking forward or turning left to avoid obstacles for as long as the program keeps running.

Now that you've completed the program and you understand how it works, run it. Check out how the robot walks. Try pointing the robot at some obstacles, like walls or pieces of furniture. Is the robot able to avoid them?

EXERCISE 4-1

1. Try reading through the walking core program in plain language, or pseudocode. To review how to do that, refer to the previous chapters.

2. Try changing the walking core program to make the robot go backward or turn right when it sees an obstacle, rather than turn left. Use the `walk backward` and `turn right by N steps` custom blocks defined in stacks D and E.

3. Can you make the robot patrol an area by walking straight, turning around, and going back the way it came? Can you make it walk around in a square? How many steps should you specify in the `turn left` or `turn right` custom blocks, input to make the robot turn 90 or 180 degrees?

building the transformer

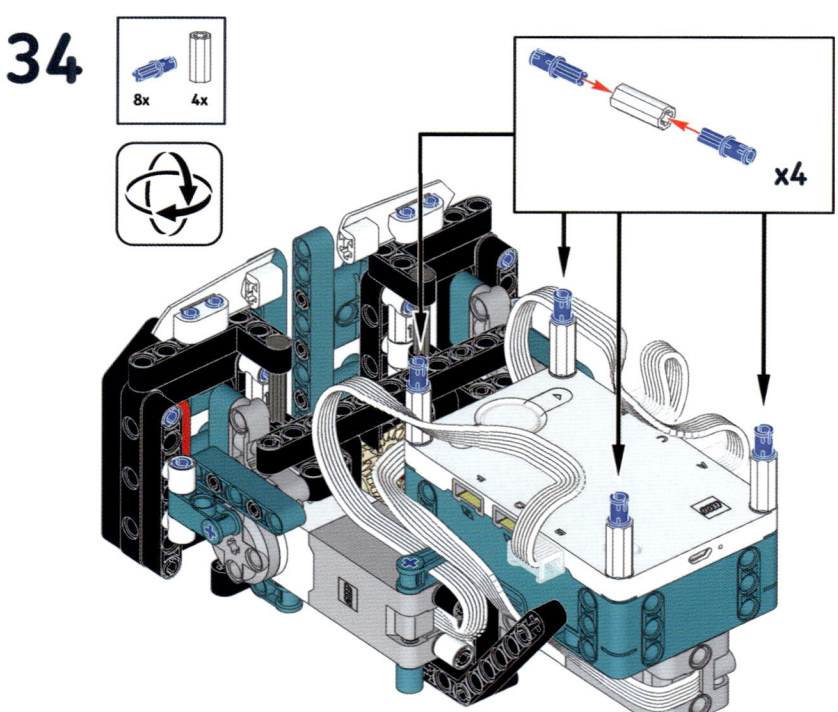

Remove the head assembly and take it apart. You'll need those parts later.

35

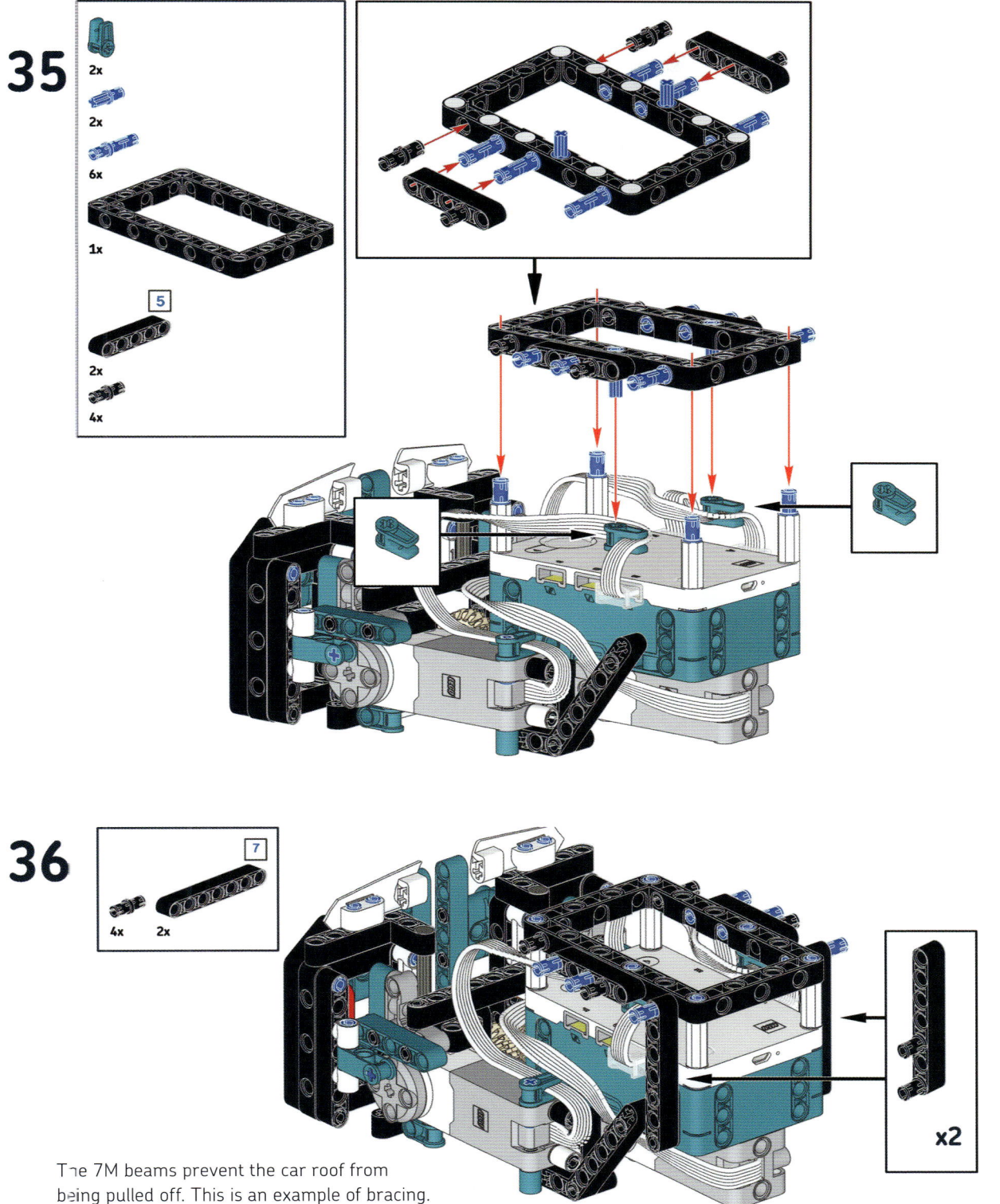

36

The 7M beams prevent the car roof from being pulled off. This is an example of bracing.

37

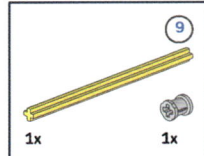

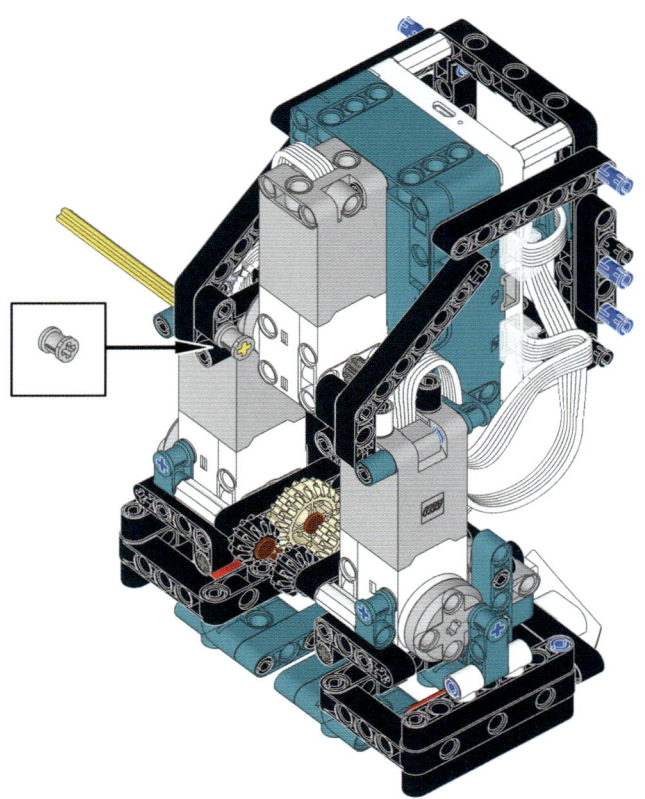

torso subassembly

38

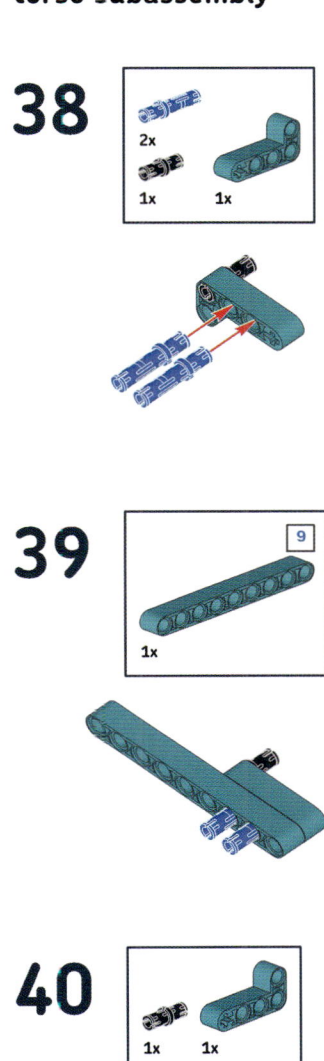

39

40

41

42

43

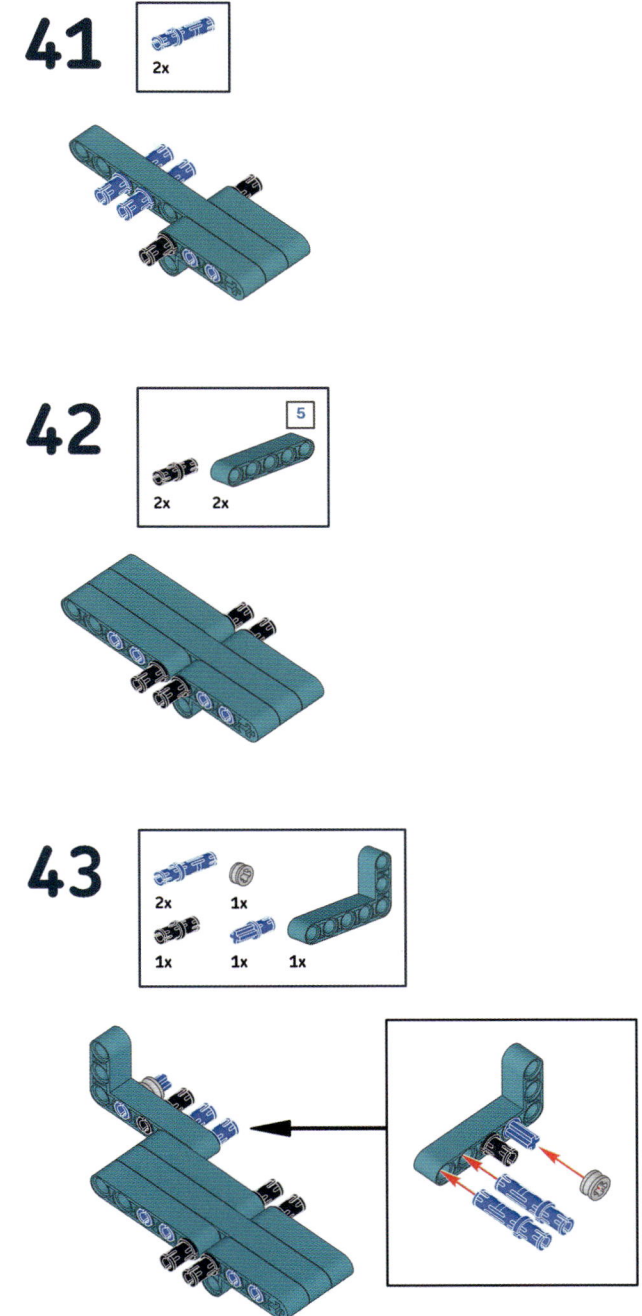

44

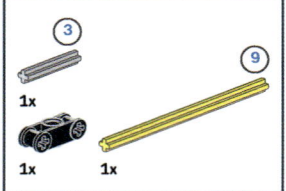

1x
1x 1x

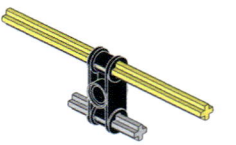

45

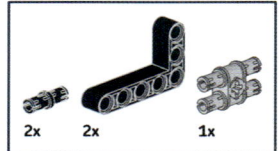

2x 2x 1x

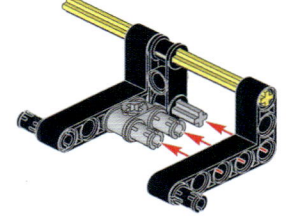

46

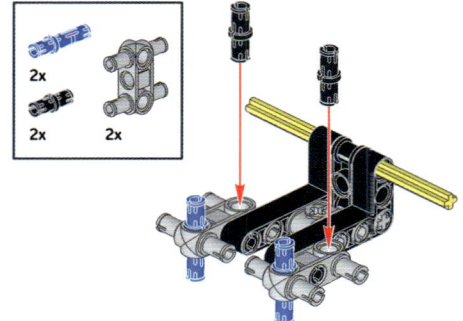

2x
2x 2x

47

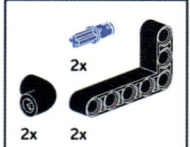

2x 2x

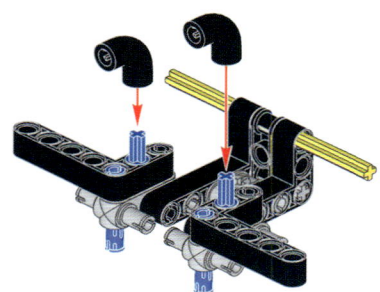

48

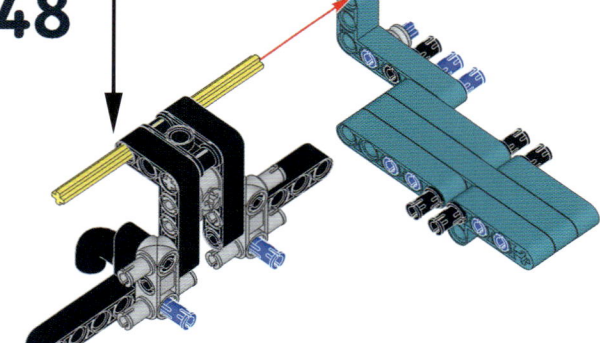

49

2x 1x
1x 1x 1x

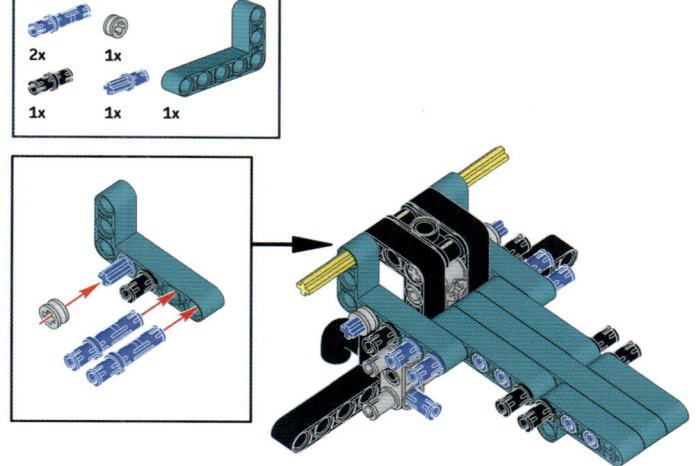

50

2x

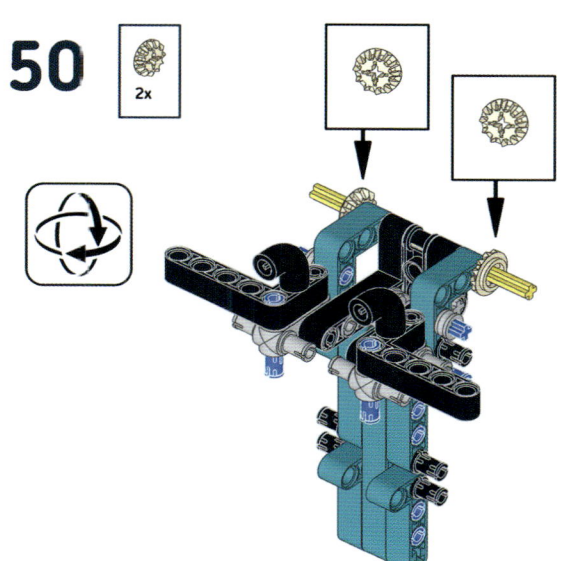

51

2x 2x ③ 2x

2x

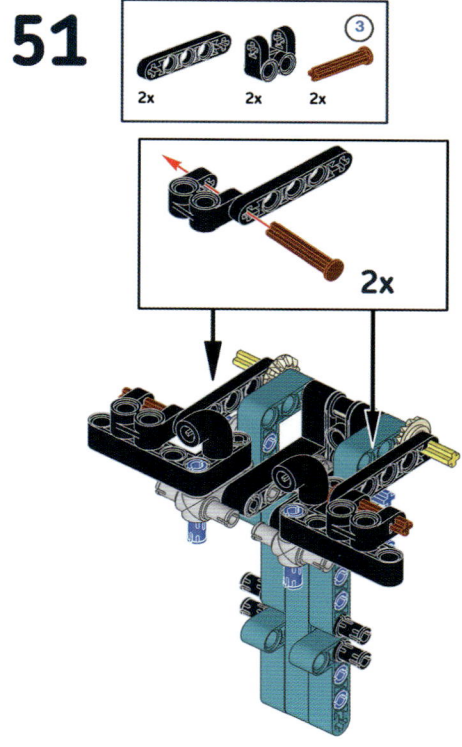

52

2x 2x

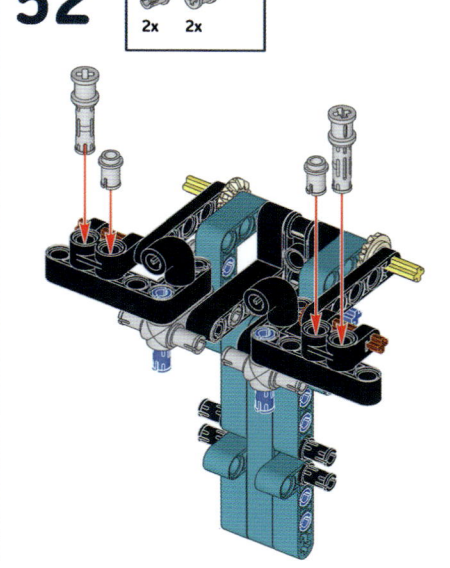

53

shoulders subassembly

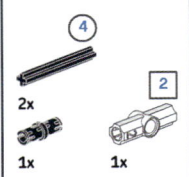

2x 1x 1x

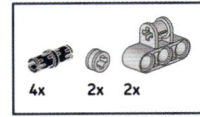

54

4x 2x 2x

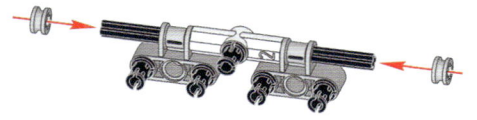

55

 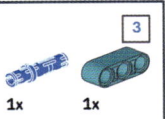

1x 1x

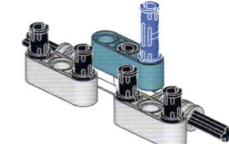

56

1x 1x

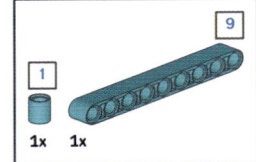

57

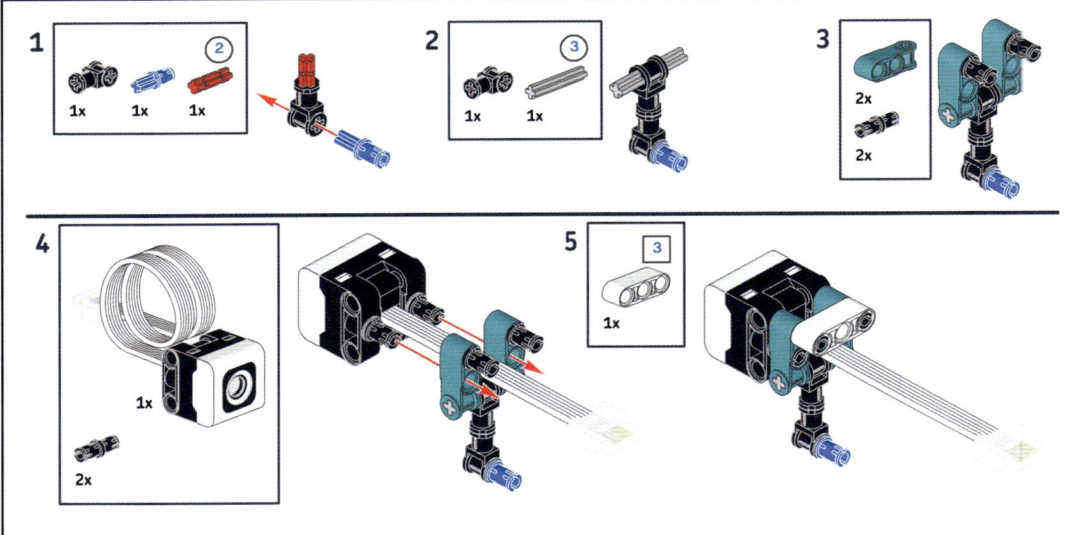

1

1x 1x 1x

2

1x 1x

3

2x 2x

4

1x

2x

5

1x

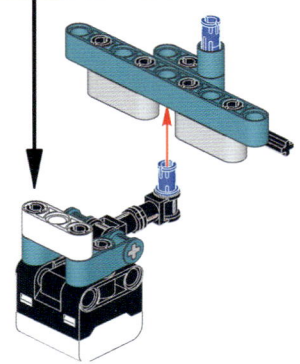

58

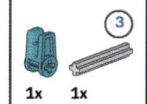

1x   1x   (3)

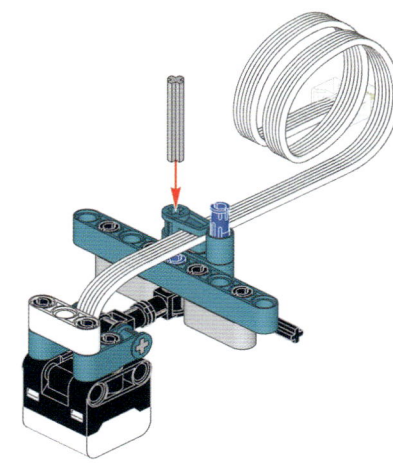

59

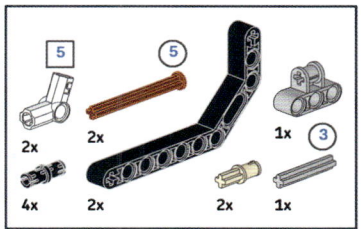

5   5   1x   (3)

2x   2x   1x

4x   2x   2x   1x

1

2

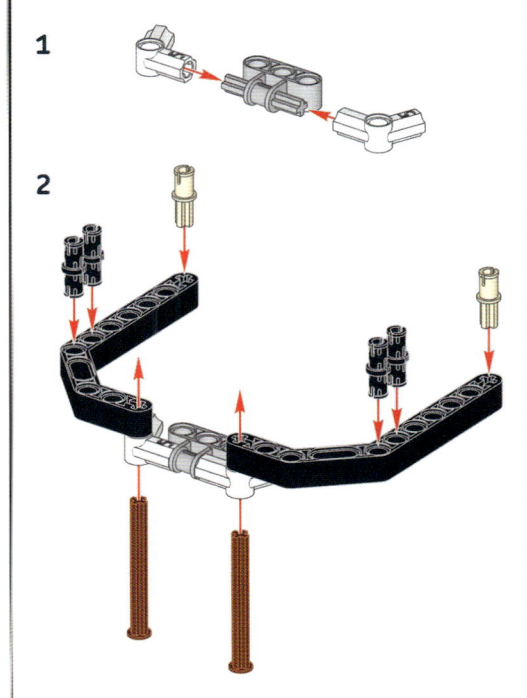

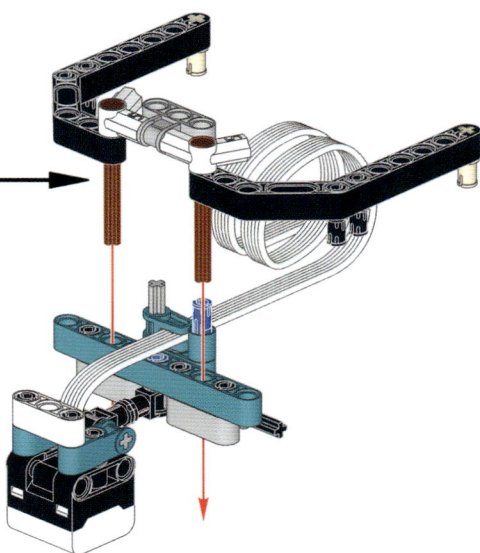

60

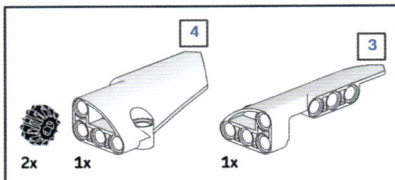

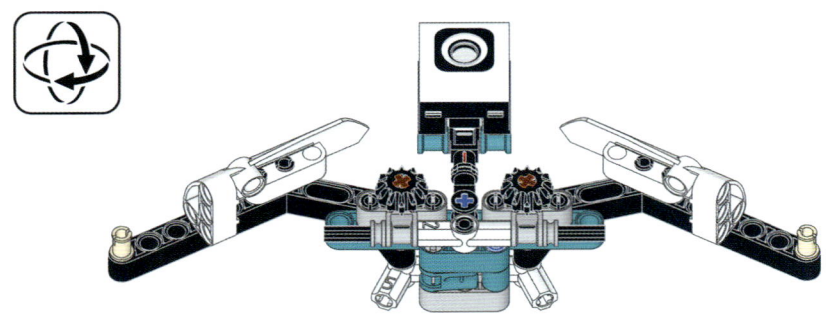

61

Align the arms correctly before engaging the gear teeth. Check out the next step.

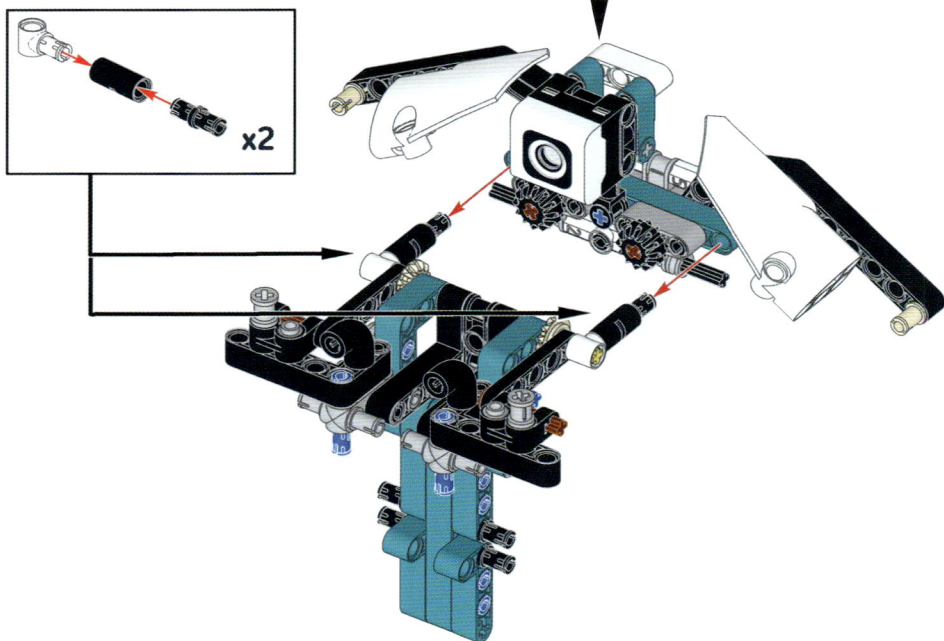

62

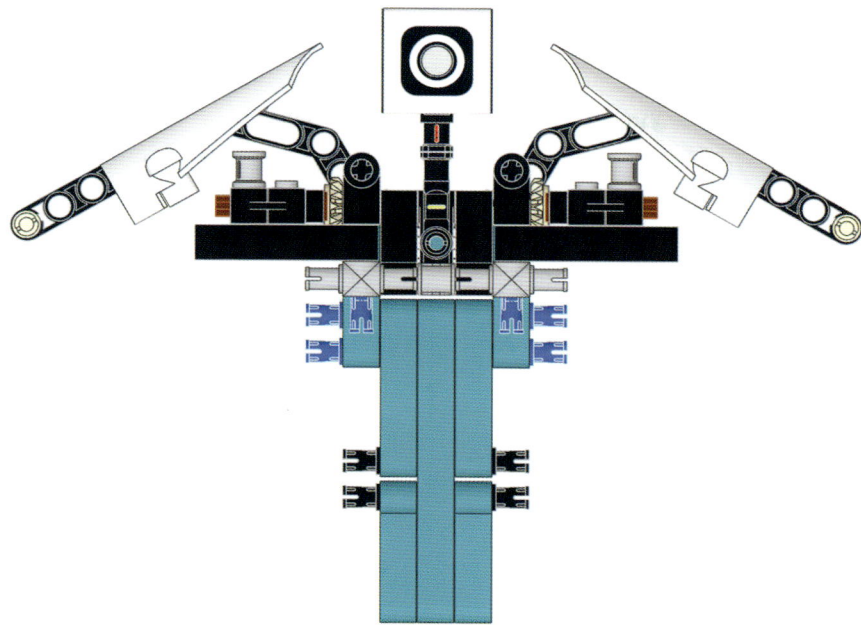

Make sure the arms are symmetrically aligned and angled as shown here when you attach them. This is how the arms will appear when the Transformer is in biped mode (chest is folded down).

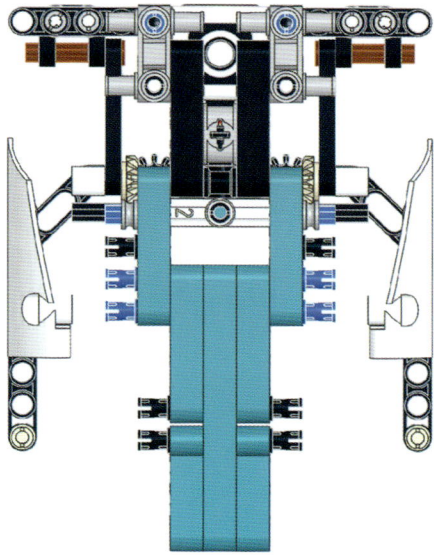

To check that you assembled the arms correctly, make sure the arms are parallel to the teal pieces of the torso when the Transformer is in car mode (chest is unfolded).

63

2x

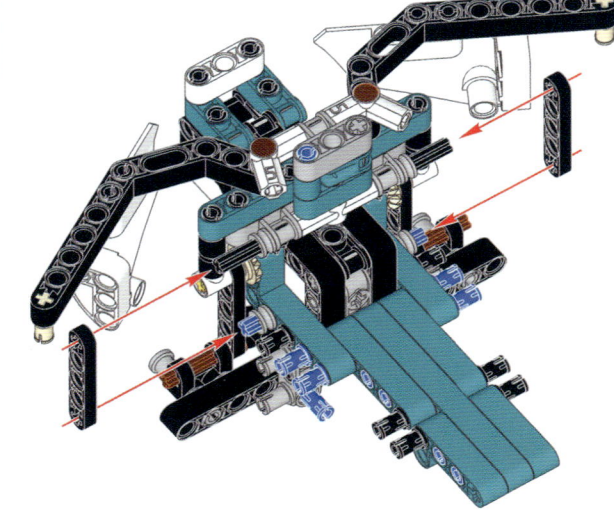

64

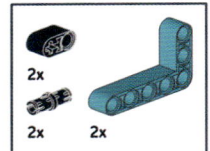

2x

2x 2x

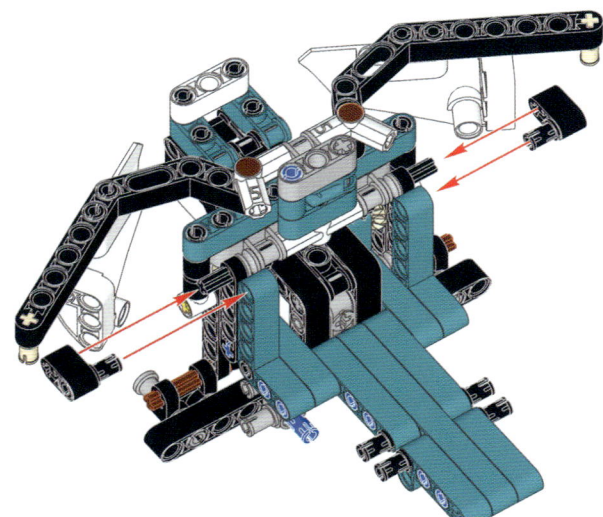

65

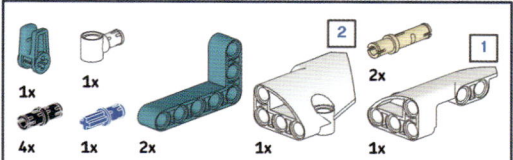

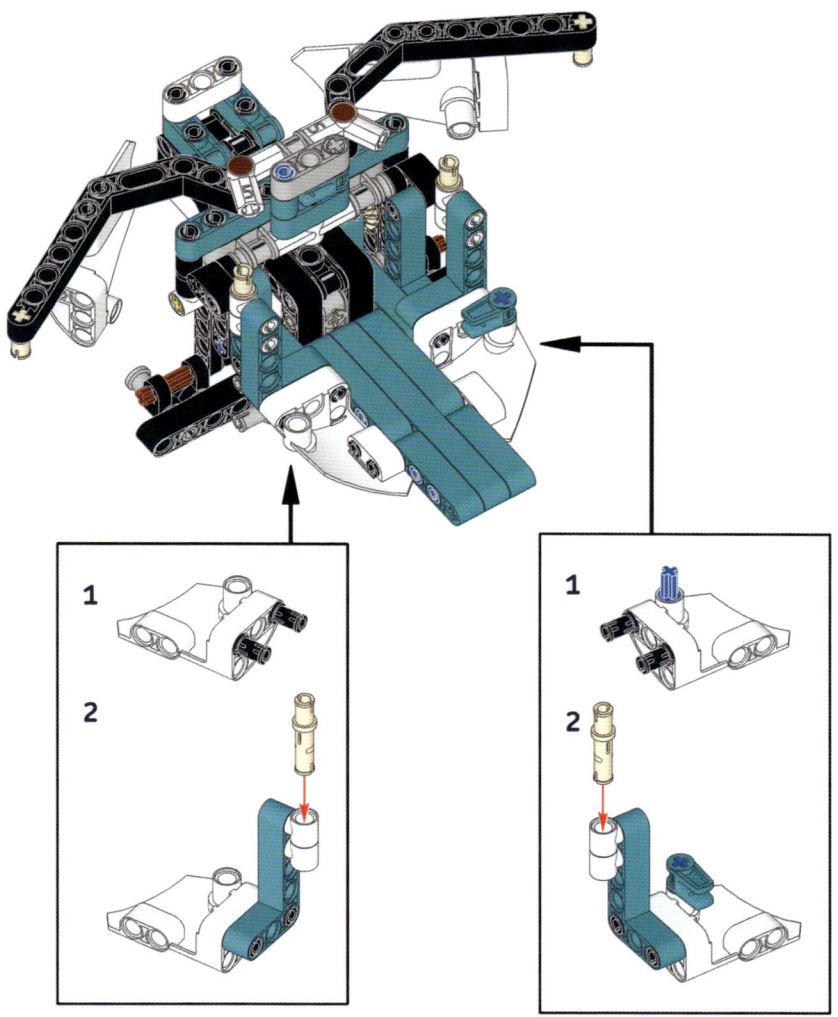

66

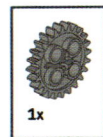

1x

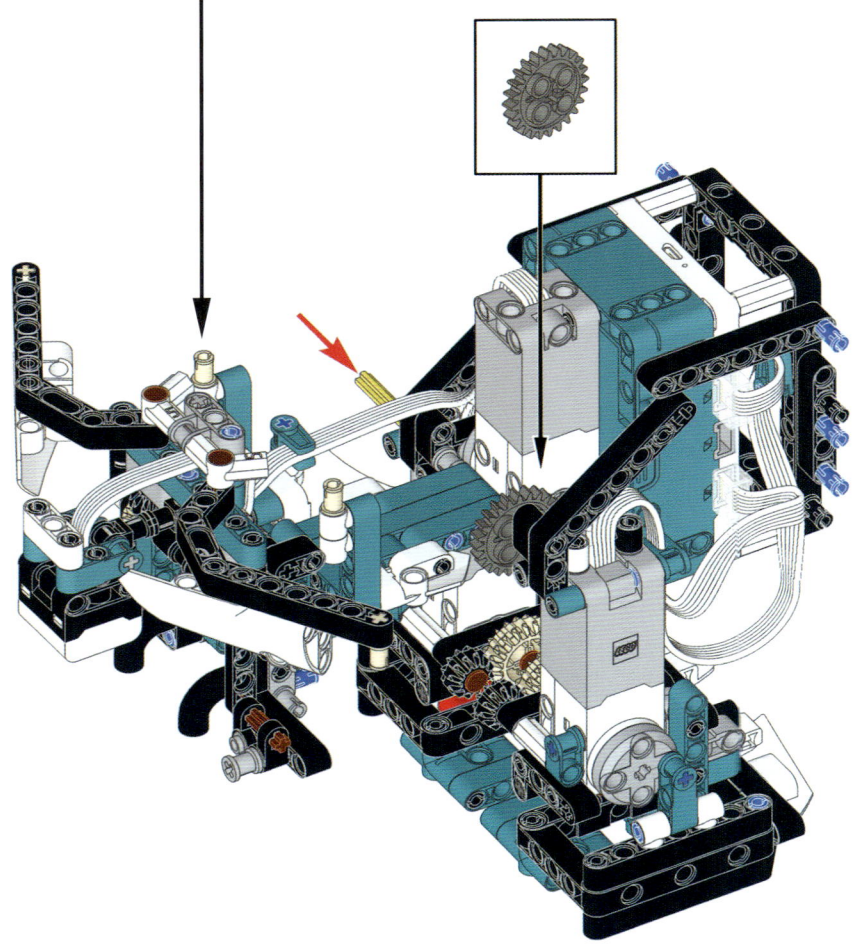

The driving gear attached to the motor shaft has 8 teeth, and the driven gear attached to the torso axle has 24 teeth. The gear ratio is 24:8 = 3:1, which means that the torso axle will spin three times slower than the motor shaft. This gearing makes the motor's torque three times stronger. *Torque* is a twisting force that, when applied to an object, makes it rotate. In this case, the gears allow the motor to lift the torso with three times less effort than if the motor were driving the torso axle directly.

67

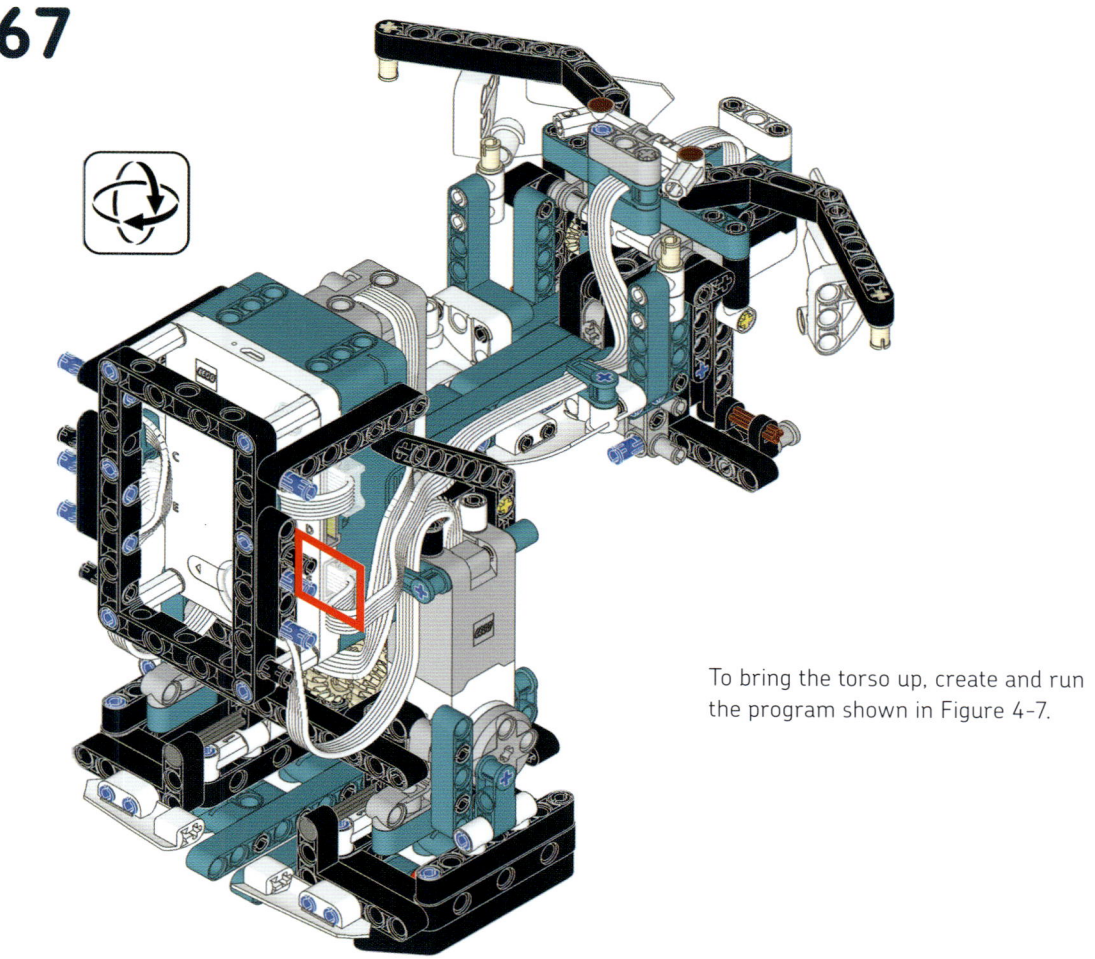

To bring the torso up, create and run the program shown in Figure 4-7.

moving the robot torso up

At step 67, you need to bring the torso up in order to keep building the robot. You could move the torso manually by applying a little force to it, but you would risk damaging the yellow axle, the gears, or the motor itself. To move the torso safely, create and run the program shown in Figure 4-7. Save the program as *calibrate_robot*, as you'll need to reuse part of it in the final program for SARKIAP-1.

First, we set the speed of the motor that moves the torso, connected to port E, to **10%**. We want to run the motor clockwise to bring the torso close to the Hub. However, when we reuse this sequence in the final Transformer program, the torso might already be close to the Hub (in car mode), in which case running the motor clockwise would stress the mechanism. Therefore, we run the motor counterclockwise for 1 second, moving the torso away from the Hub. Then we start running the motor clockwise as we want.

After waiting for 1 second to give the motor time to start running, we use a **wait until** block to let the motor keep running until its speed becomes 0. This will happen when the torso beams touch the motor body and the torso can't move anymore. At that point, we stop the motor. Then we use two **motor go to position** blocks to align the robot's feet. Finally, we use the **stop** block to end the block stack and exit the program.

Run the program, and the torso should move up. Now you can keep building from step 68.

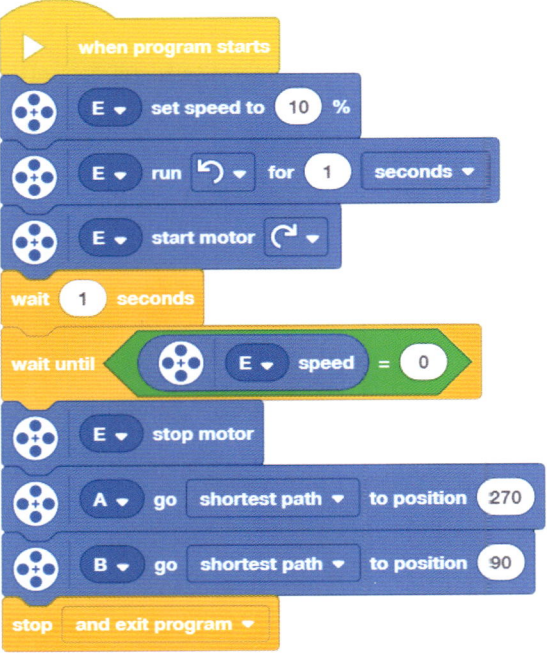

Figure 4-7: This sequence brings the robot's torso up, in the process of transforming into a car. You need to execute this program to keep building the robot.

finishing the transformer

68

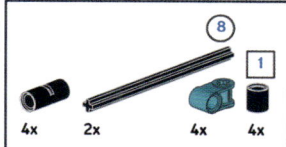

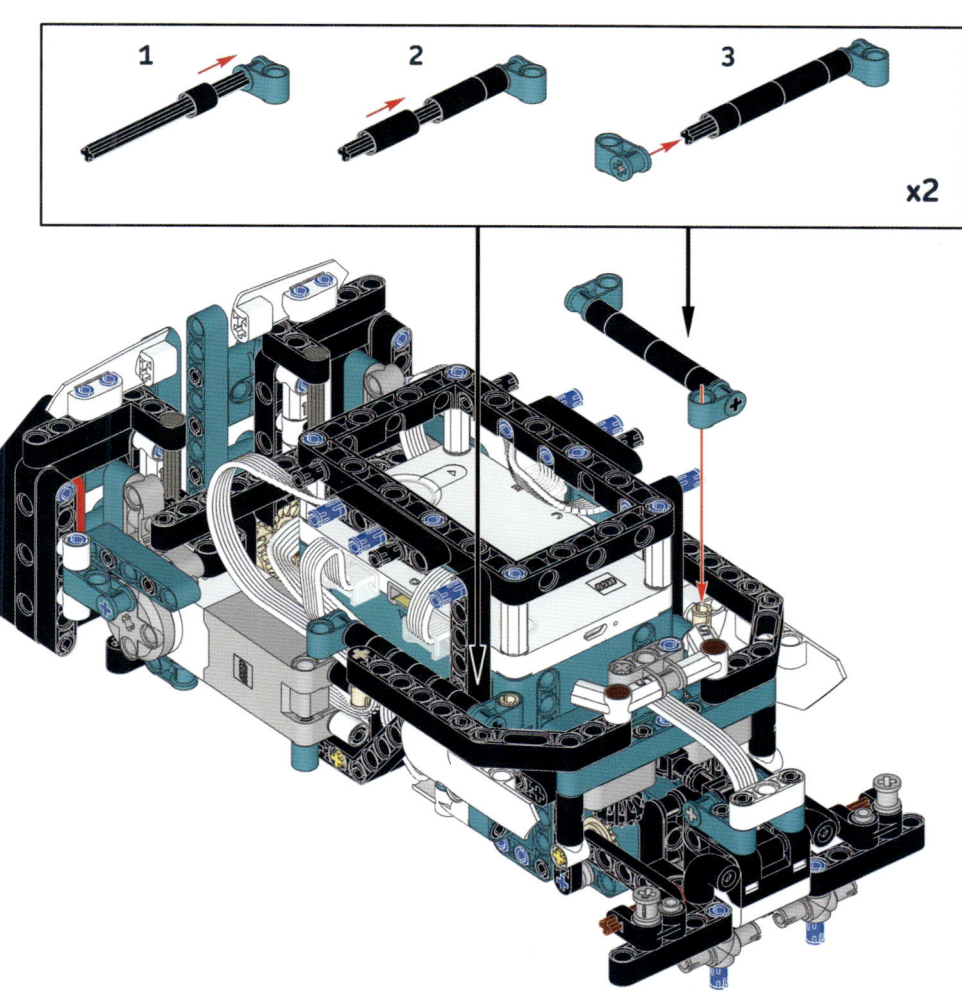

69

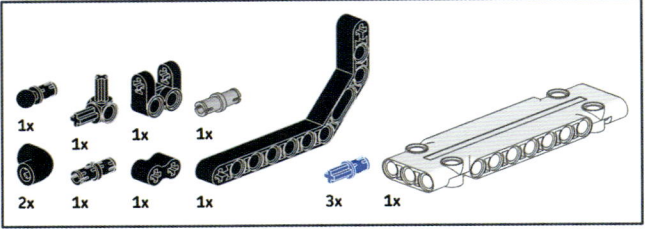

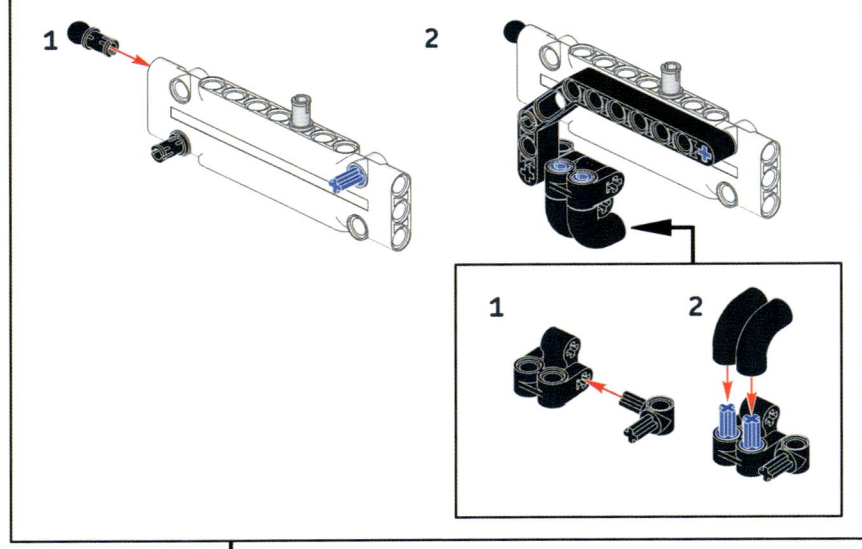

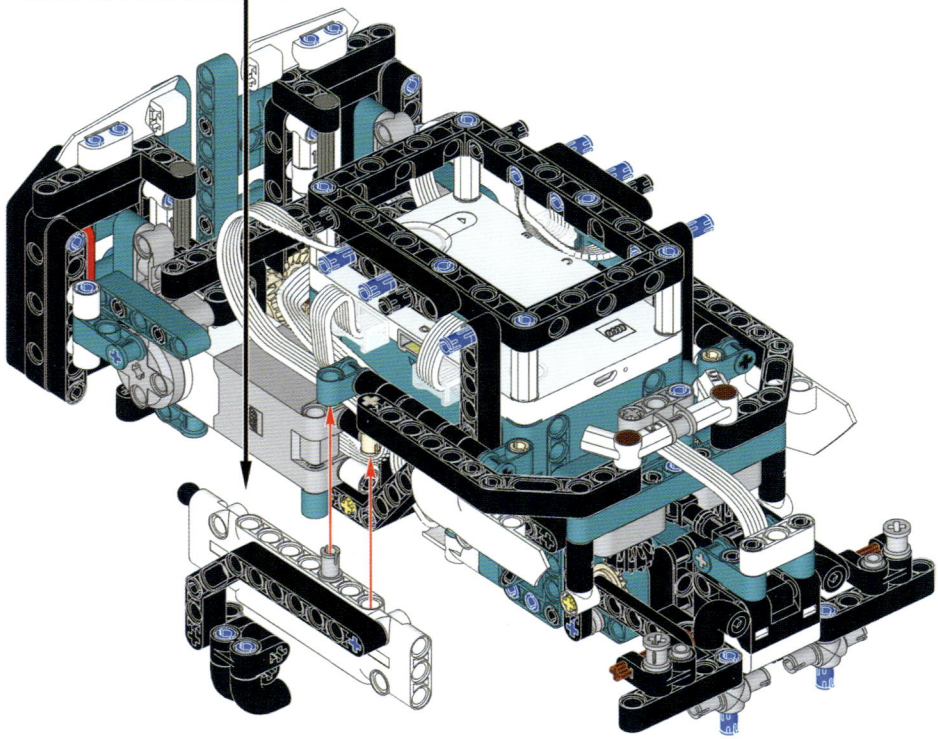

70

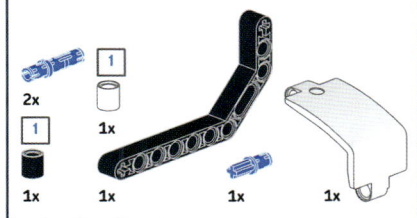

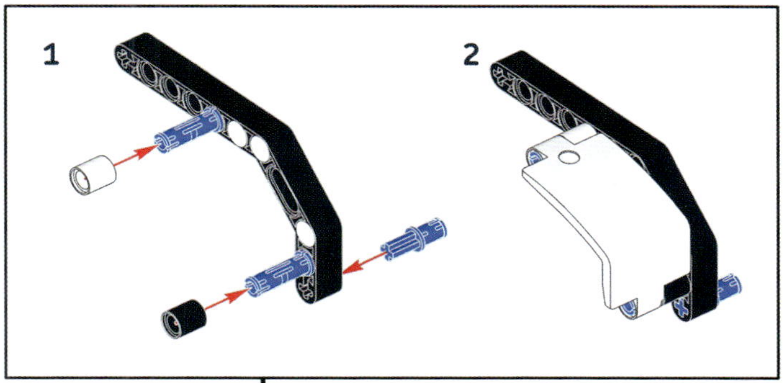

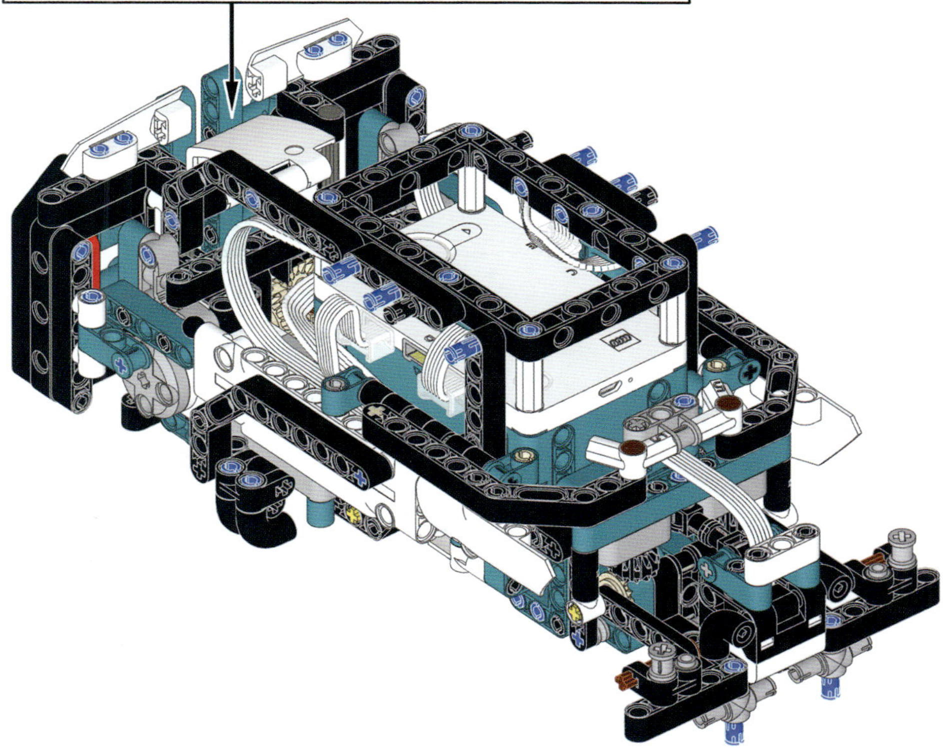

2x
1x
1x
1x
1x
1x

1

2

71

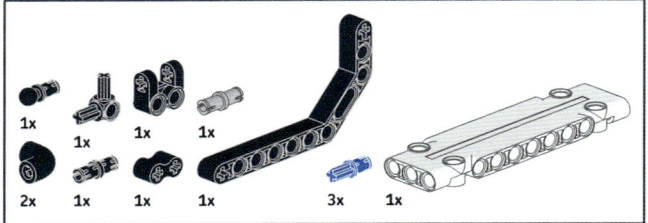

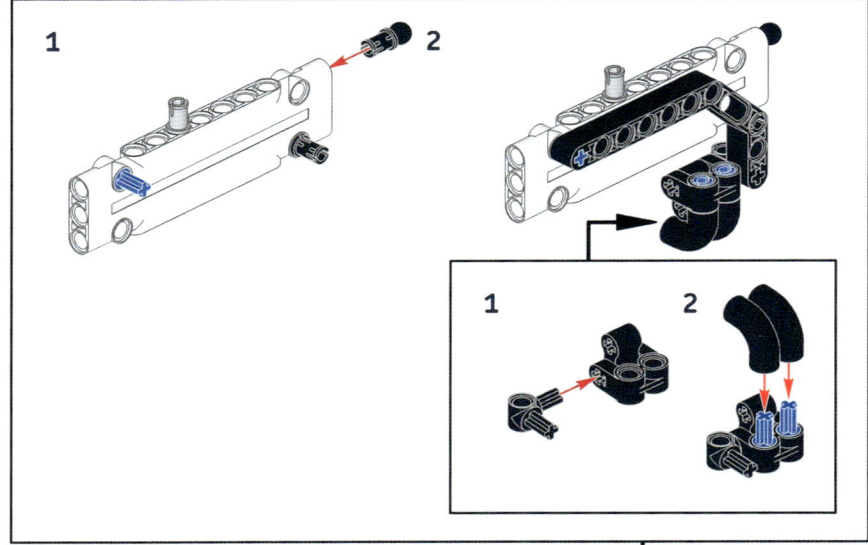

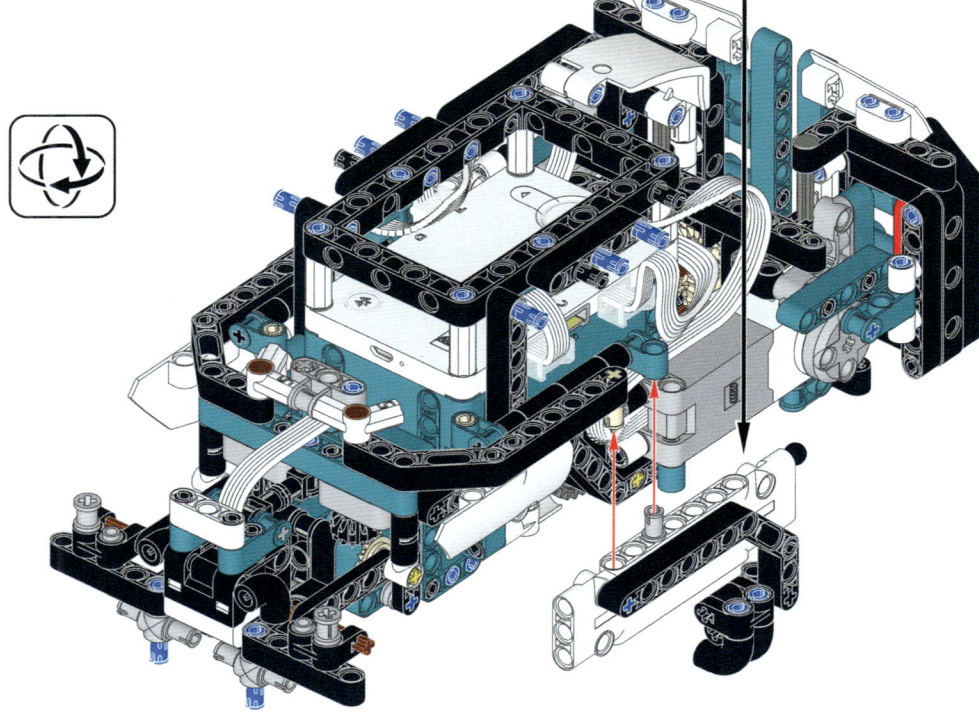

70

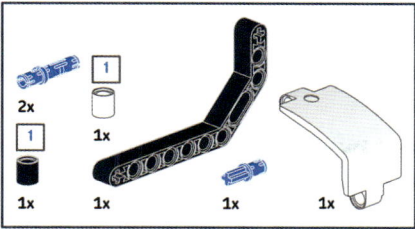

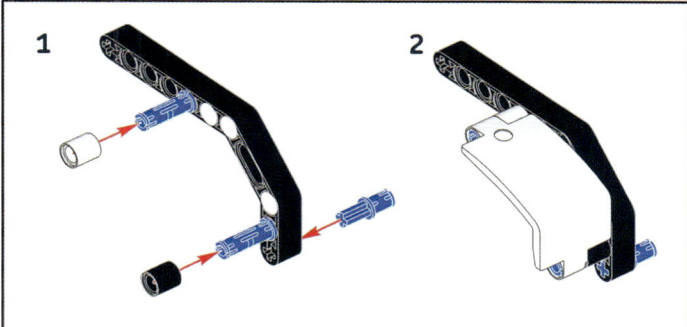

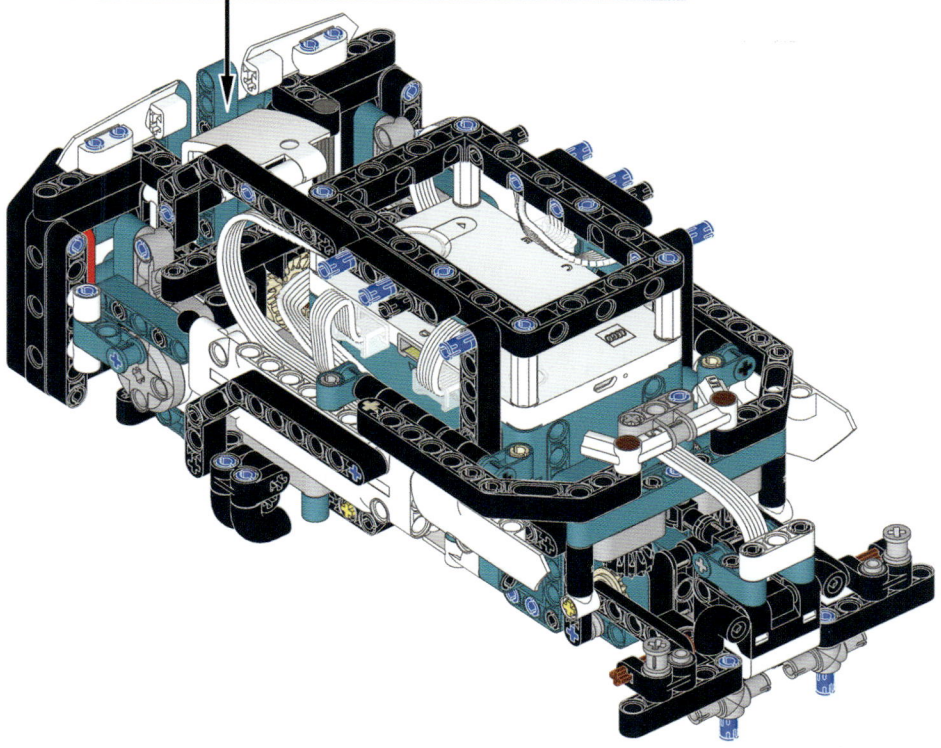

73

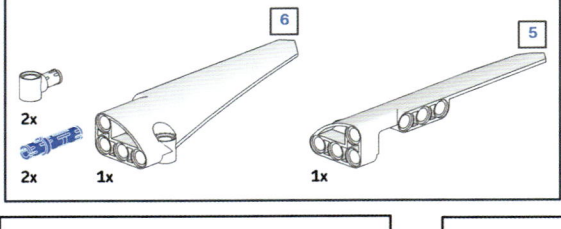

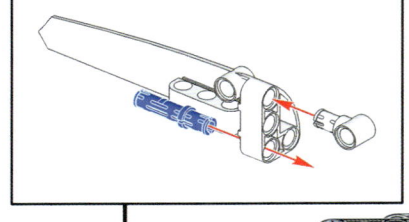

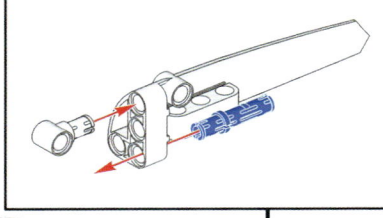

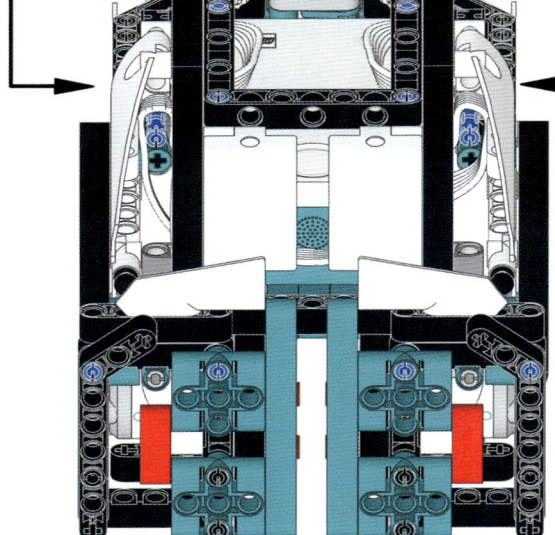

Attach the side panels to the black and blue pins on the side of the roof frame. From this view, you can see how to pass the motor cables around the long blue pins.

74

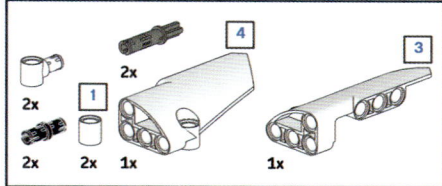

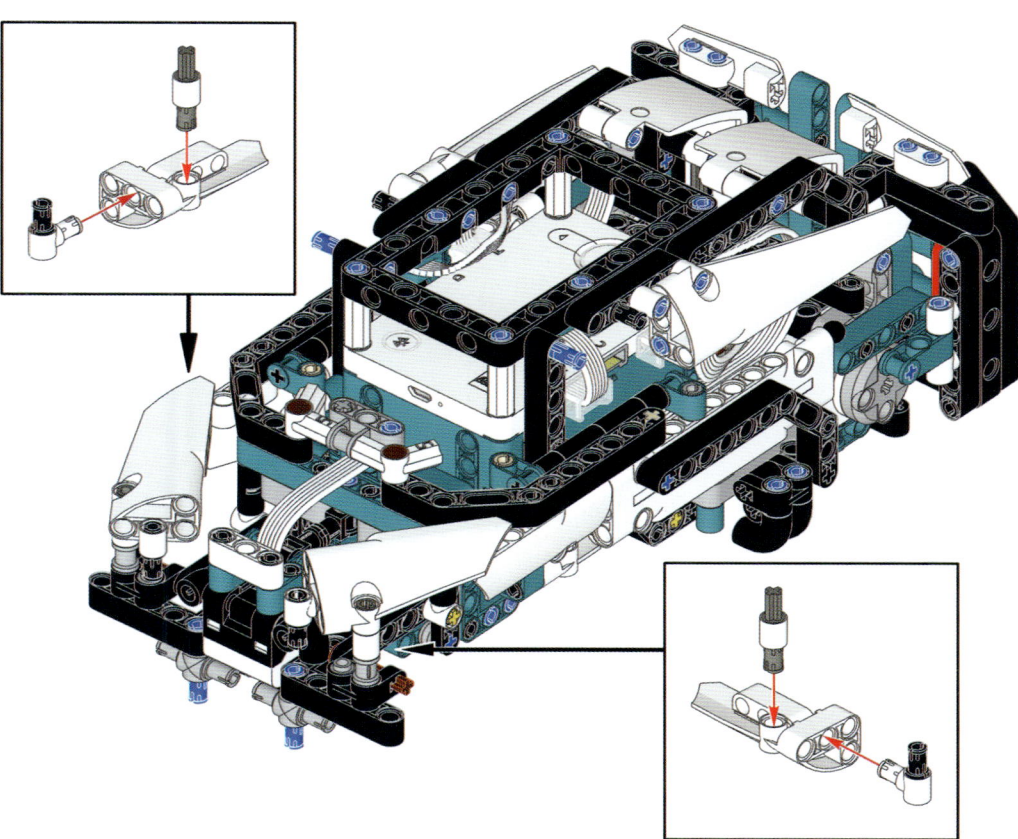

windscreen subassembly

75

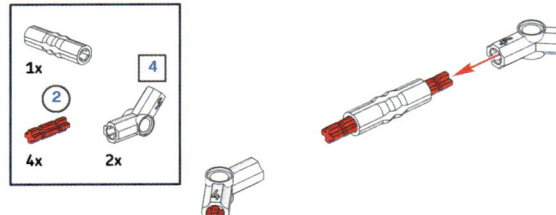

1x
②
4x   2x   4

76

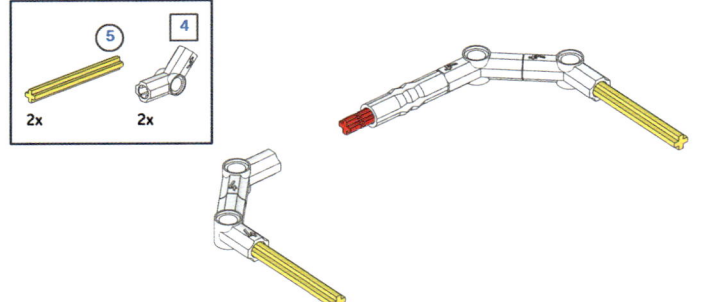

⑤
2x   2x   4

77

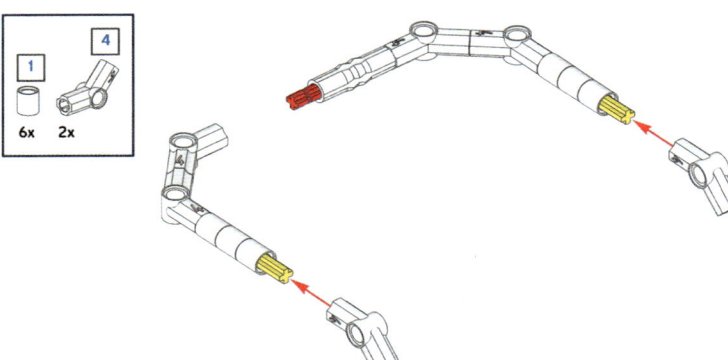

①
6x   2x   4

78

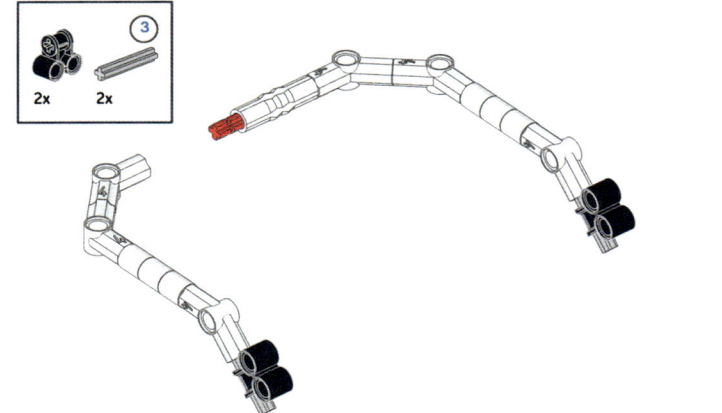

③
2x   2x

79

1x

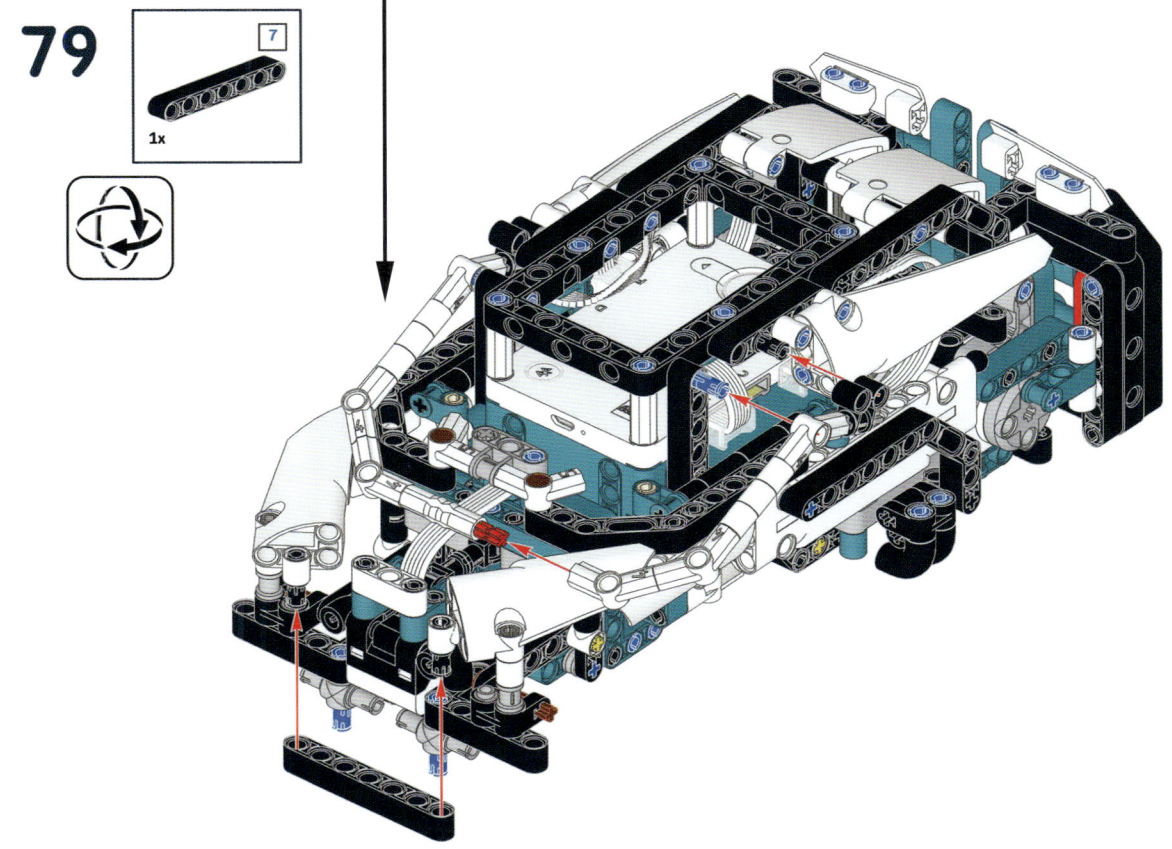

80

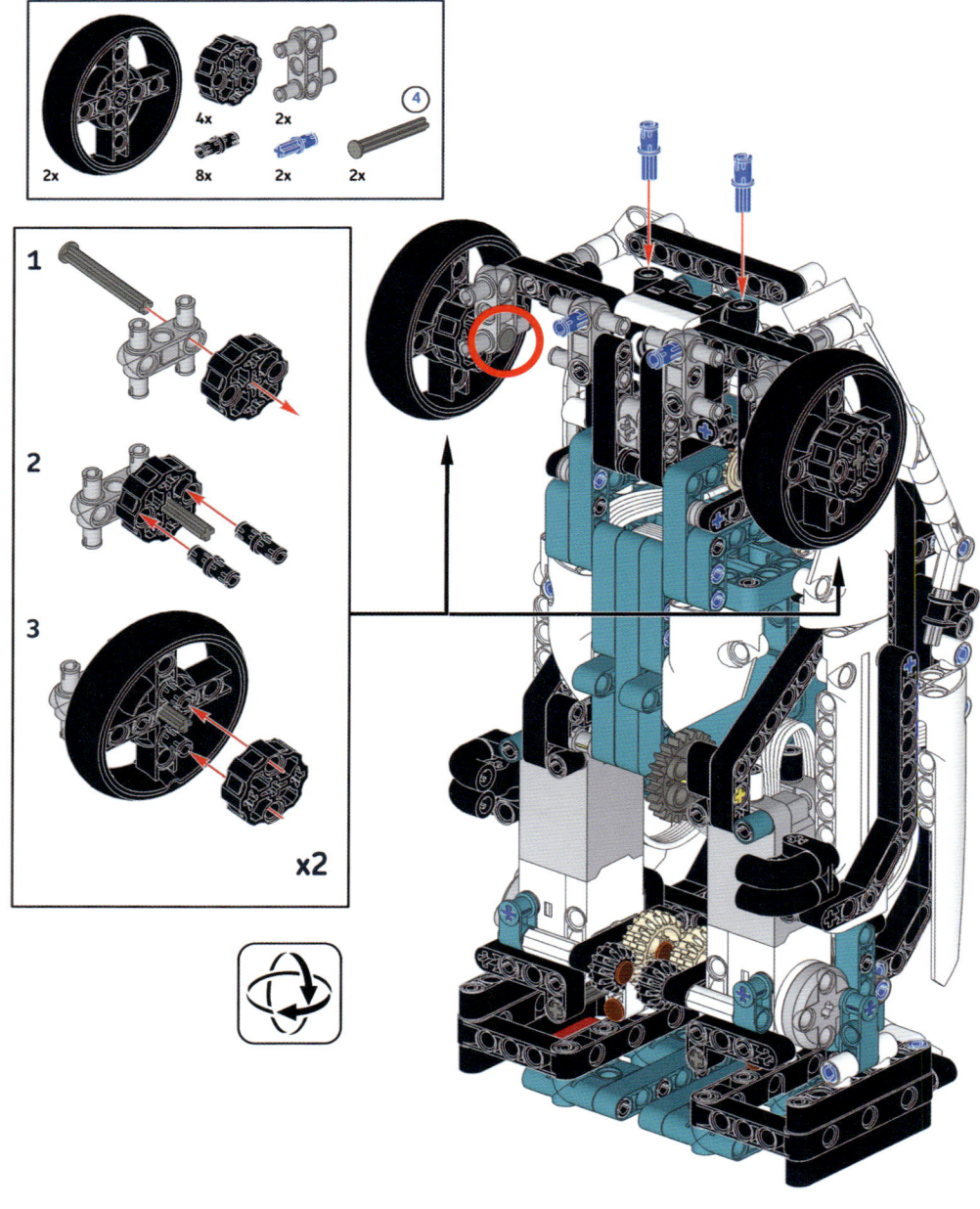

81

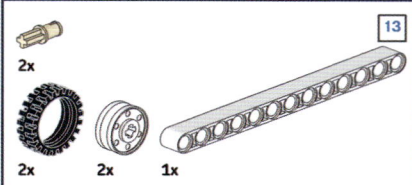

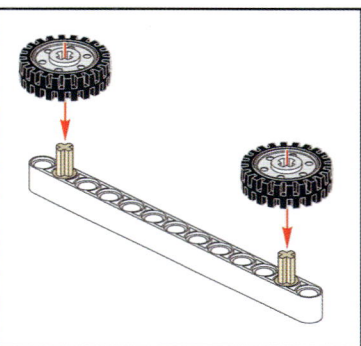

The small wheels prevent the main wheels from shaking when the robot is walking.

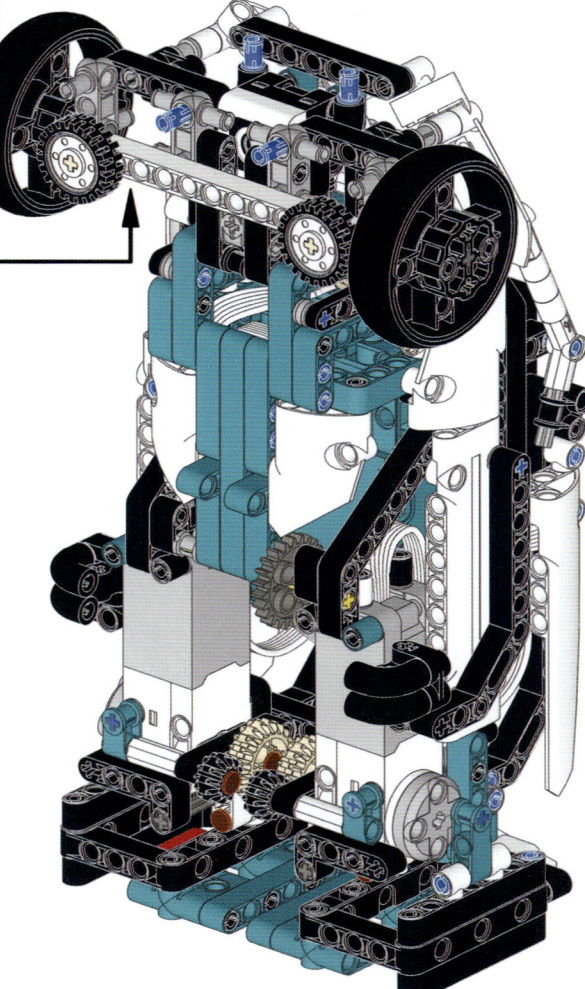

82

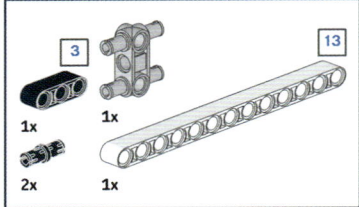

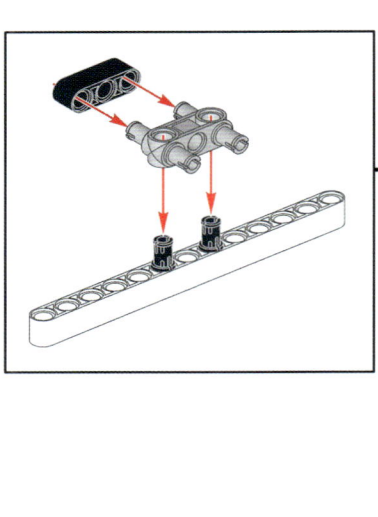

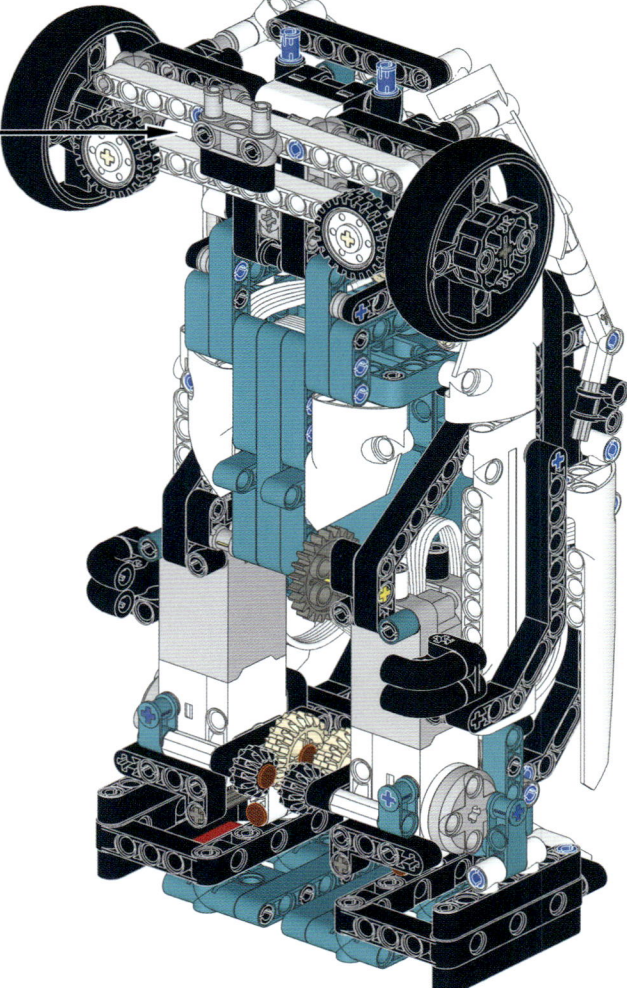

front bumper subassembly

83

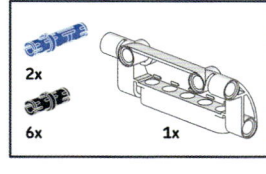

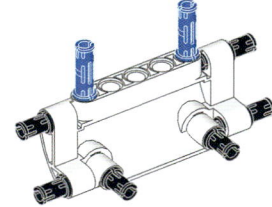

84

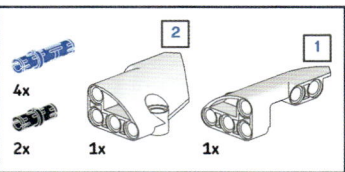

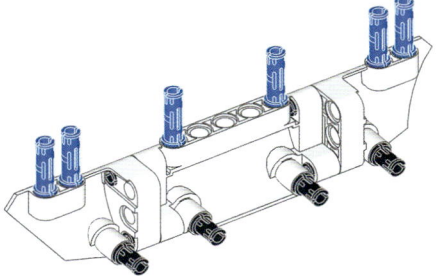

85

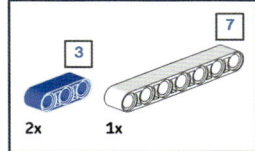

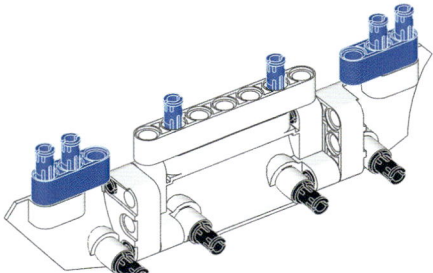

86

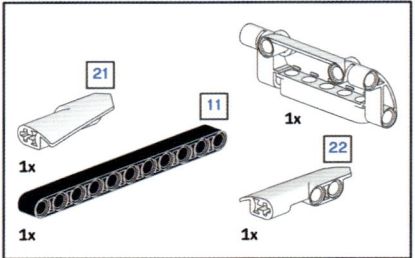

21
11
22
1x
1x
1x
1x

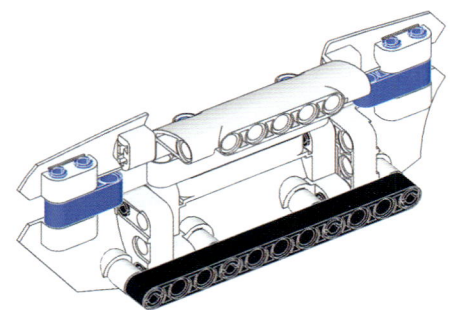

87

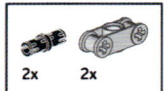

2x 2x

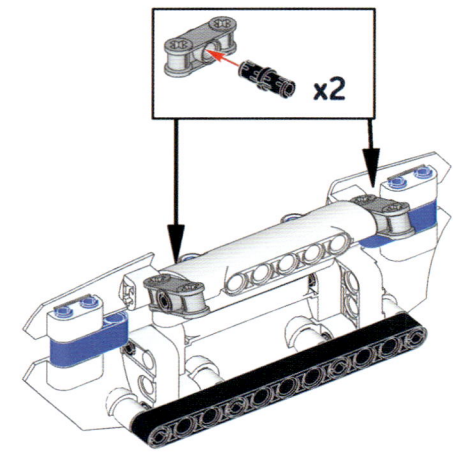

x2

88

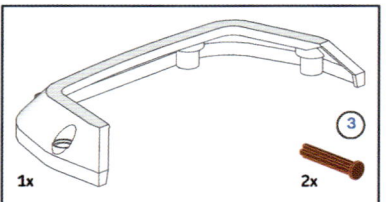

1x 2x ③

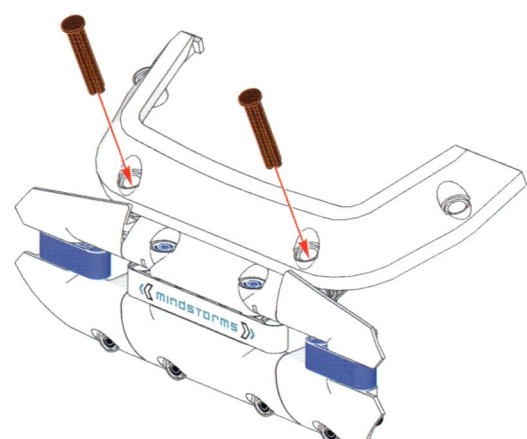

89

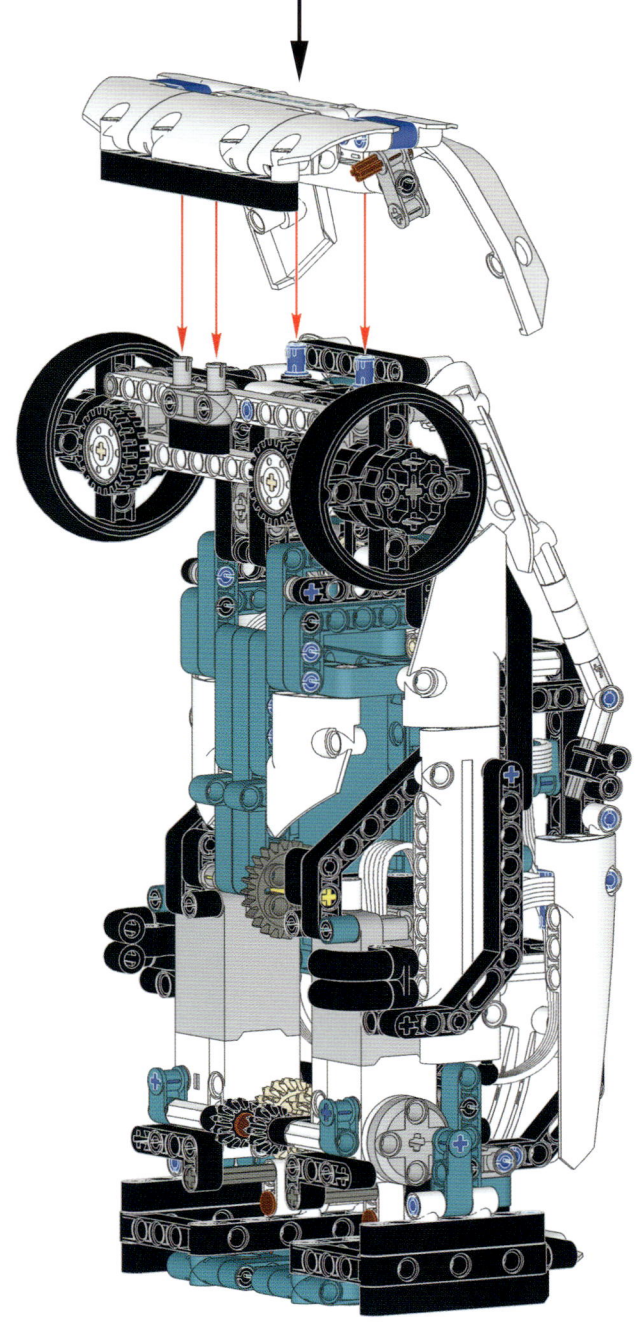

90

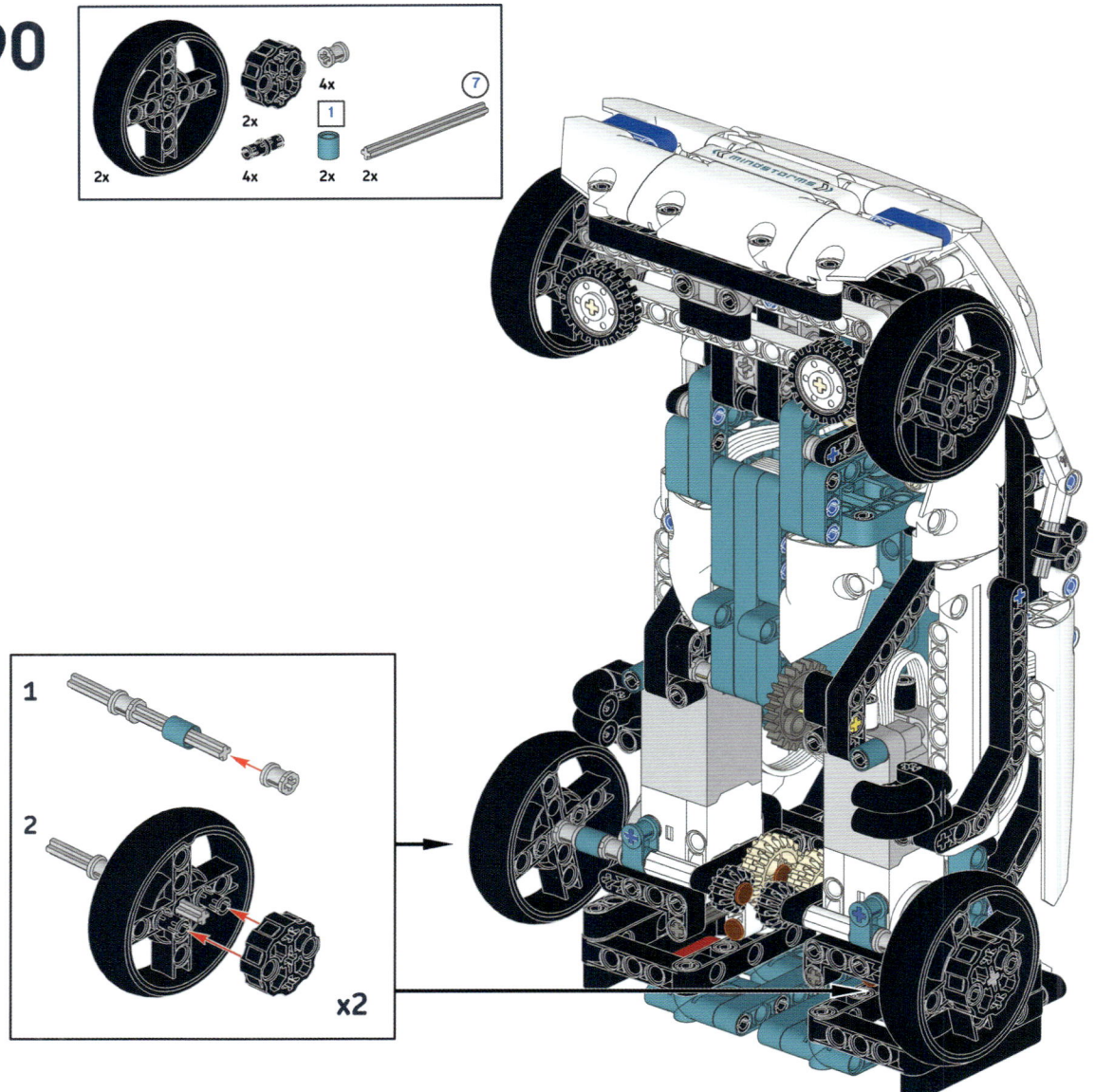

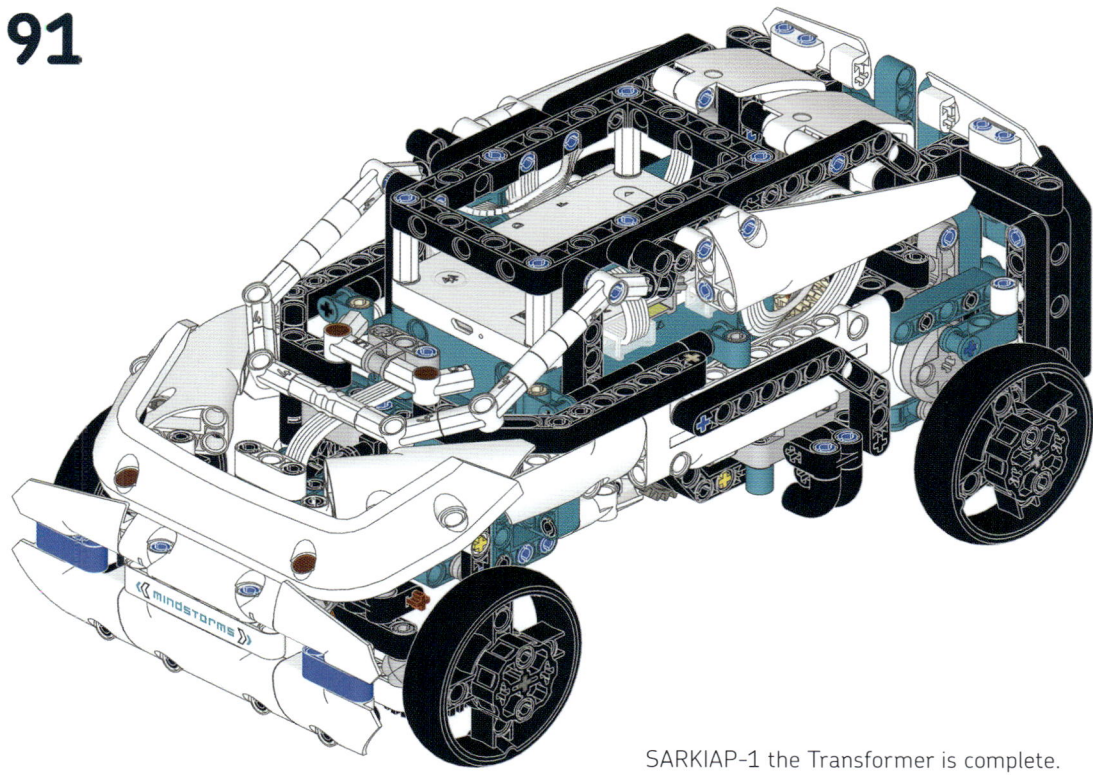

SARKIAP-1 the Transformer is complete.

creating a custom remote-control interface

Before you can start creating the final program for SARKIAP-1, you need to create a custom remote-control interface using the app's built-in editor. Using the interface, you'll be able to control the Transformer's movements—both as a humanoid robot and a car. Duplicate the *walking_robot* program and rename it *transformer_RC*. Then follow the steps on the next page to create the custom remote control, using the screenshots as reference.

1. Click the **Remote Control** button, located in the middle-right side of the programming area. An empty remote-control interface should appear.

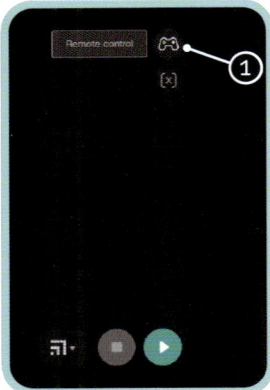

2. Click the **Edit** button at the top-right corner to start building the interface. The Remote Control Editor should appear.

3. Click the **Add Widget** button (the green circular button with a plus sign).

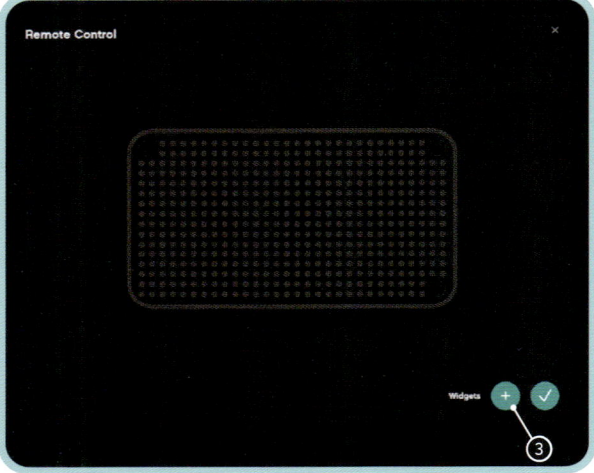

4. The Add Widget box contains Controls and Sensors. Pick **Joystick**.

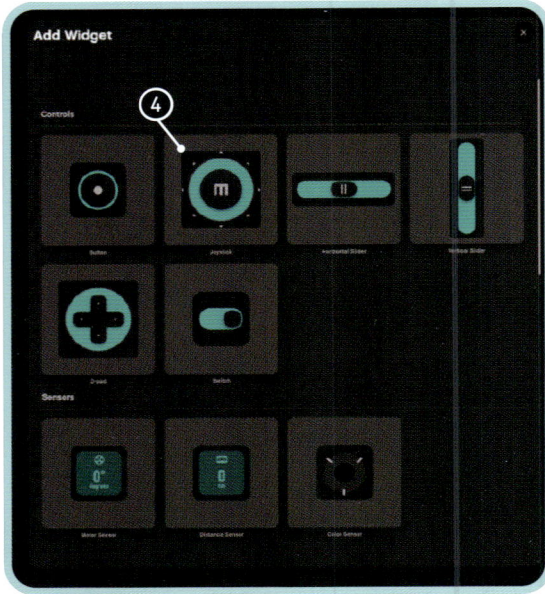

5. Configure the joystick using the three-dot menu in its top-right corner. Change its color to yellow and leave the default name J1. We'll use this joystick to steer the robot.

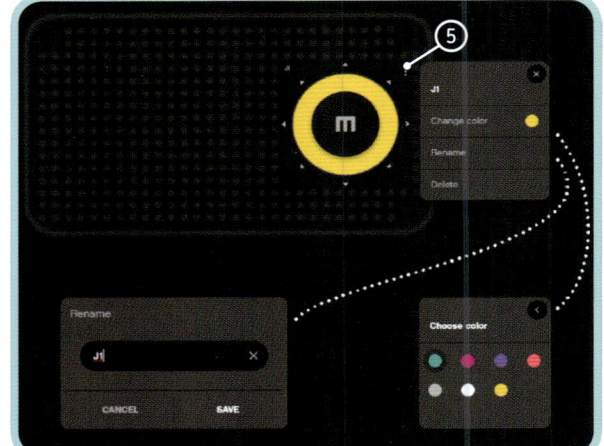

6. Using the Add Widget box, create three buttons. Use each button's configuration menu to change its color and rename it. Name one button **B_car**, make it white, and place it in the bottom-left of the controller. Pressing this button will make the robot transform into car mode.

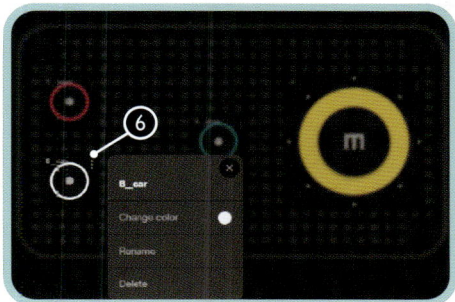

7. Name the second button **B_biped**, make it magenta, and place it in the top-left of the controller. Pressing this button will make the robot transform into biped mode.

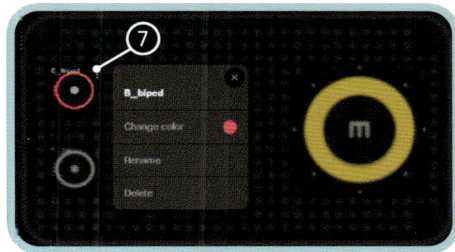

8. Name the third button **B_drive**, make it green, and place it in the middle of the controller. This button will make the car take a short run on its own.

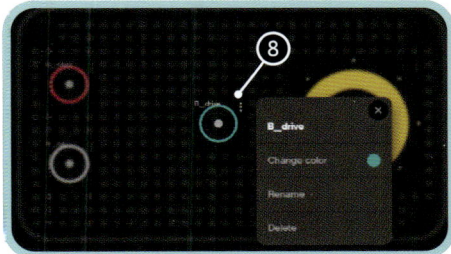

NOTE It's a good habit to use a prefix to identify the widget (*B* for *button*, *J* for *joystick*, and so on).

9. Click the round button with a check mark to close the Remote Control Editor.

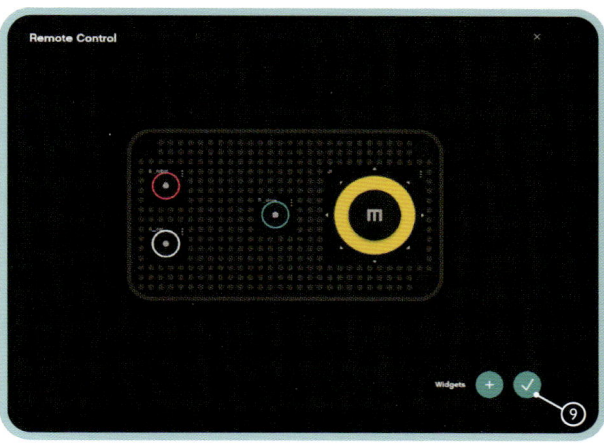

10. You can resize the controller window by dragging the icon in the bottom-right corner. To hide it, click the **Remote Control** button in the programming area.

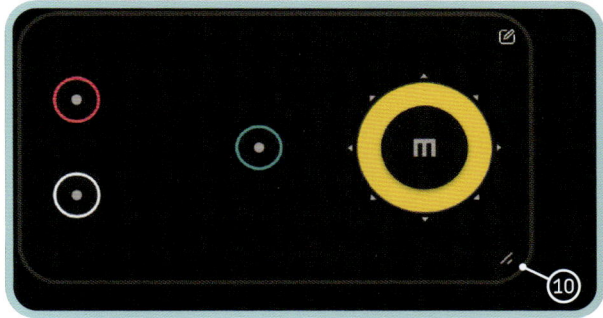

For you to use the remote control, the program's storage position must be set to Streaming mode. On smart devices, the app will ask you to rotate the screen horizontally to use the remote control. The Controller window will show a Close button in the top-left corner (to hide it) and a Play button in the bottom-right corner to start the program, as shown in Figure 4-8. On a computer screen, the Controller window will float on top of the programming area, and you can interact with the widgets with your mouse.

NOTE In the controller you just created, you added buttons and a joystick. The Remote Control Editor also lets you add switches, sliders, and widgets that report the readings coming from the sensors.

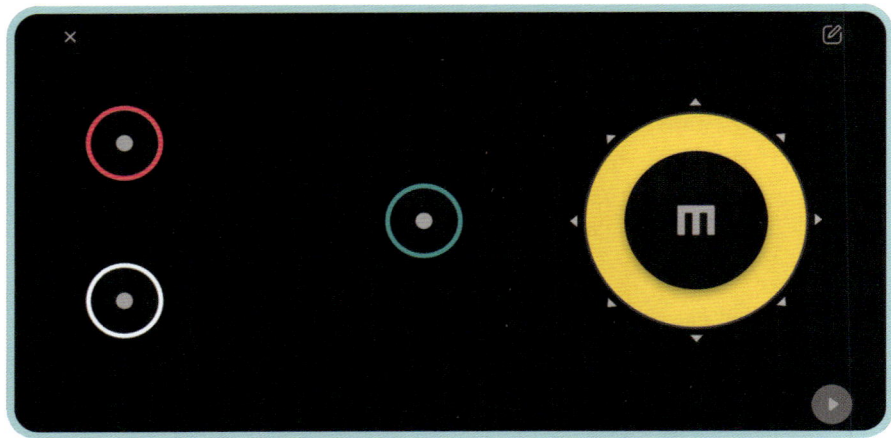

Figure 4-8: The remote controller looks a bit different on smart devices with small screens.

reacting to remote-control inputs

Now that you've created the remote control, you'll want to add new blocks to the Remote Control palette to manage the events and readings generated when you use the remote-control widgets. Each type of widget should have its own blocks. For example, you can see the new blocks that allow you to manage the joystick inputs in Figure 4-9.

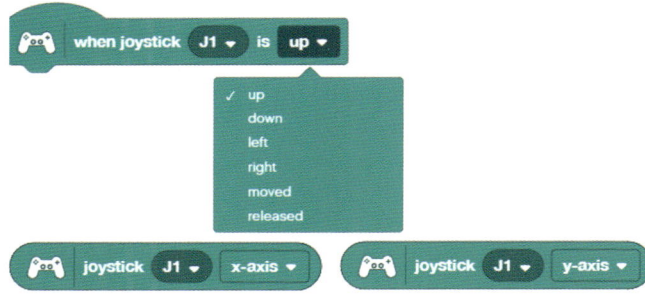

Figure 4-9: The programming blocks that report the position of the joystick of the remote controller

The when joystick is hat block executes the blocks attached to it when the specified joystick widget is in the specified state. This block will trigger only if the specified event occurs and won't retrigger if the joystick is still in the same state after the attached blocks have executed. You can set this block to trigger if the joystick is up, down, left, right, moved in any direction, or released. If you have multiple joysticks in your controller, you can specify the name of the joystick you want to react to.

The joystick axis block reports the current position of the specified joystick widget along the specified axis; you can specify x-axis for the horizontal movement or y-axis for the vertical movement. The block has round sides, meaning it reports numeric values as input to whatever block it's plugged into. The x-axis values range from -100 for the leftmost position to 100 for the rightmost position. Likewise, the y-axis values range from -100 for the bottom position to 100 for the top position. Both axes are at 0 when the joystick is released.

As you'll see in stacks L and M of the final Transformer program in Figure 4-17, to effectively use the value of the joystick position in your programs, you should plug joystick axis blocks into comparison blocks to test whether the joystick has moved past a certain position. The comparison blocks will give you logic values you can use as conditions for control structures like if then, wait until, and while blocks. For example, you might write code that says, "While the joystick's y-axis value is greater than 30, move forward."

The Remote Control palette should also have two new blocks for managing button inputs, as shown in Figure 4-10.

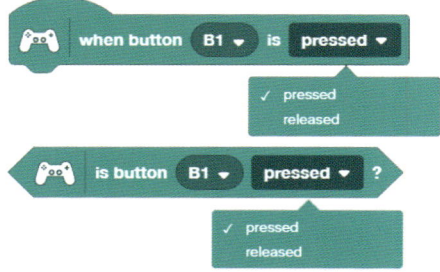

Figure 4-10: The programming blocks that report the state of the buttons of the remote controller

The `when button is` hat block triggers the attached blocks when the specified button is pressed or released. The stack won't repeat if the button is still pressed (or released) after the blocks have executed. The `is button` block has pointy sides, meaning it reports logic values: `true` if the specified button is in the specified state or `false` if it isn't. For example, it can be configured to check whether the button named B1 is pressed or released.

reading the orientation of the hub in space

As mentioned in Chapter 1, the Hub has a built-in sensor that measures the Hub's orientation in space. The `hub angle` block shown in Figure 4-11 reports readings from this sensor. As you can see in the figure, changes to the *pitch angle* mean the Hub's USB port is pointing up or down, changes to the *roll angle* mean the Hub's motor and sensor ports are pointing up or down, and changes to the *yaw angle* mean the Hub's USB port is pointing left or right with respect to the ground. The yaw angle can be reset to 0 with the `set yaw angle` block.

In the Transformer program, you'll use pitch angle readings to determine whether the robot is in car mode (Hub lying down) or in biped mode (Hub standing up). Based on the pitch angle, the program will react to the remote-control commands in different ways. For example, if you press the button to transform into a car but the Hub's pitch reading tells the program the robot is already in car mode, nothing will happen.

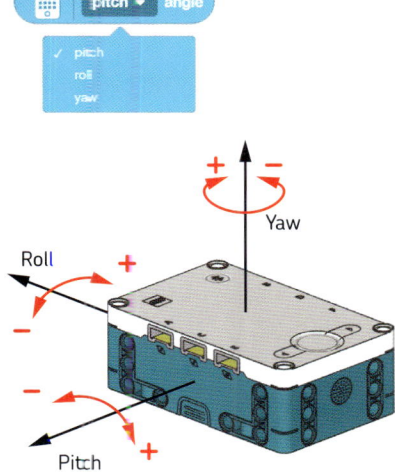

Figure 4-11: The Hub has an internal sensor that measures its angle along three axes.

starting multiple stacks running together

In our final Transformer program, we sometimes need to move motors A and B to align the robot's feet. It would be great to move both feet at the same time, but since the motors need to move to different positions, we'll need to use a separate block for each motor. Fortunately, the `broadcast message` and `when I receive message` blocks shown in Figure 4-12 give us a way to execute multiple blocks at the same time. You'll find these blocks in the Events palette.

The `broadcast message` block sends a message you specify (for example, `alignFeet`) to all the blocks in your program. The `when I receive message` block can be set to listen for the specified message. When the message is received, any blocks attached to the `when I receive message` block will run. By setting multiple `when I receive message` blocks to listen for the same message, you can make multiple block stacks run at the same time.

After the message has been sent, the next block in the stack after `broadcast message` will be executed immediately. If you use the `broadcast message and wait` block instead, the program will wait until all of the stacks attached to the `when I receive message` blocks have finished before proceeding.

programming the transformer

Now we're ready to create the final remote-control program for the Transformer. The complete program is distributed across Figures 4-13 through 4-17. Continue adding these blocks to the *transformer_RC* program where you created the custom remote control.

When the program starts, stack A in Figure 4-13 handles some important setup. First we call the `align feet` custom block, defined in stack B, which moves both feet to the same position. As we've discussed, the motors controlling the

Figure 4-12: The broadcast message blocks work with the when I receive message block to allow you to start running multiple stacks of blocks together.

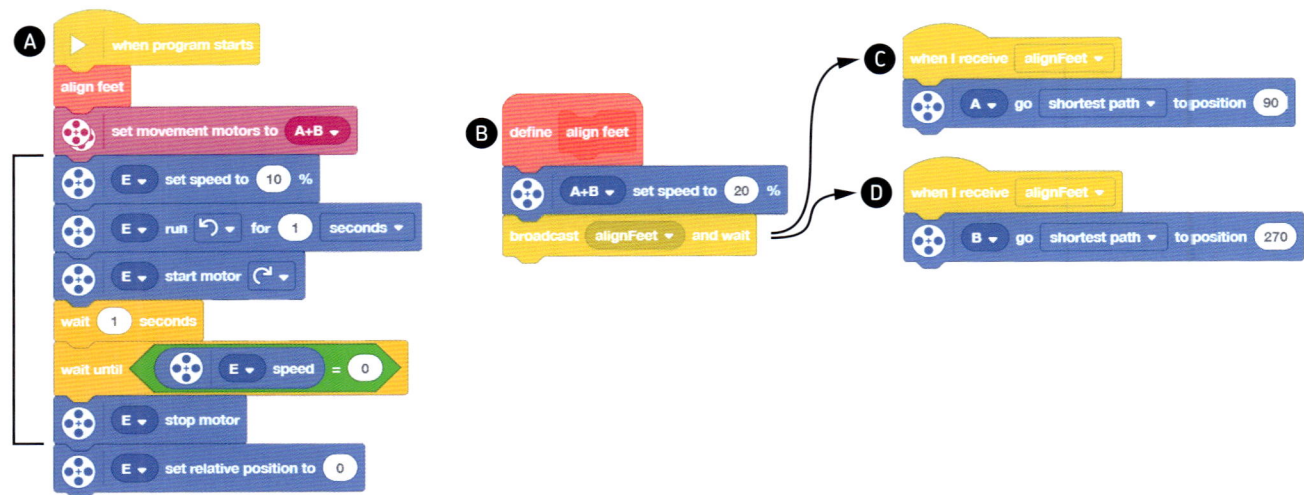

Figure 4-13: Final Transformer program, part 1: main stack and align feet

feet need to rotate to different positions, so we can't use a single motor go to position block to move them together. To run the two motor go to position blocks *in parallel*, we broadcast the alignFeet message from stack B with the broadcast message and wait block. Stacks C and D both begin with a when I receive message hat block. The two stacks execute together when they receive the alignFeet message, moving both motors into position at the same time.

The align feet custom block in stack B ends only when both stacks C and D have finished their execution. Then stack A continues with the set movement motors block, which sets motors A and B as the driving motors for all the Movement blocks in the program. The next sequence of blocks, marked with a bracket, is the same as in the *calibrate_robot* program you already created (compare with Figure 4-7) to rotate the torso of the robot against the motor body.

After the torso has been brought into position, the set relative position block (from the More Motors extension palette) resets motor E's relative position to 0. This way, we can say the motor is at position 0, regardless of its actual position, and we can measure other motor positions relative to that baseline. We'll do this in the become biped and become car custom block definitions (stacks E and F, shown in Figure 4-14), where we use motor go to relative position at speed blocks to rotate the motor relative to the 0 position we've set.

become biped and become car custom blocks

The custom blocks become biped and become car, defined in stacks E and F (Figure 4-14) both use motor E to transform the model into biped mode and car mode, respectively. The transformation sequences (depicted in Figure 4-1) are really fun to watch. Later in the program, we'll see how these stacks are triggered in response to inputs from the remote control.

Notice that the align feet custom block we first used in stack A is reused in both stacks E and F. This is an example of why defining a custom block with a clear, self-contained purpose is so convenient: we can reuse the block whenever we want.

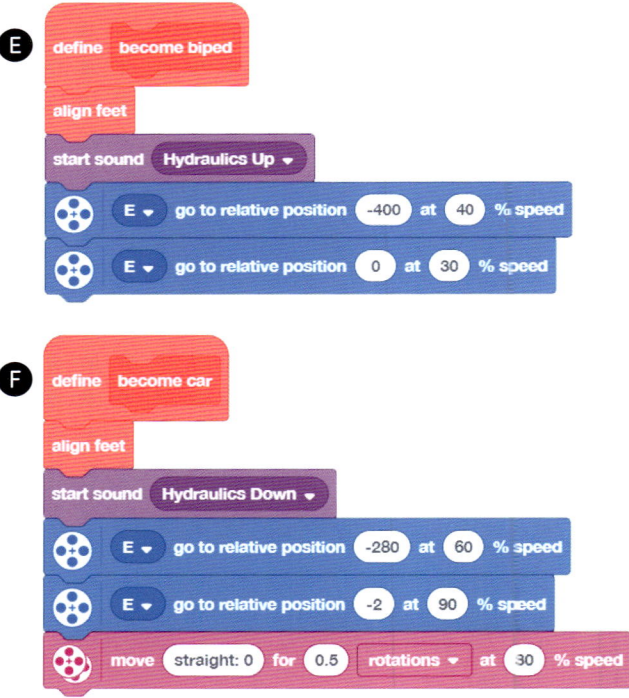

Figure 4-14: Final Transformer program, part 2: become biped and become car custom block definitions

NOTE The Hydraulics Up and Hydraulics Down sounds are in the M.V.P. section of the Sound Library.

custom blocks to make the robot walk and turn

The custom blocks defined in Figure 4-15, which allow the robot to walk and turn when it's standing up, are adapted from the ones we made for the walking core program in Figure 4-2. Update those original custom blocks and definitions as shown in the figure. Instead of walking forward and backward for an indefinite amount of time, or repeating the turning sequence for a fixed number of steps, we now use `repeat until` blocks to continue moving in the correct direction until a particular condition becomes true. We get those conditions from the joystick on our remote control.

For each `repeat until` condition, we use a comparison block to check the value of one of the joystick axes against a specified value. For example, if the `walk forward` custom block is triggered, the robot will continue walking forward until the joystick's y-axis value is less than 70—that is, until you're no longer pressing the joystick forward on the remote control. By checking the joystick position against a high value like 70, we create a *dead zone* around the joystick's resting position, making the robot easier to control. With the dead zone, you have to really move the joystick to make the robot walk the way you want. Without the dead zone, small, accidental movements of the joystick could make the robot walk in unwanted directions.

car RC and biped RC custom blocks

Whether the Transformer is in car mode or biped mode, you'll be moving it with the same joystick on the remote control. We need to make a `car RC` custom block to remote-control the Transformer when it's in car mode and another custom block, called `biped RC`, to remote-control the Transformer when it's in biped mode. You can see the definitions for these blocks in Figure 4-16.

In stack L, the `car RC` stack, the `if then else` block creates another dead zone in the middle of the joystick's x-axis. This way, the car will drive straight unless you move the joystick far from the center. We use the `is between` operator block to check whether the joystick's x-axis (horizontal) reading is in the range -30 to 30. If this is the case, we start the motors to drive straight at a speed that's proportional to the joystick's y-axis (vertical) position. We take the y-axis reading and multiply it by -0.8 to set the speed. The farther forward or backward you move the joystick, the faster the car will drive forward or backward. You'll find the `multiply` (*) block in the green Operators palette.

NOTE We multiply the joystick's y-axis position by a negative number because the gears turn the car's wheels in the opposite direction of the motors. Negative speeds therefore make the car drive forward.

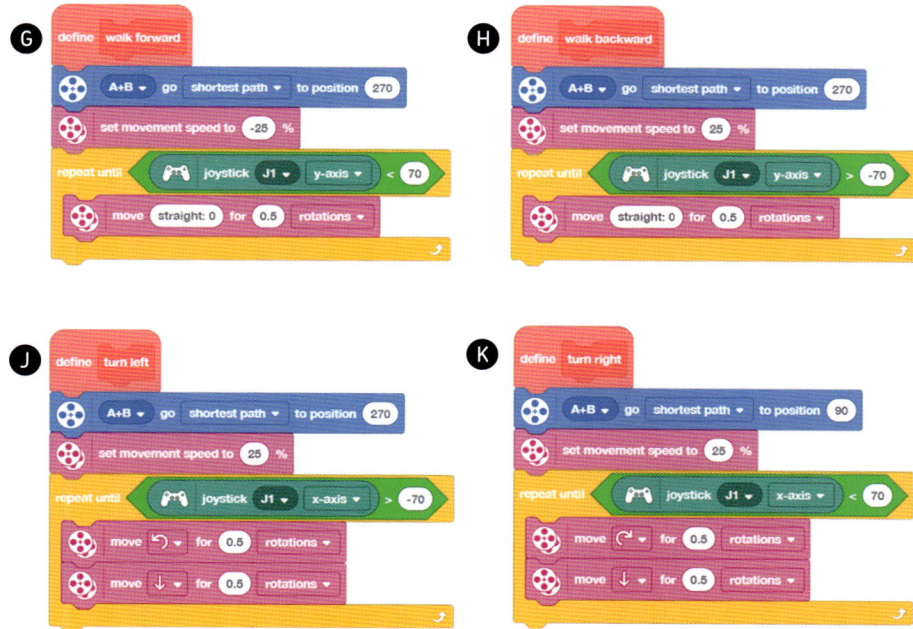

Figure 4-15: Final Transformer program, part 3: walk forward, walk backward, turn left, and turn right custom block definitions

If the joystick's horizontal position is outside the -30 to 30 range, the start moving with steering at speed block in the else space makes the car curve to the left or right. This Movement block makes the car turn by running one motor faster than the other. For example, to curve to the left, the right motor will spin faster than the left one. The front wheels are mounted on a *passive steering* mechanism, which naturally follows the curve imposed by the powered rear wheels.

The start moving with steering at speed block automatically sets the motors to different speeds based on the steering value you specify, which controls how sharp the turn will be. In this case, the block gets a steering value by multiplying the joystick's x-axis position by 0.4. This calculation takes a joystick x-axis value ranging from –100 to 100 and produces a steering value between –40 and 40. We multiply by 0.4 because higher steering values would make the front wheels get stuck in a weird position.

In the biped RC custom block definition (stack M), a series of if then blocks check the position of the joystick and make the robot walk in the appropriate direction when it's standing up. This is where we call the custom blocks we defined in Figure 4-15. For example, the first if then block checks whether the joystick's vertical position is greater than 60, and if that's true, calls the walk forward custom block.

program stacks that manage the widgets events

The final part of the program, shown in Figure 4-17, contains the stacks that are triggered by the Remote Control hat blocks when you interact with the controller widgets.

When you press the **B_drive** button on the controller, stack N will execute, making SARKIAP-1 perform a series of maneuvers only if it's in car mode. To see if it's in car mode, we check the Hub pitch angle. If the pitch angle is less than

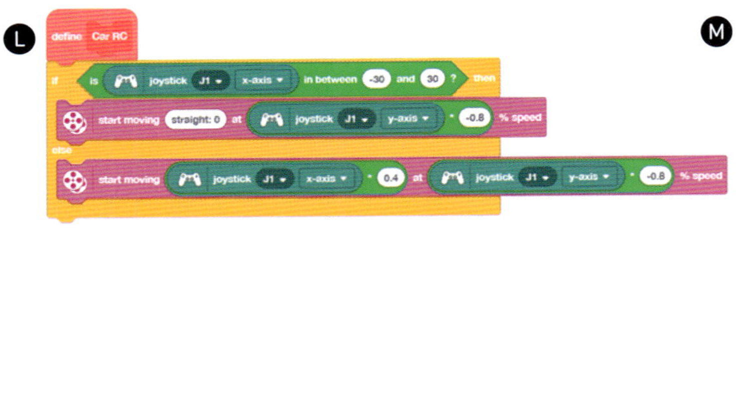

Figure 4-16: Final Transformer program, part 4: custom blocks to manage remote-control steering commands: car RC and biped RC

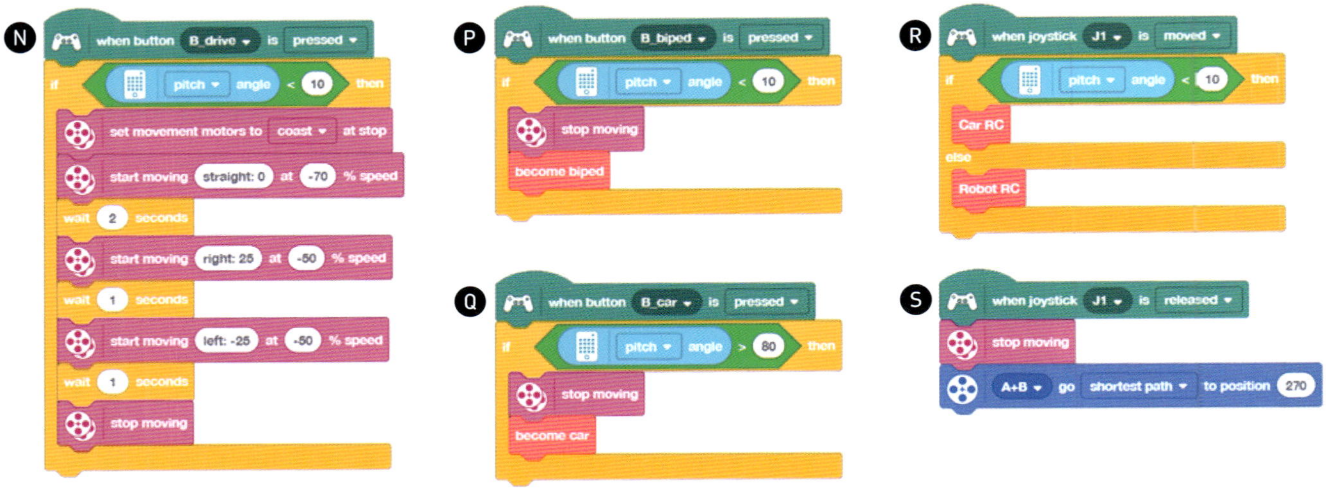

Figure 4-17: Final Transformer program, part 5: remote-control widget event-driven stacks

10 degrees, the Hub is horizontal, and all the wheels are touching the ground.

When you press the **B_biped** button (stack P), if the Hub's pitch angle is less than 10 degrees, the Transformer is in car mode. The motors are stopped with the `stop moving` block, and the `become biped` custom block performs the sequence of actions needed to transform the car into a humanoid. If SARKIAP-1 is already standing up, the Hub pitch angle will be much greater than 10, the comparison block will return `false`, and no actions will be taken.

A similar process happens in stack Q, which is executed when you press the **B_car** button. If the Hub's pitch angle is greater than 80 degrees, the Transformer is standing up in biped mode, and it should be brought back to car mode by the `become car` custom block.

Stacks R and S are very important, because they decide what happens when the joystick is moved, or released, respectively. You should always use a `when joystick is moved` and a `when joystick is released` block in a program for a robot controlled by a joystick.

When you move the joystick in any direction, stack R is executed. The `if then else` block checks the Hub pitch angle to see whether the Transformer is in car or biped mode. It then executes the `car RC` block or `biped RC` block accordingly (see Figure 4-16). When you release the joystick, stack S is executed, the driving motors are stopped, and the feet are aligned.

When you're done building the program, test it out! Remember that the storage position must be set to Streaming mode for the remote-control blocks to work. Place the Transformer on its wheels and watch it go through the calibration procedure when the program starts. Move the joystick to drive it around. Then, try pressing the button **B_biped** and see SARKIAP-1 shape-shift into a walking humanoid robot. If you move the joystick all the way up, SARKIAP-1 should start walking forward. Now, press the button **B_car** and see the robot roll back into a car.

what you've learned

While building SARKIAP-1, you learned some interesting building techniques, such as how to make a robot walk and how to use gears to speed up the rotation of an axle or to let a motor lift a load using less effort. While programming the Transformer, you learned how to use many new programming blocks, how to create custom blocks, how to make a remote-control interface, and how to broadcast messages to make multiple stacks run in parallel.

In the next chapter, you'll build a cute and shy pet turtle that you can interact with. You'll program it to have interesting behavior, as if it has a mind of its own.

shelly the turtle

The world rides through space on the back of a turtle.
—Terry Pratchett, foreword to *The Color of Magic*

LEGO MINDSTORMS sets aren't just for building alien robots and remote-control vehicles. With the same LEGO elements, you can also build cute creatures like Shelly, a pet turtle you can pamper, feed, and take for a walk. But be careful—she's very shy! Even a gentle tap on her back might cause her to hide her head in her shell and shiver with fear.

In this chapter, you'll build Shelly and discover the mechanisms to make her walk, turn, and move her head in and out. You'll also learn how to make a program to give her interesting behaviors, with a bit of unpredictability.

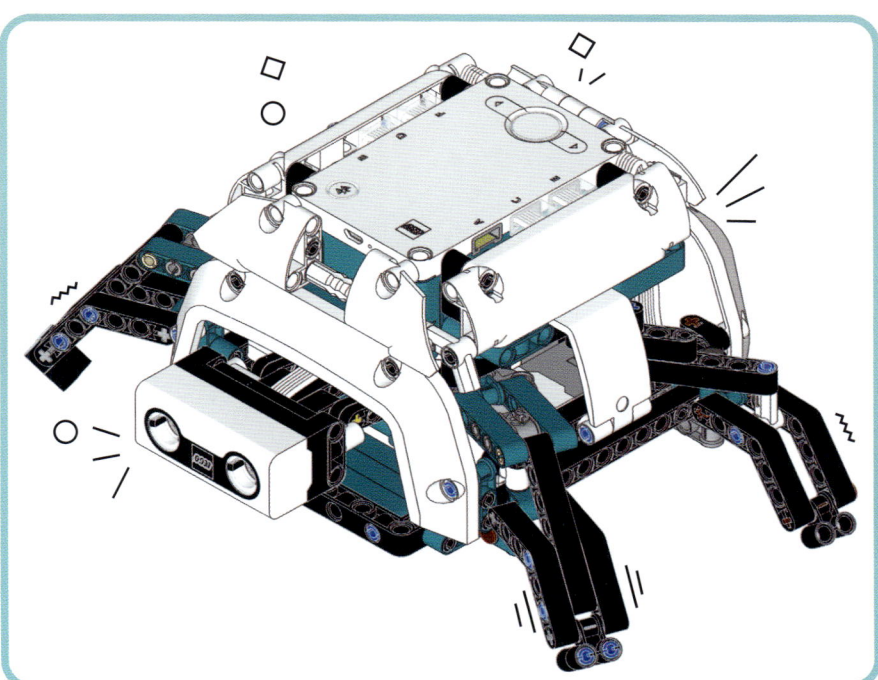

Figure 5-1: Shelly is a shy robotic pet turtle.

building the turtle

In this section, you'll find step-by-step instructions to build Shelly the Turtle. Along the way, you'll find some notes describing key design choices and interesting techniques.

1

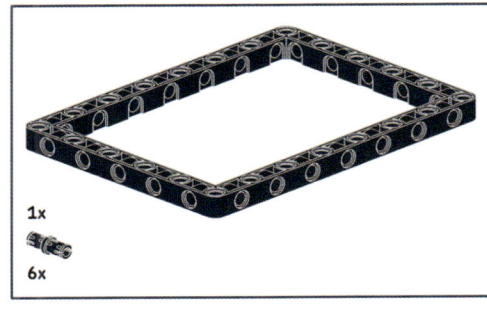

1x

6x

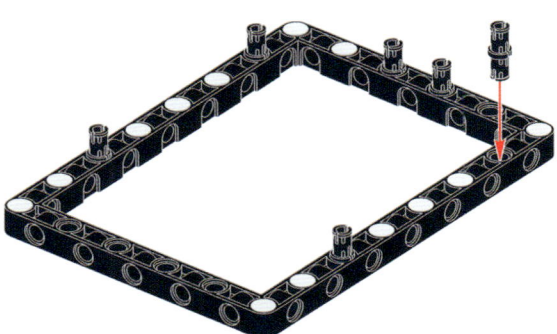

The white dots should help you insert the pins in the correct holes.

2

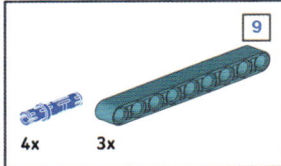

9

4x   3x

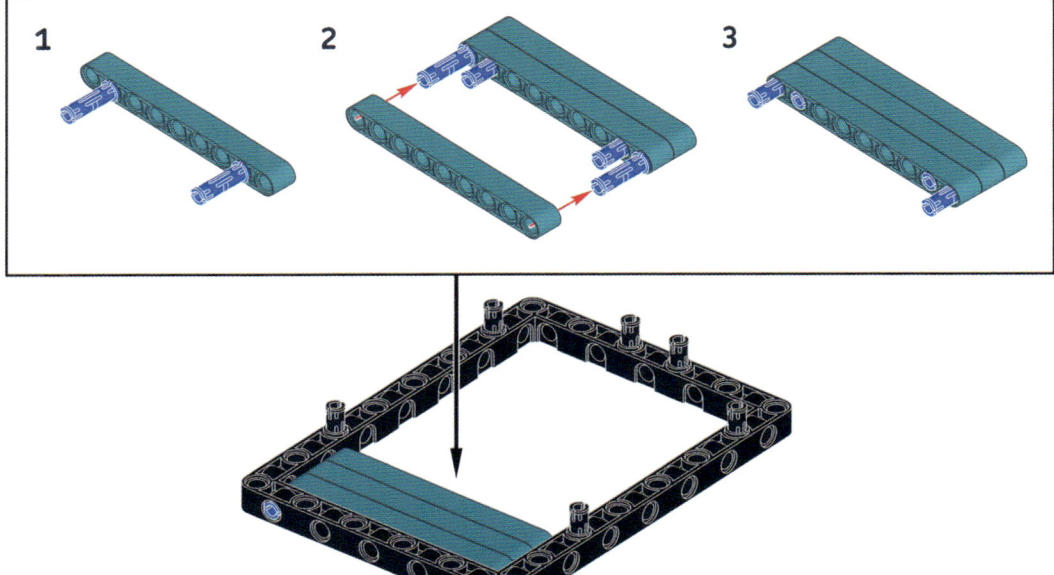

3

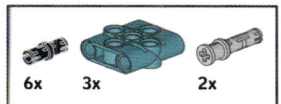

6x   3x   2x

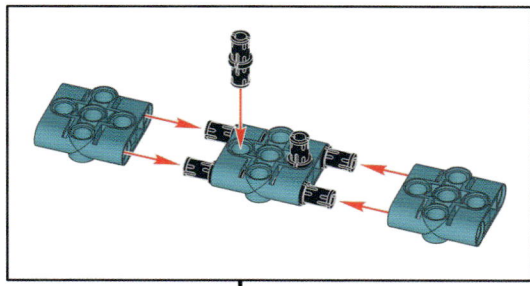

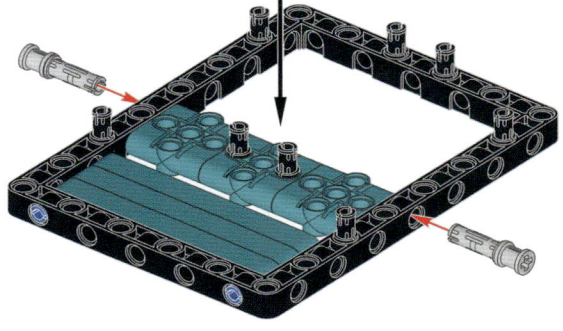

4

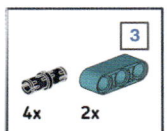

4x   2x   [3]

x2

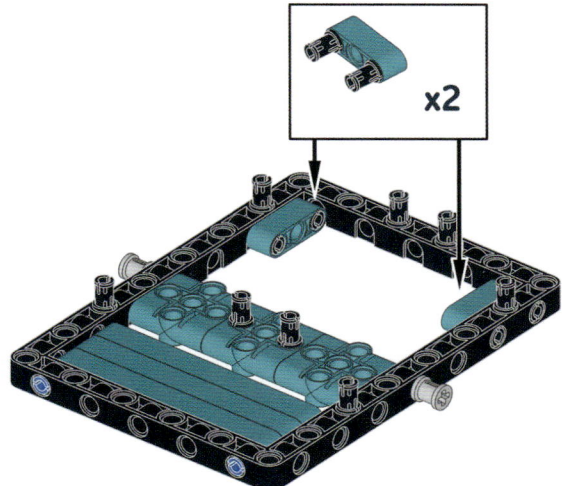

5

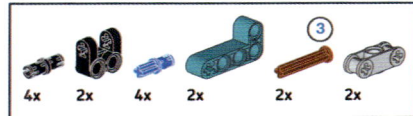

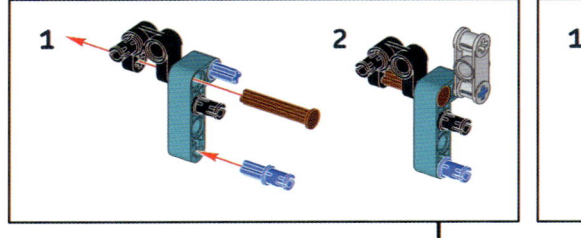

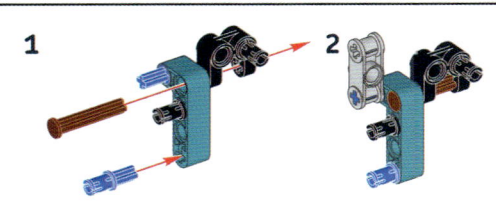

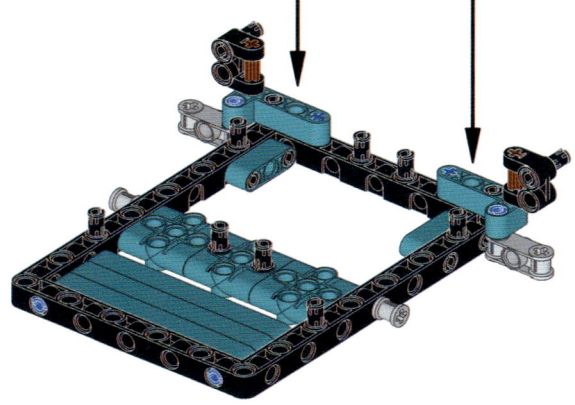

6

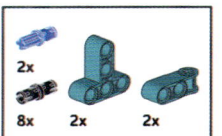

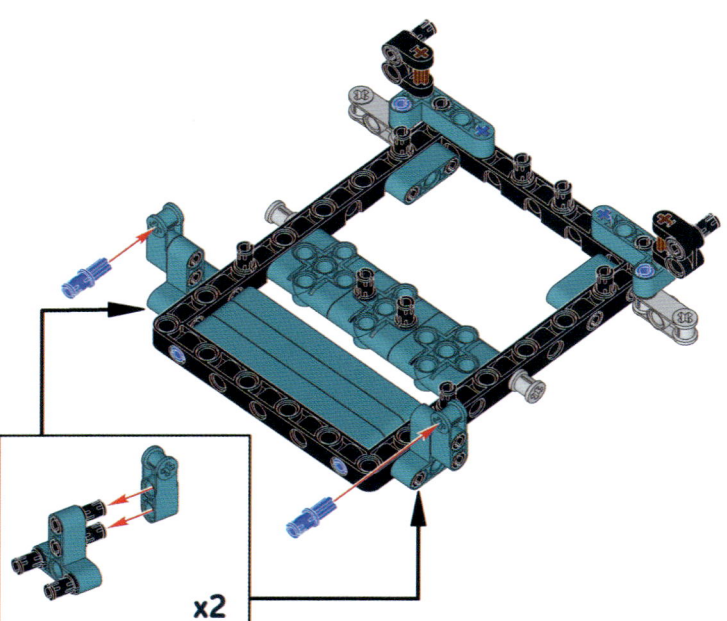

x2

left motor subassembly

7

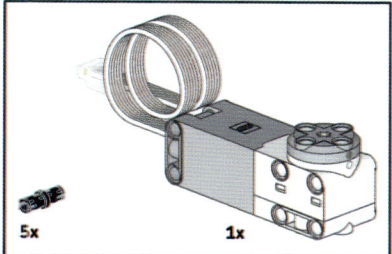

5x   1x

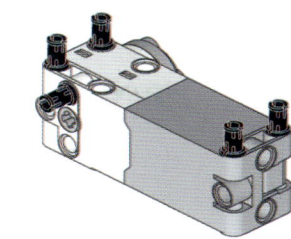

8

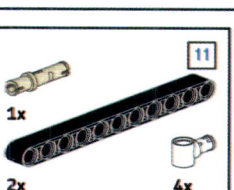

1x   11

2x   4x

The 3M tan pin should go in the marked hole of the motor shaft.

9

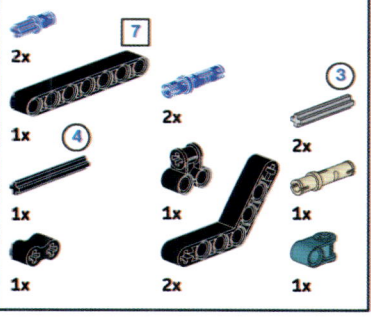

2x   7

1x   2x   3

4   2x

1x   1x   1x

1x   2x   1x

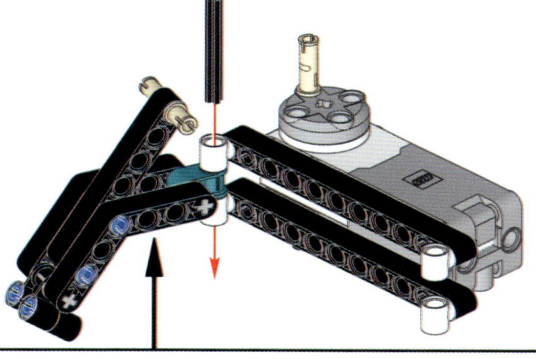

The rubber beam adds grip to the legs when Shelly walks.

1

The teal cross block should be free to rotate on the grey 3M axle. Insert the axle through the teal block's round hole, not its cross hole.

2

3

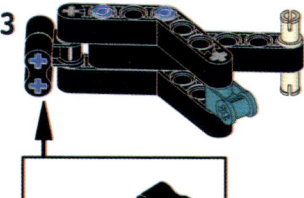

10

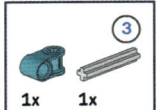

1x 1x

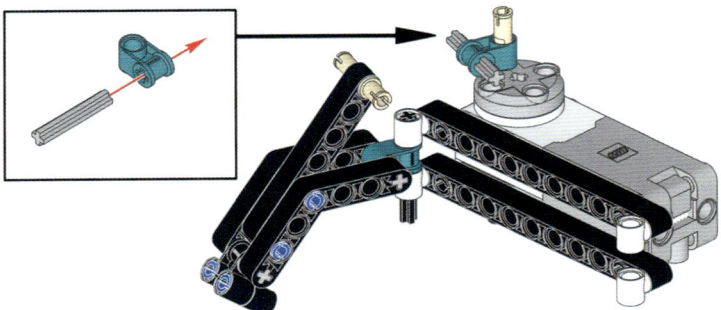

11

1x

2x

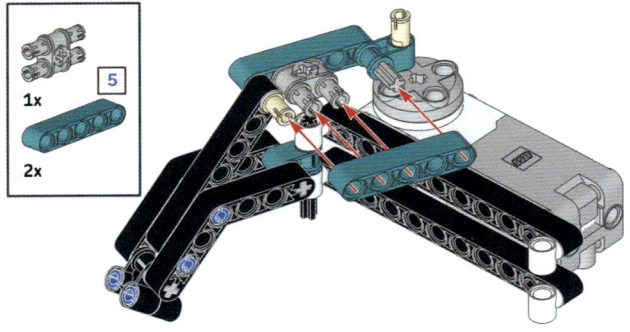

---

right motor subassembly

12

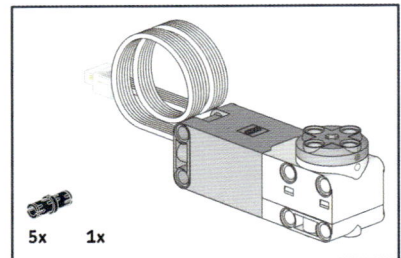

5x 1x

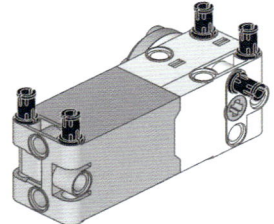

13

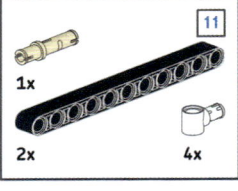

1x

2x 4x

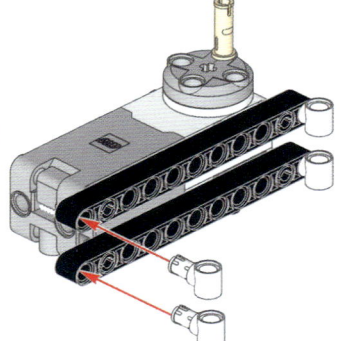

The 3M tan pin should go in the marked hole of the motor shaft.

14

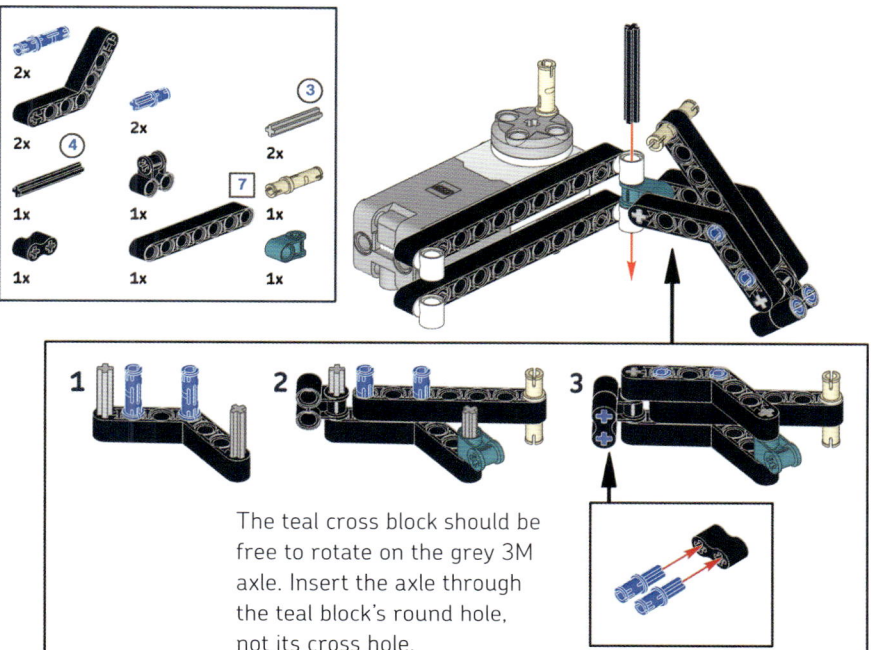

1 **2** **3**

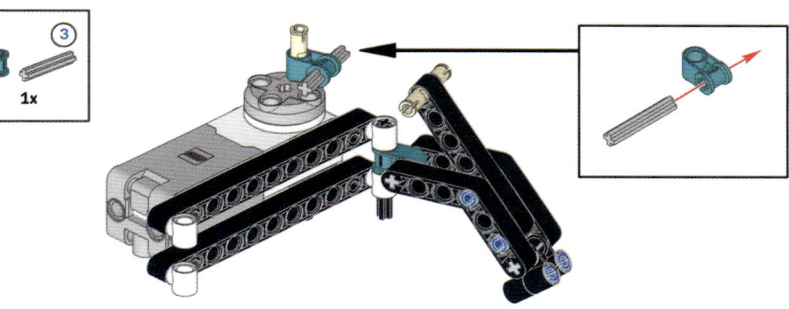

The teal cross block should be free to rotate on the grey 3M axle. Insert the axle through the teal block's round hole, not its cross hole.

The rubber beam adds grip to the legs when Shelly walks.

15

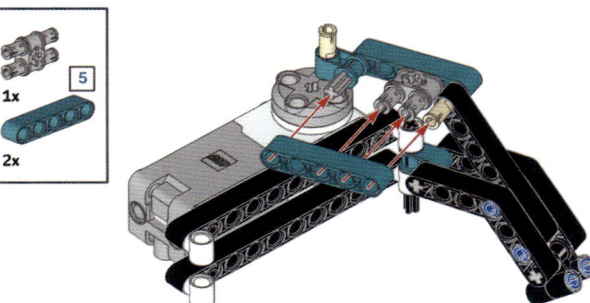

16

17

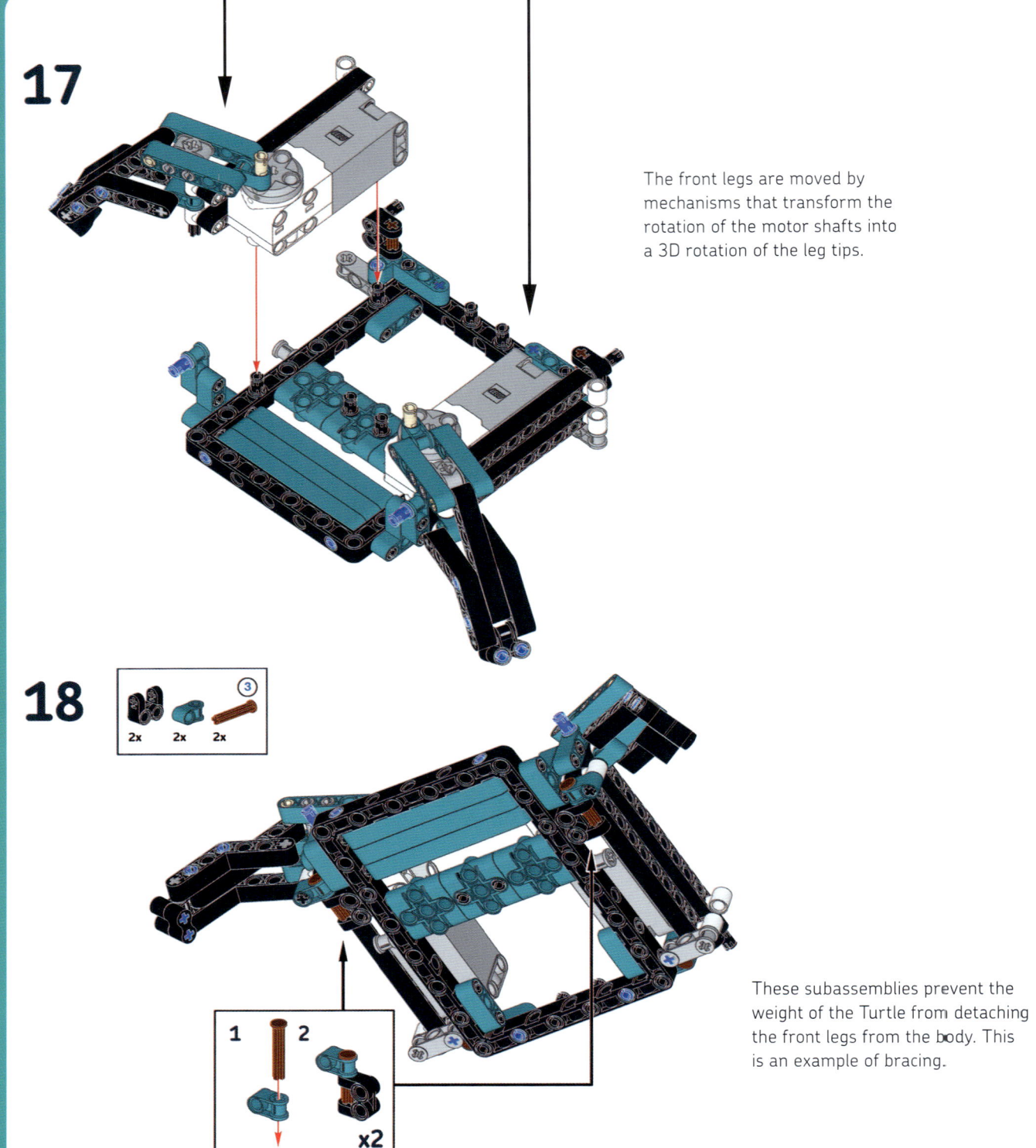

The front legs are moved by mechanisms that transform the rotation of the motor shafts into a 3D rotation of the leg tips.

18

2x   2x   2x   ③

1   2   x2

These subassemblies prevent the weight of the Turtle from detaching the front legs from the body. This is an example of bracing.

19

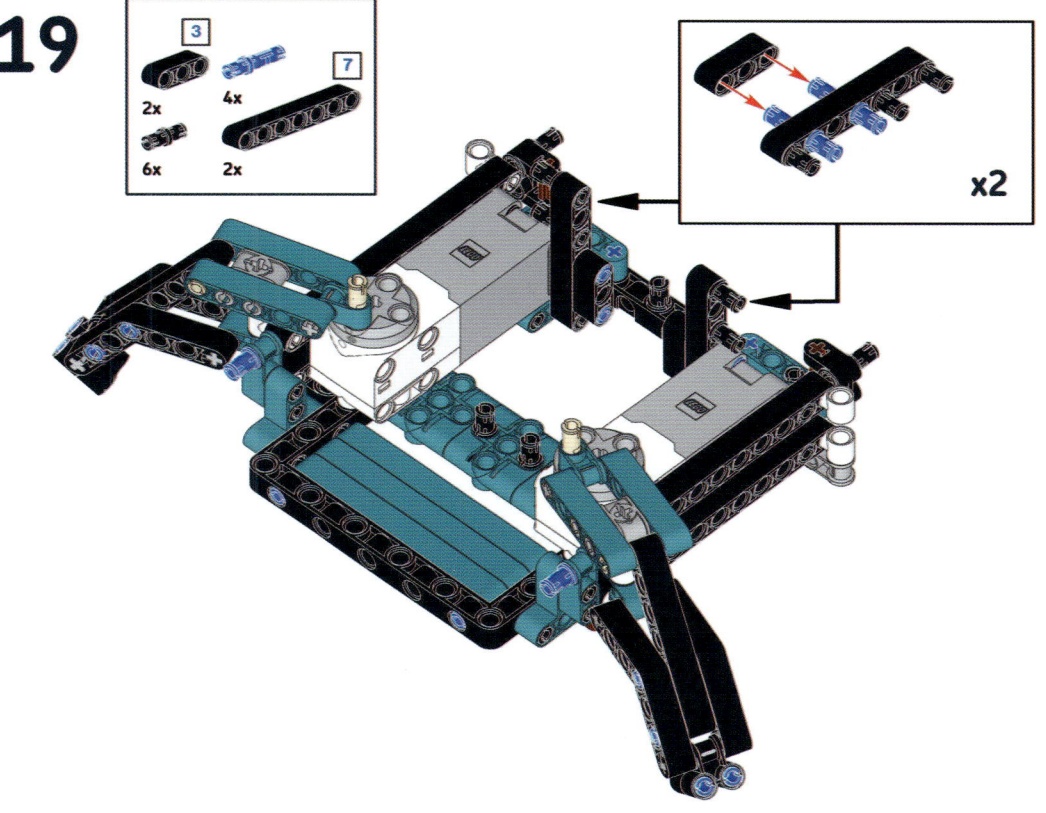

These beams will lock the motors into place so they can't be pulled away from the body. This is another example of bracing.

x2

20

If you want, you can make the back legs symmetrical by reversing the direction of the brown axles on one of the legs.

21

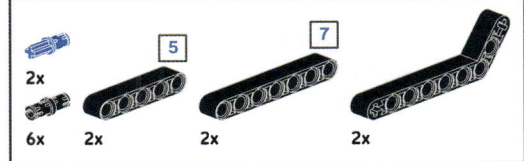

2x

6x 2x 2x 2x

These linkages transmit the motion of the motor to the back legs so that when a front leg is moved forward, the back leg is moved backward, and vice versa. These linkages could be straight, but I chose bent beams to hide the links inside the shell as much as possible.

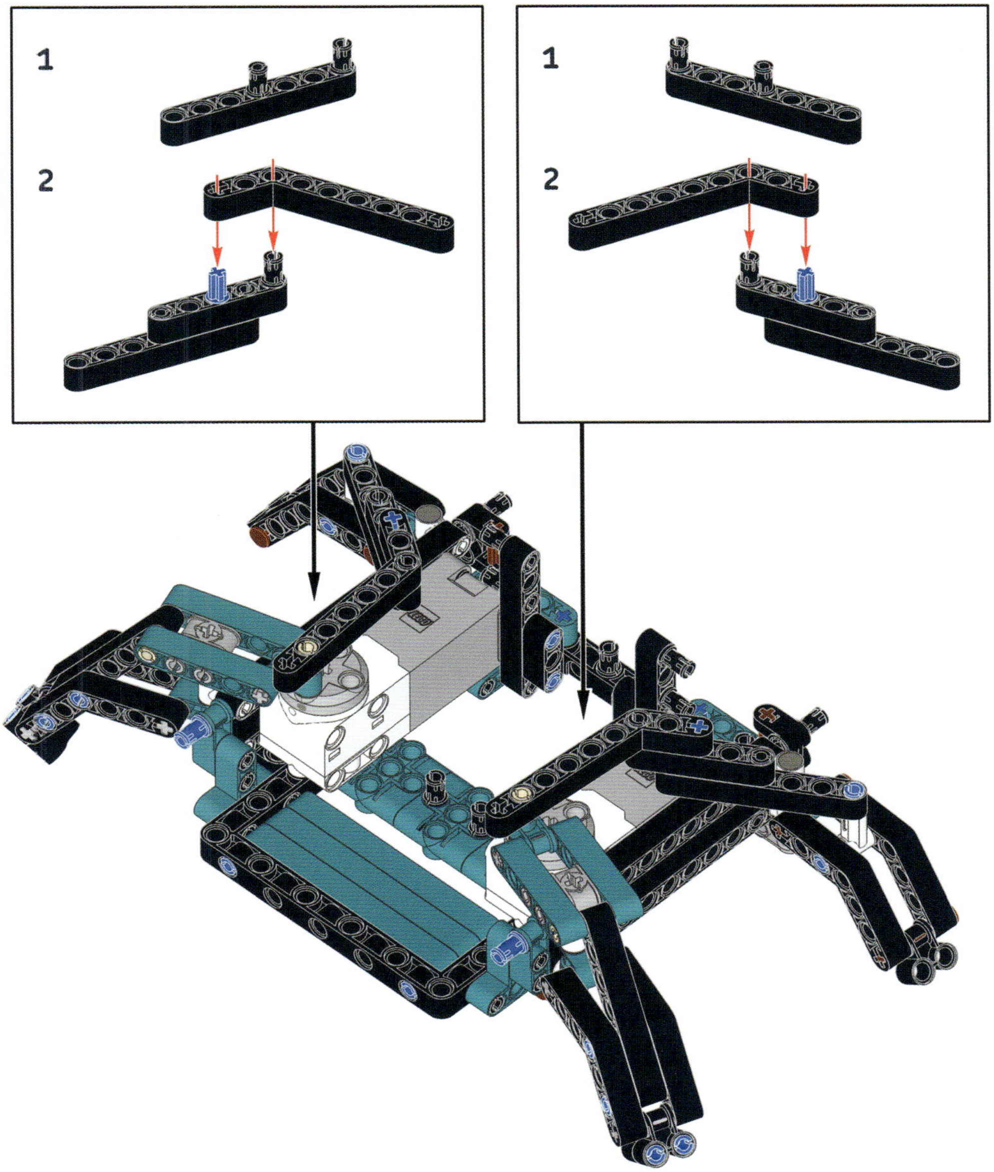

22

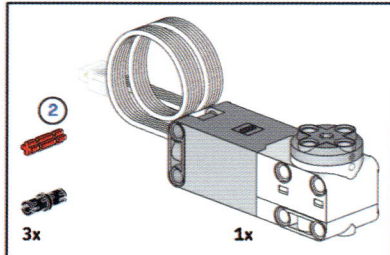

2

3x　　1x

The black pin goes in the marked hole of the shaft.

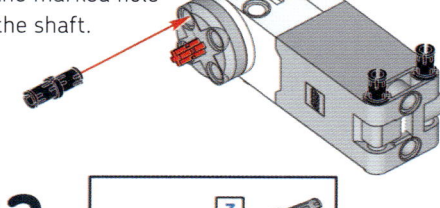

23

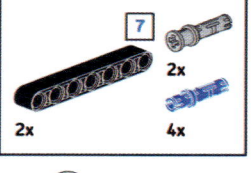

7

2x　　2x

2x　　4x

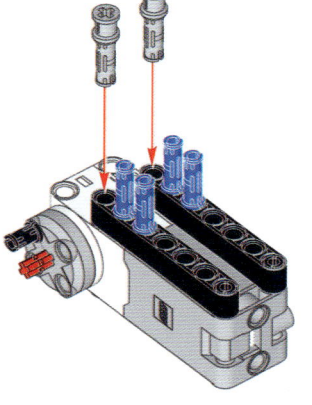

24

3

2x　　2x

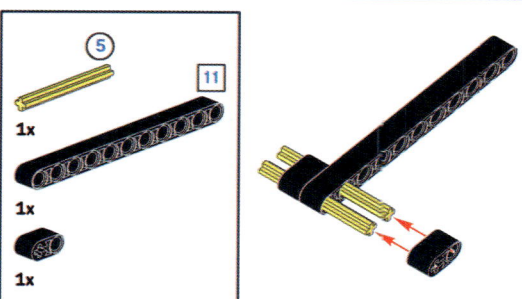

25

5

1x　　1x

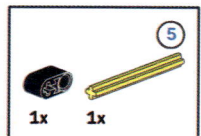

26

5

11

1x

1x

1x

27

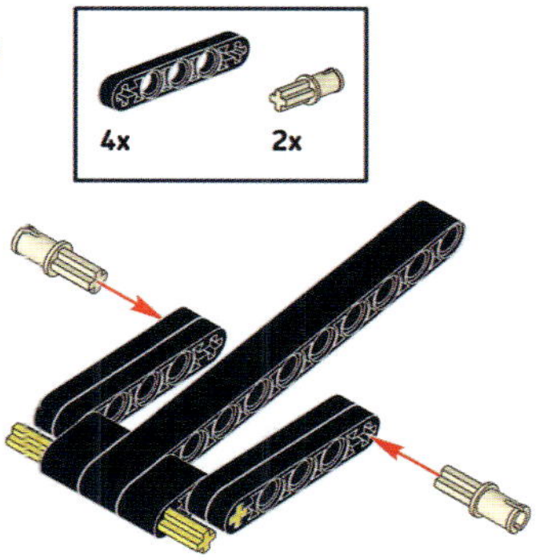

4x 2x

28

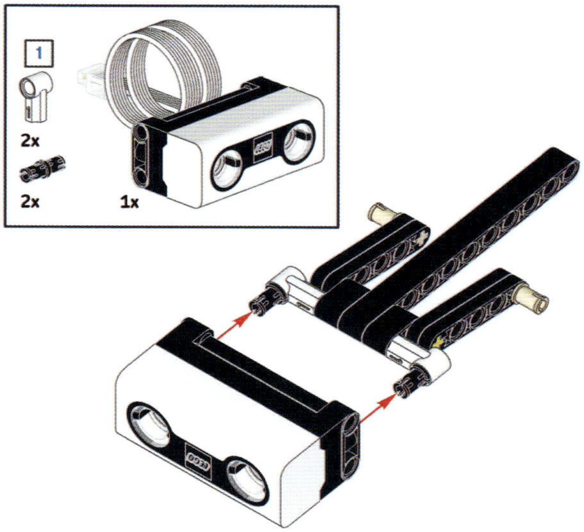

1
2x

2x 1x

29

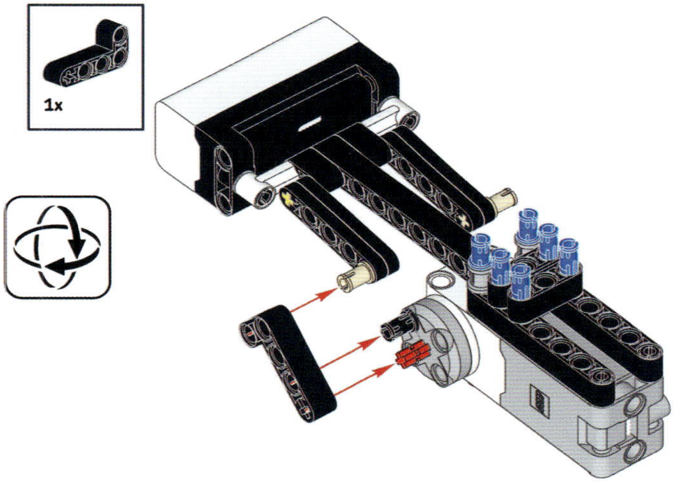

1x

30

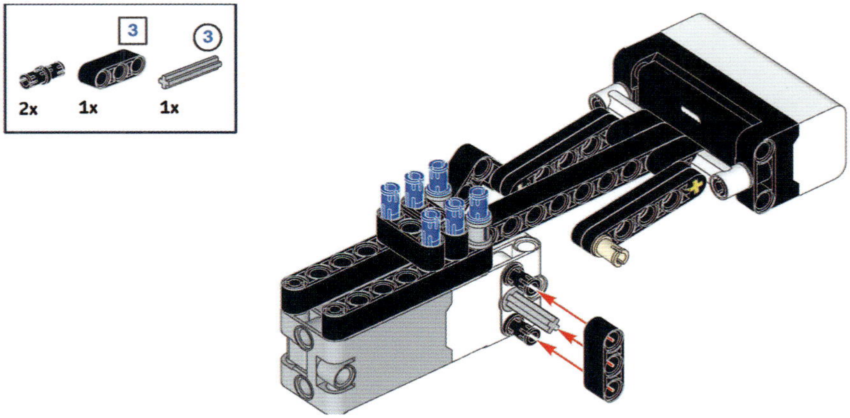

3 ③
2x 1x 1x

31

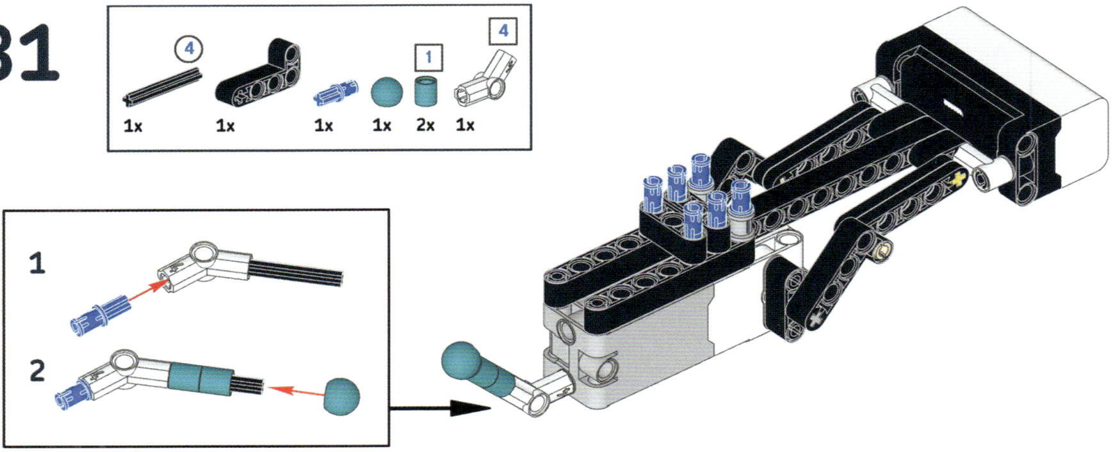

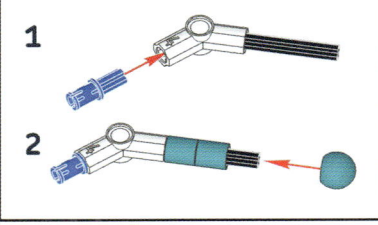

1x 1x 1x 1x 2x 1x

1
2

32

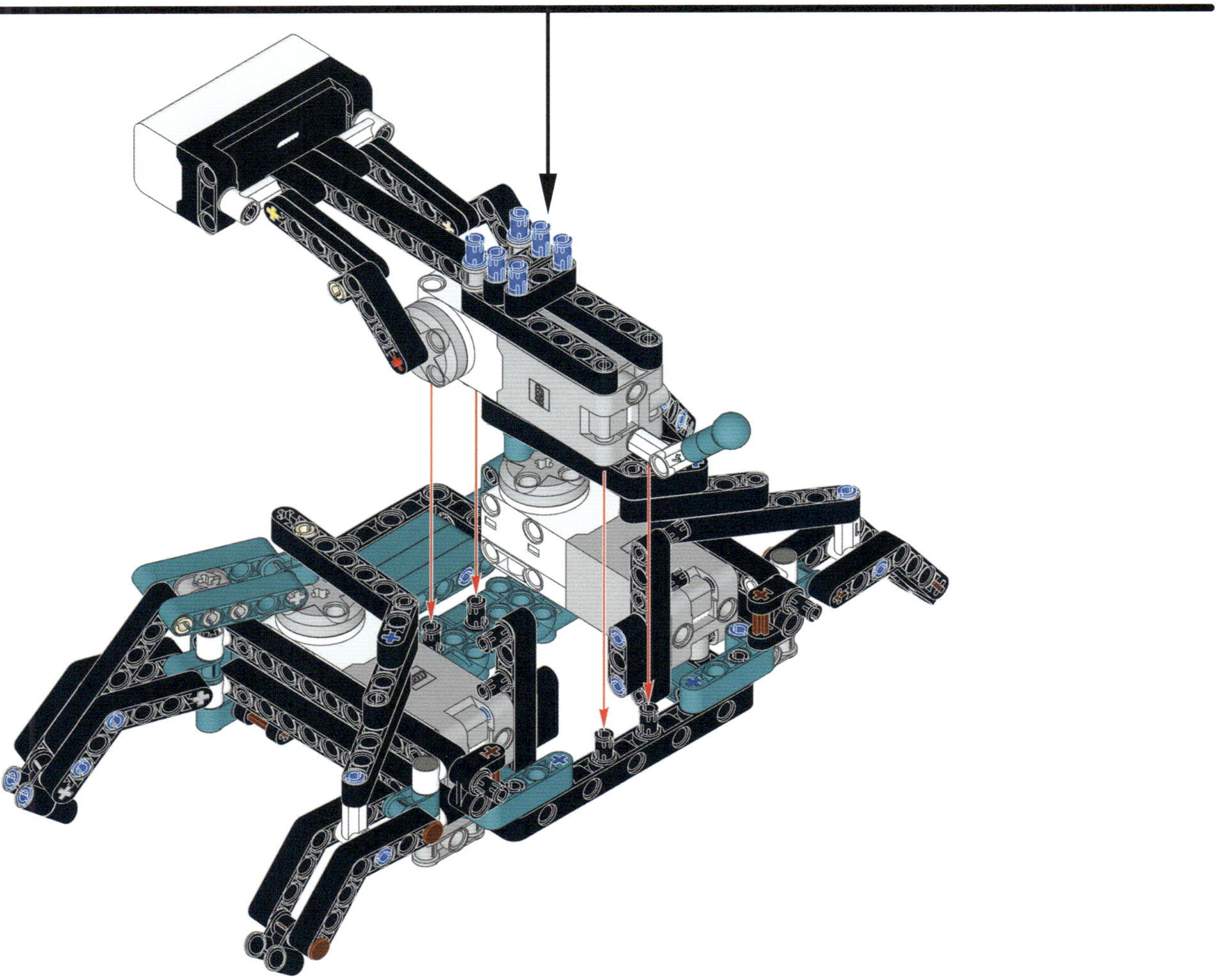

33

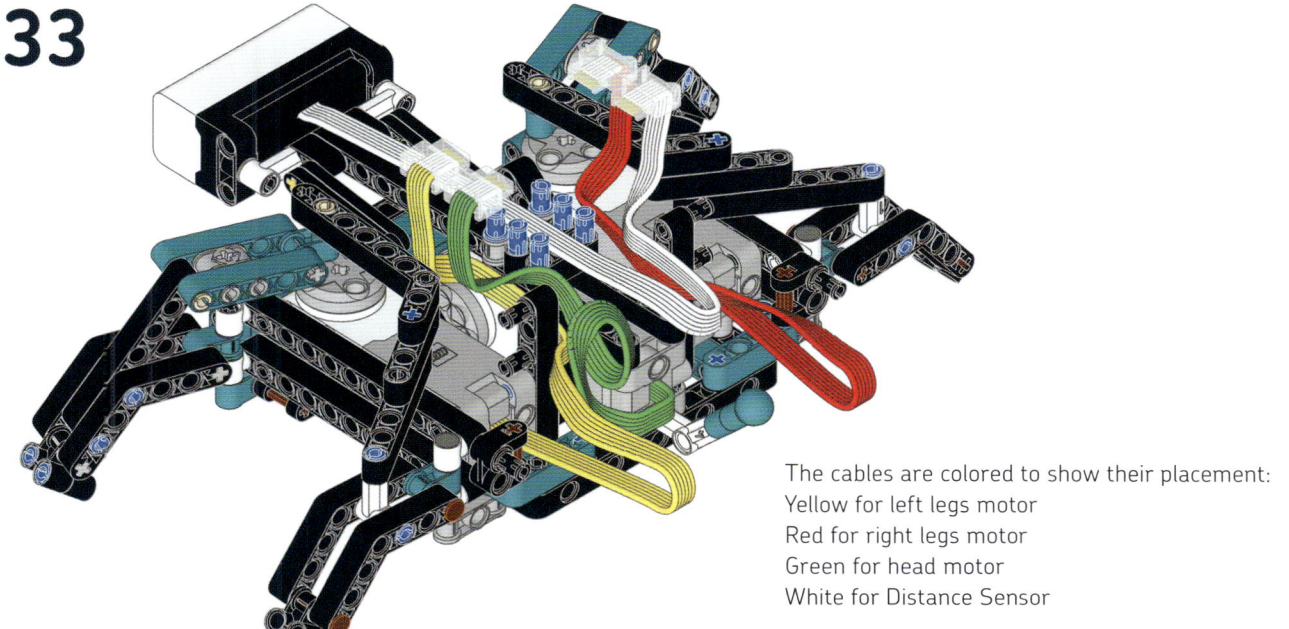

The cables are colored to show their placement:
Yellow for left legs motor
Red for right legs motor
Green for head motor
White for Distance Sensor

34

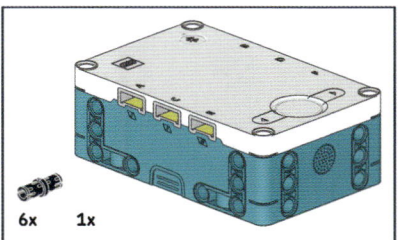

6x   1x

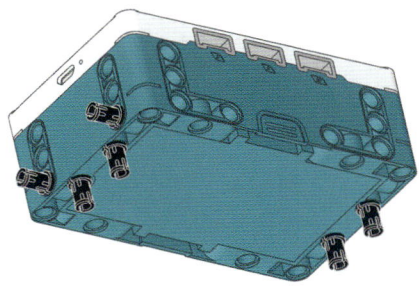

35

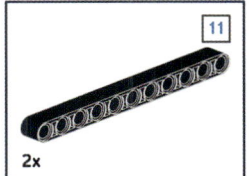

11

2x

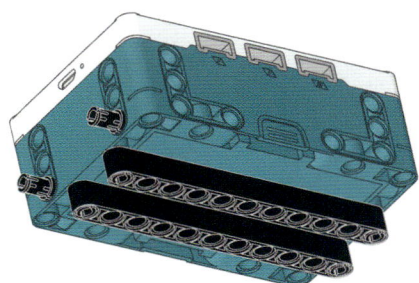

36

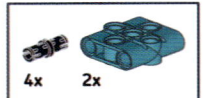

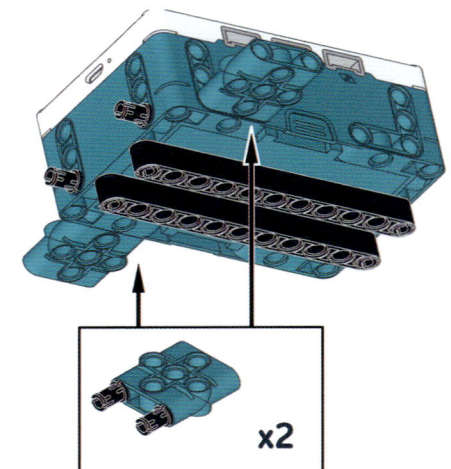

x2

37

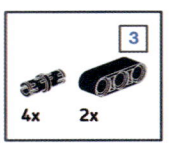

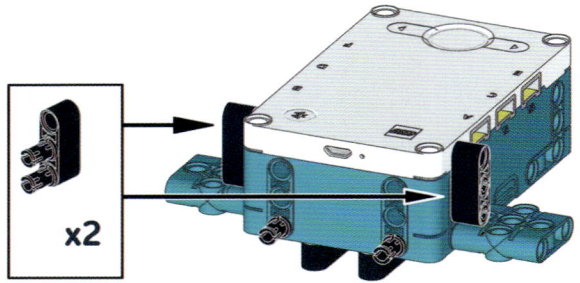

x2

38

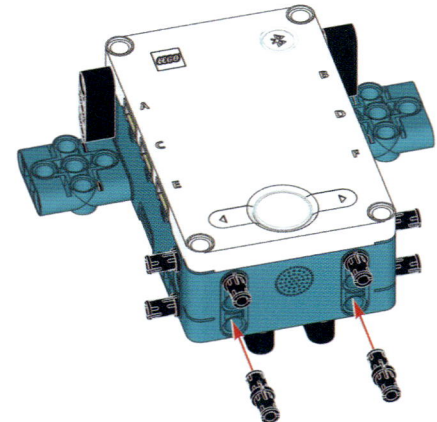

39

2x

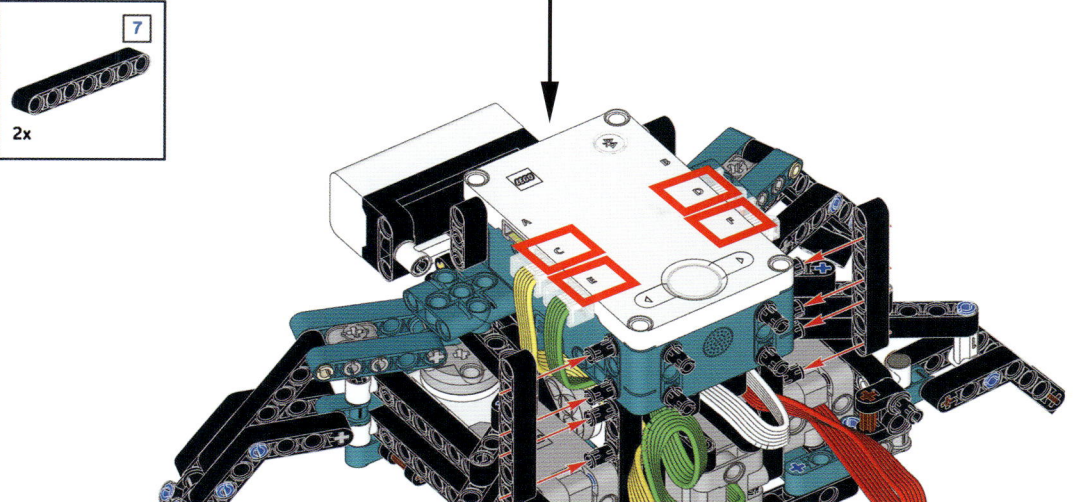

40

1x 2x 4x 2x 1x

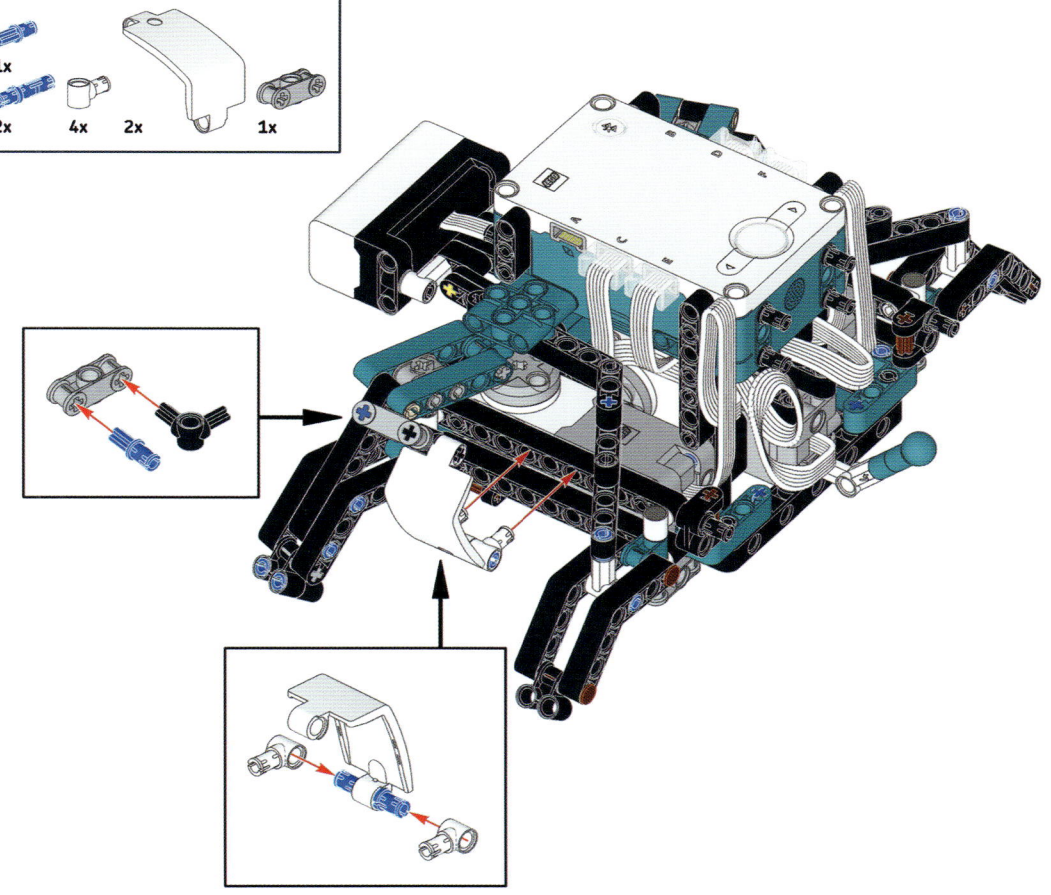

41

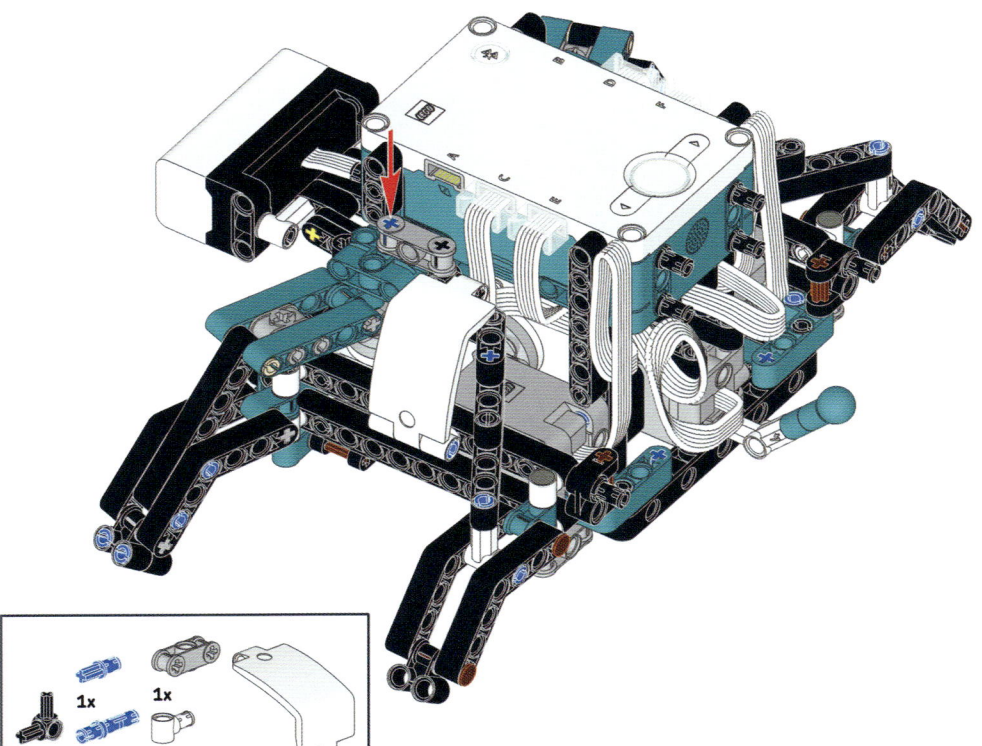

42

1x **1x**
1x **1x** **2x** **1x**

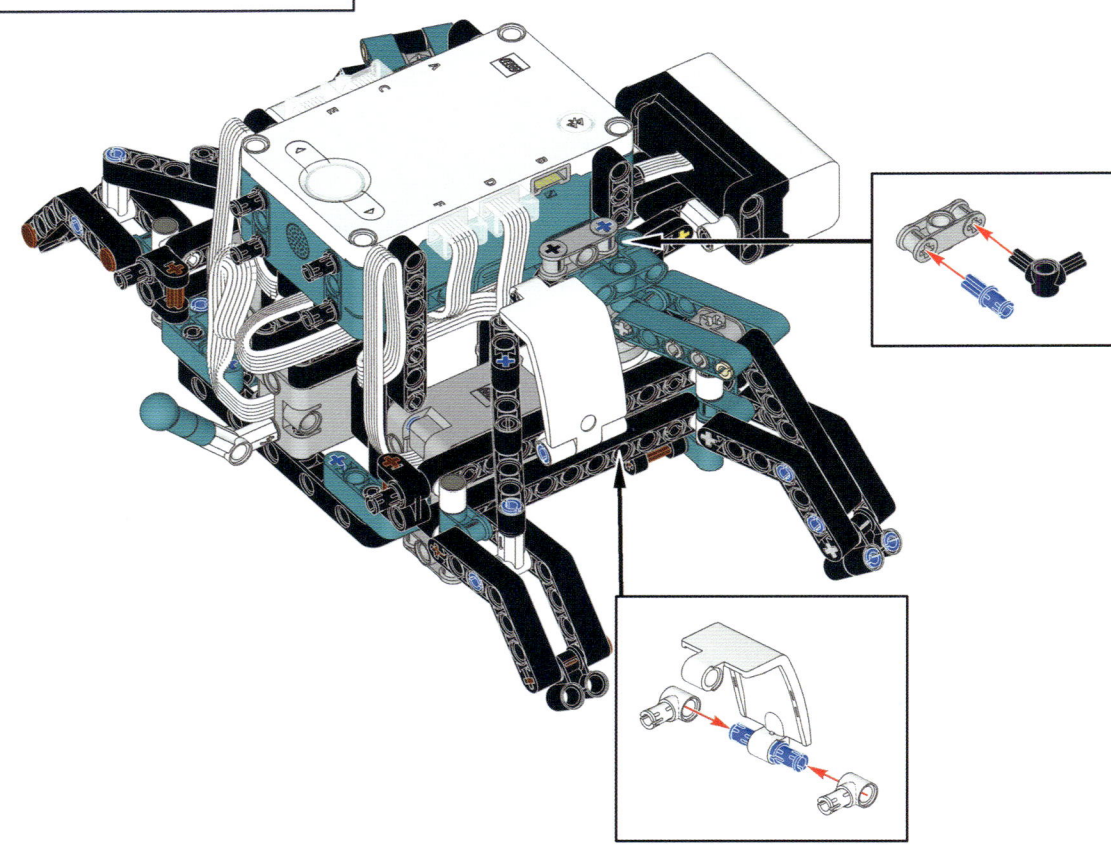

43

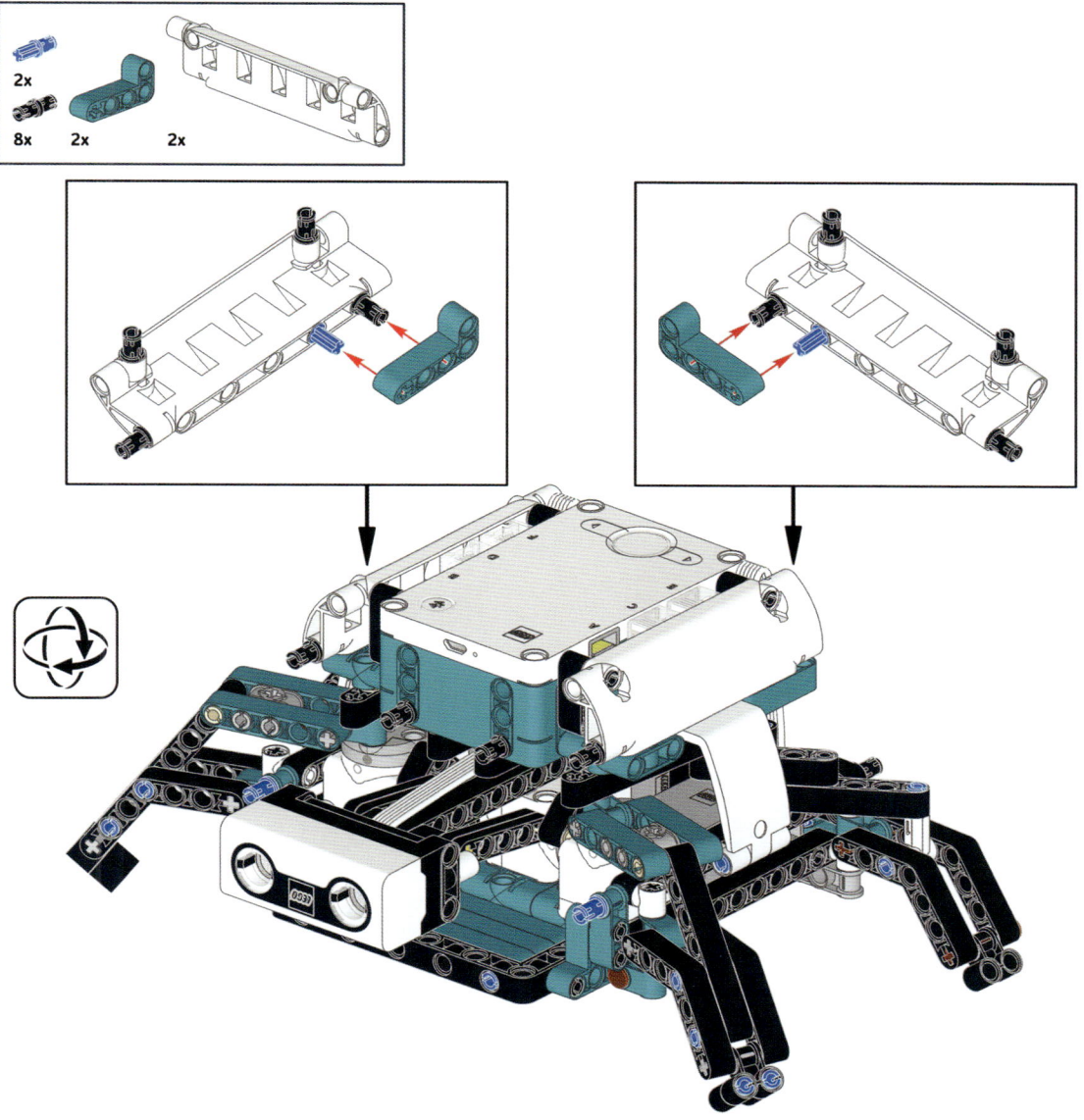

44

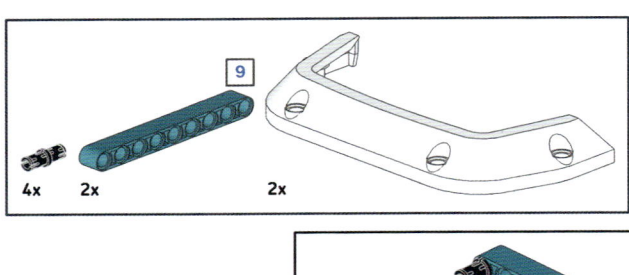

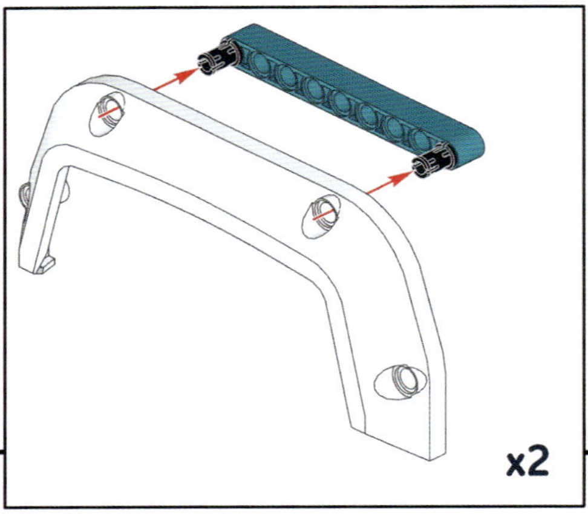

These "mudguard" pieces are decorative but also brace the hub against the bottom of the shell, making the Turtle sturdy.

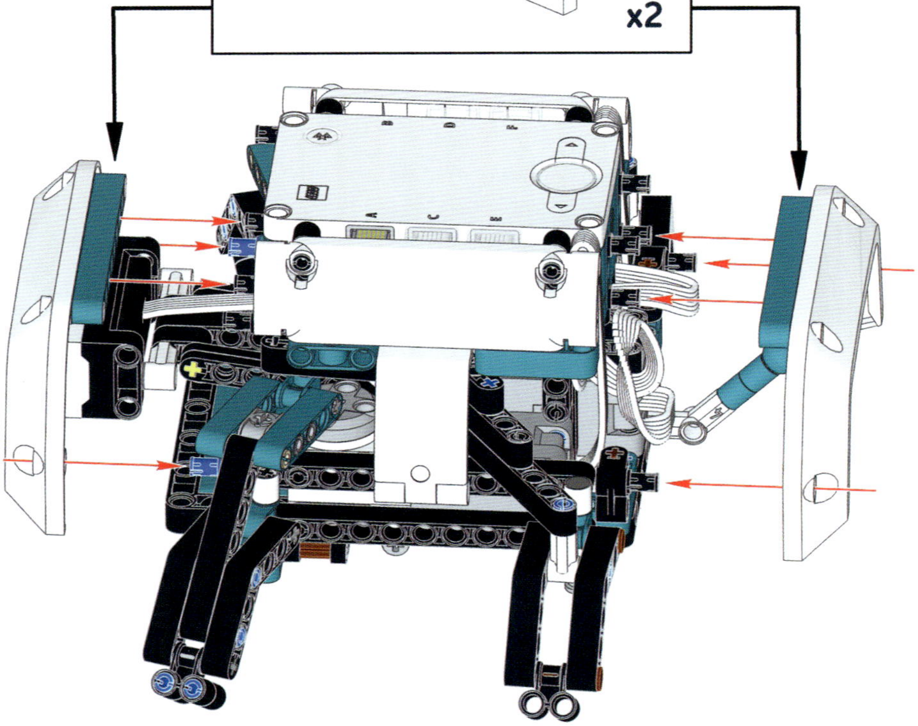

x2

45

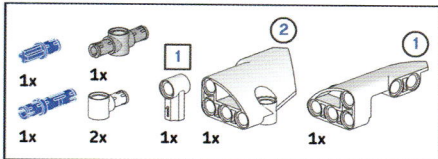

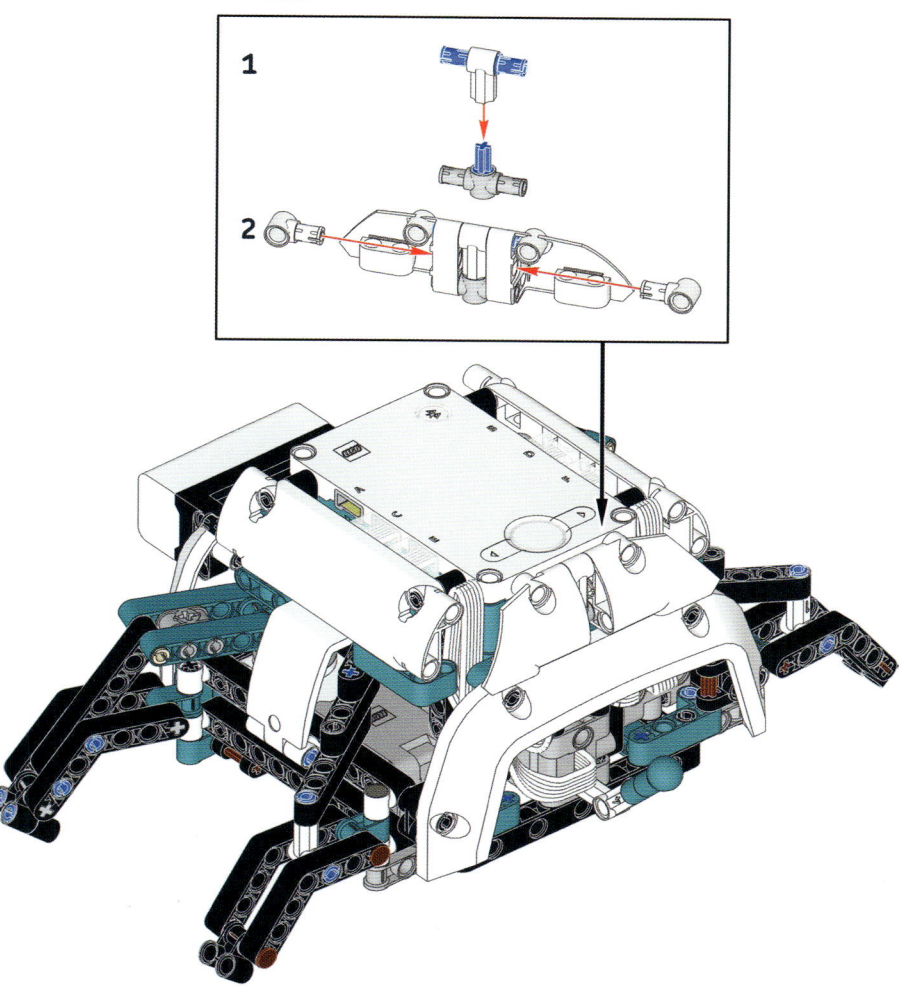

46

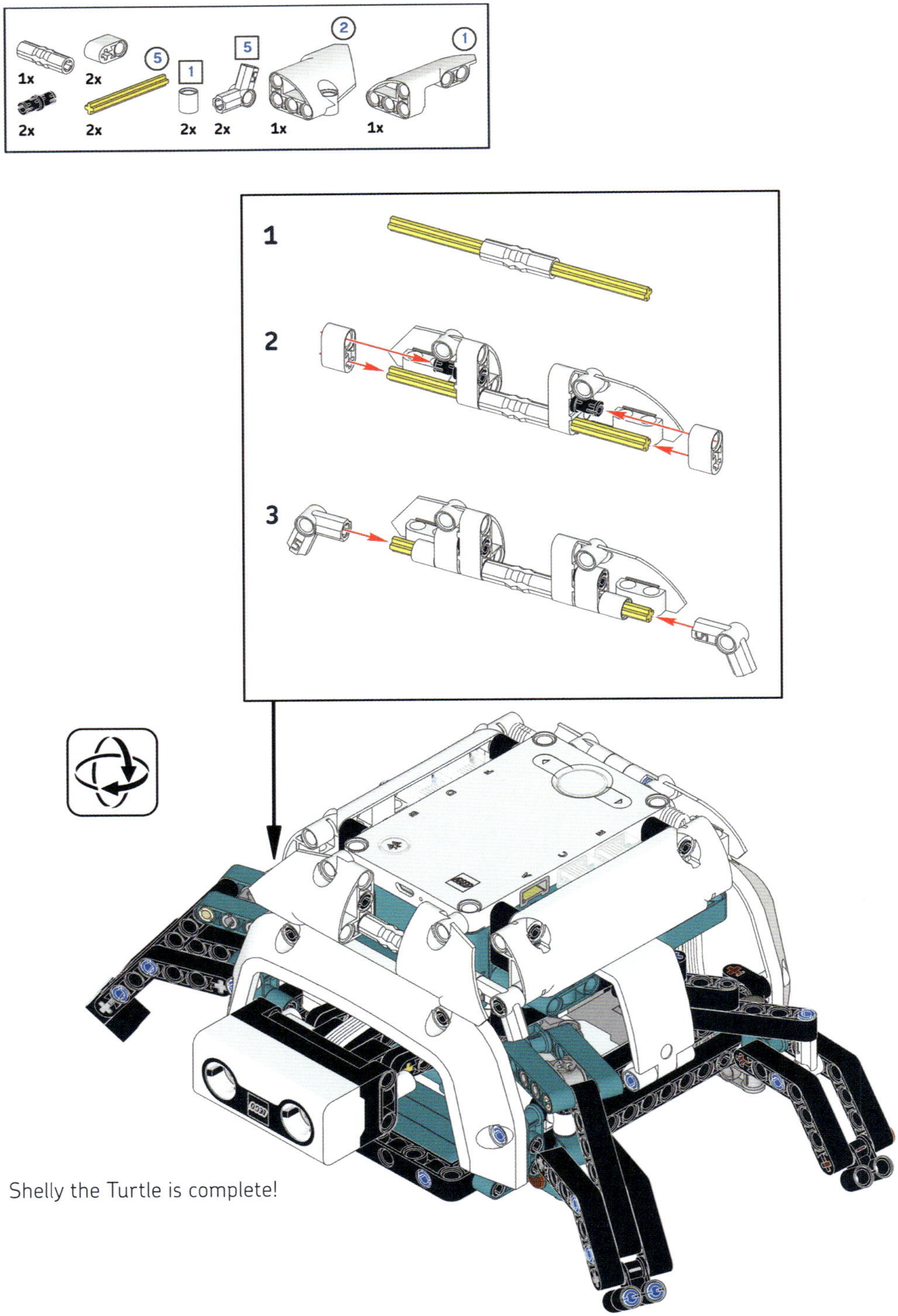

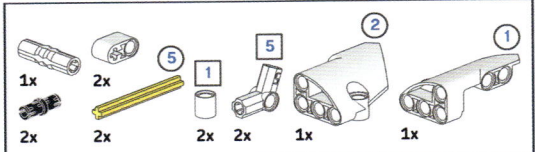

1x 2x
2x 2x 2x 2x 1x 1x

1

2

3

Shelly the Turtle is complete!

programming the turtle

Now that you've built Shelly, it's time to create a base program to make her walk, move her head in and out, and react to disturbances. We'll program the model to have a shy personality, like a real turtle. When she's walking, she should stop and quickly pull her head into her shell as soon as she sees an object that comes too close. Thanks to the Hub's built-in sensor, we can also make Shelly react to other annoyances, such as being knocked on the shell or flipped over.

To start developing programs for my robots, I find it helpful to sketch their behavior using natural language first. Here's a plan for our first program:

```
When program starts
Move head out
Repeat forever:
    Start walking forward
    Wait until you see an object or you're flipped over
    Stop moving
    Hide head
    Tremble with fear until you are finally left alone
    Show head
    Randomly turn right or left a random number of steps
Go to the beginning of the loop
```

This plan is implemented in the base program shown in Figure 5-2. Apart from a few extra setup blocks at the beginning, reading through the word blocks is exactly like reading through the plan.

To create the program, first make the custom blocks, defined in Figures 5-3 and 5-4, and then proceed with the main stack shown in Figure 5-2. See Chapter 4 to review how to create custom blocks. The steps to make the complex logic conditions used by the wait until block ❷ in stack A and the if then block ❹ in stack G are highlighted inside the gray frames. When you're done, save the program as *turtle_base_program*.

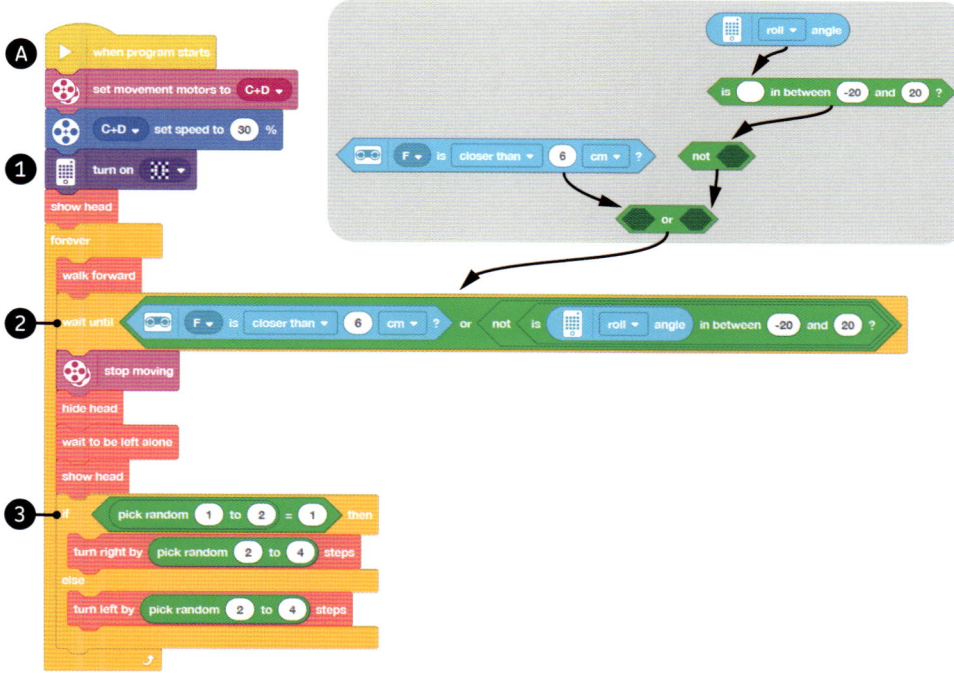

Figure 5-2: The main stack of the Turtle base program

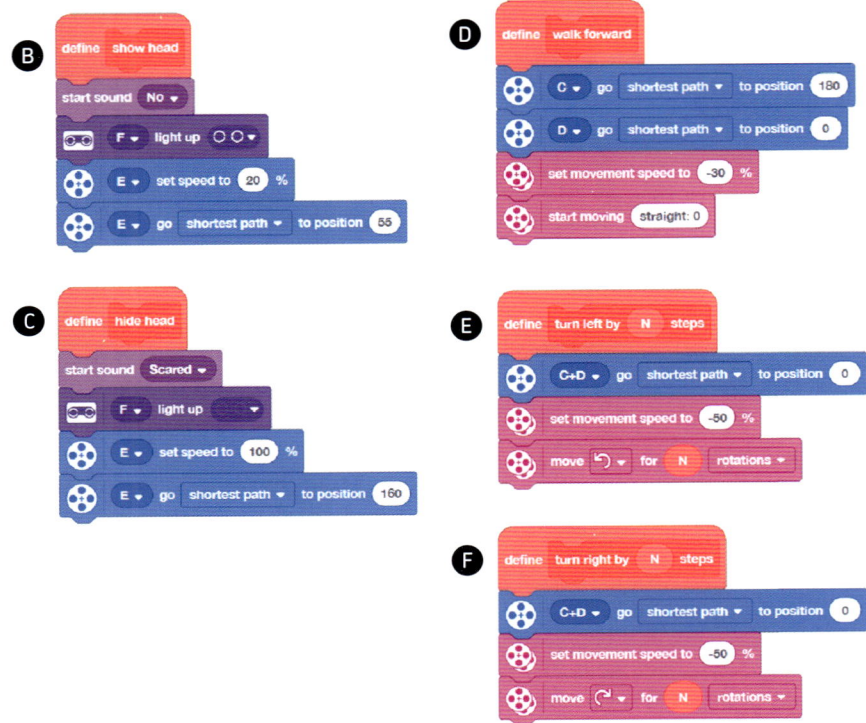

Figure 5-3: The custom blocks that manage Shelly's movements: moving the head in and out, walking, and turning

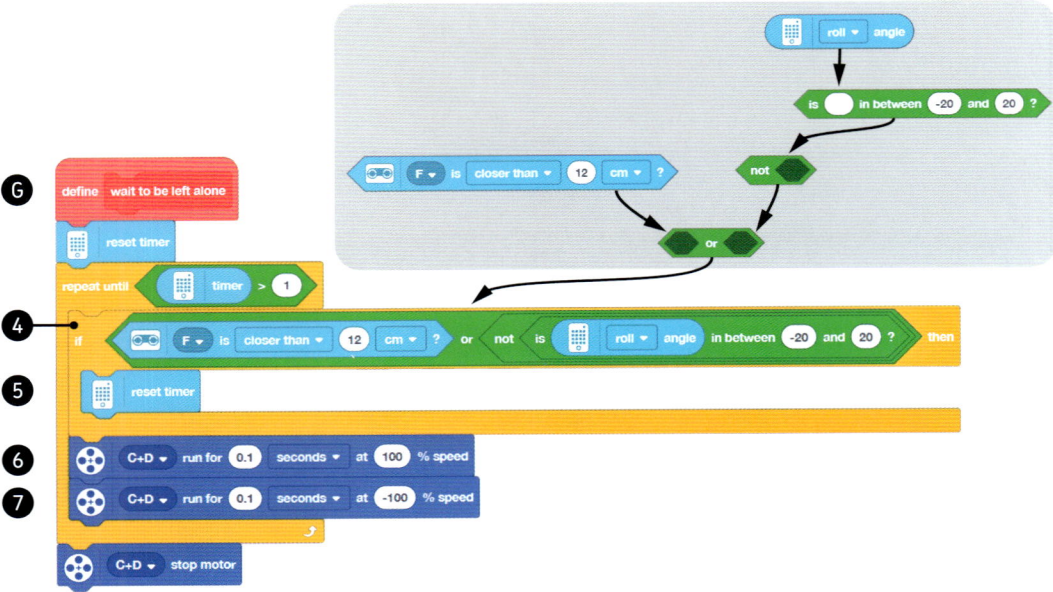

Figure 5-4: The custom block that makes Shelly tremble with fear until you leave her alone

creating custom patterns on the Hub display

The `turn on light matrix` block in stack A (❶ in Figure 5-2) allows you to light up your own pattern on the Hub display. The pattern remains lit until the display is told to do something else by another Light block or the program is stopped. We use this block to create a turtle shell design on Shelly's back.

Use the drop-down menu on the block to build the pattern you want to display. The menu lets you set each light in the Hub's 5×5 display separately. First choose a brightness level with the slider on the right; then click a square in the 5×5 grid to set a light to that brightness level. Click the square again to turn that light off. You can turn all the lights off or on with the two buttons at the bottom.

Figure 5-5: A custom light pattern for our Turtle's shell

For our Turtle program, try reproducing the pattern shown in Figure 5-5. Or you can design a pattern of your own!

creating complex logic conditions

We've already used logic operators to create composite logic conditions when we built the Gobbler in Chapter 3. Let's now look closely at the long logic expressions in our Turtle program. The first is used by the `wait until` block ❷ of stack A shown in Figure 5-2. You can read it like so:

Wait until the distance read by the Distance Sensor is less than 6 cm OR the value of the roll angle of the Hub is NOT between -20 and 20 degrees.

A Distance Sensor reading of less than 6 cm indicates an object is getting close to Shelly. The `NOT` operator negates the result of the `is between` block, so the second condition checks whether the Hub's roll angle value is *outside* the specified range, indicating Shelly has been flipped over. The `wait until` block will let Shelly keep walking forward until one of the two conditions combined with the `OR` block becomes true. As soon as one of the conditions is met, the rest of the program continues, starting with the `stop moving` block.

The logic expression used by the `if then` block ❹ in stack G, shown in Figure 5-4, is similar to the previous one and can be read in the same way:

If the distance read by the Distance Sensor is closer than 12 cm OR the value of the roll angle of the hub is NOT between -20 and 20 degrees, then . . .

As we'll see, this condition makes Shelly keep trembling with fear until the object in front of her goes away or she's no longer flipped over.

understanding the blocks to walk and move the head

Figure 5-3 shows the custom block definitions that make Shelly move her head, walk forward, and turn. The `show head` block definition (stack B) starts by playing the No sound from Charlie's sound library. Then it turns on the Distance Sensor lights, sets a low speed for the motor, and tells it to turn to the specified angle, moving the head out from under the shell. The `hide head` block definition (stack C) is similar. It plays the Scared sound from Charlie's sound library, turns off the Distance Sensor lights, sets the maximum speed for the motor, and turns the motor to pull the head inside the shell.

WARNING If the head does not come fully out, or you hear the motor whistling and not moving, you might not have assembled the crank that moves the neck slider correctly. Double-check the assembly: the black pin in step 22 of the building instructions should go in the marked hole of the motor shaft.

The `walk forward` custom block definition (stack D) starts by aligning motors C and D so they're half a turn off from each other, meaning the left and right legs are at opposite ends of their movement loop. Then we set the speed of the driving motors (negative because of the way the motors are facing) and start running the motors to make Shelly walk straight.

Stacks E and F differ only by the steering input of the final Movement block in each stack. We move both motors to 0 to place the legs in mirroring positions, and we set a medium speed for the motors. Then a `move for duration` block moves the legs in opposite directions for a certain number of rotations to make Shelly turn left (stack E) or right (stack F). We control the number of rotations with the custom block input N, which is set by a random number generator when the custom block is called in stack A. We also use a random number generator inside an `if then else` block to decide whether Shelly should turn left or right (❸ in Figure 5-2).

using the timer

The `wait to be left alone` custom block definition (stack G), shown in Figure 5-4, makes Shelly shake her legs back and forth (blocks ❻ and ❼) to express fear (I know real turtles don't do that!) until she's left alone. She must be standing upright without any objects in sight for more than 1 second for the program to proceed. Let's see how we use the Hub's built-in timer to achieve that.

The Hub has a timer that keeps track of how much time has passed since the program started, down to the nearest millisecond—like a stopwatch. The `timer` block reports the current reading from the timer. You can start the timer over from zero at any point in the program by using the `reset timer` block. That's exactly what we do at the beginning of stack G.

Once the timer is reset, if the complex condition of the `if then` block ❹ returns `false`, the `repeat until` loop will end after 1 second:

`repeat until timer value is greater than 1 second`

However, if either the Distance Sensor sees an object closer than 12 cm or the roll angle indicates Shelly is flipped over, the timer is reset ❺, preventing it from reaching 1 second. The motor blocks ❻ and ❼ continue shaking Shelly's legs, and the `repeat until` loop starts again.

For the loop to end and the motors to stop, the `if then` condition must remain false—that is, Shelly must be standing upright and not see a nearby object—for more than 1 second. Otherwise, the `if then` block will execute the `reset timer` block, and this will happen again and again.

creating more interesting behavior

Now let's create a new program to make Shelly's behavior more interesting. Our program will define three possible moods for Shelly: Inactive, Scared, and Sleepy, with a different behavior for each mood. We'll also create two actions, Wander and Eat, and we'll define the logic for when each mood or action should take place. When we run the program, Shelly will seem to have a mind of her own. In other words, she'll be *autonomous*.

To create an autonomous robot, you should always start by planning its behaviors with pencil and paper. If you rush into programming without a clear plan, you won't get far. Figure 5-6 shows my plan for Shelly's behaviors. There are rectangles for each mood or action and arrows showing how each mood or action can lead to another. For example, Shelly will eat only after being scared, but she might wander after being either scared or inactive.

The text next to each arrow explains the *event* that causes that change in behavior. The same events could lead to different actions depending on Shelly's current mood. For example, if Shelly sees your hand when she's inactive, she'll become scared, but when she's scared and you keep showing your hand while patting her shell, Shelly will trust you and eat. A tap when she's sleeping will wake her up, but when she's awake, the same tap will freak her out!

Our program to implement the behavior sketched in Figure 5-6 will use the same custom blocks from our first Turtle program. To get started, create a copy of the original program and save it as *turtle_behavior*. (To review how to save a copy of a program with another name, see page 29 in Chapter 3.) Then delete everything but the five custom block stacks shown in Figure 5-3.

using variables

Unlike the programs we've written so far, our new program needs a way to *remember*. It must be able to set Shelly's current mood, store what that mood is, and look up the mood to decide how Shelly should behave. We can do this with a variable.

A *variable* is a place in the Hub's memory to store a piece of data so your program can look it up later. Variables are like labeled boxes: they can contain a value, which is the data you

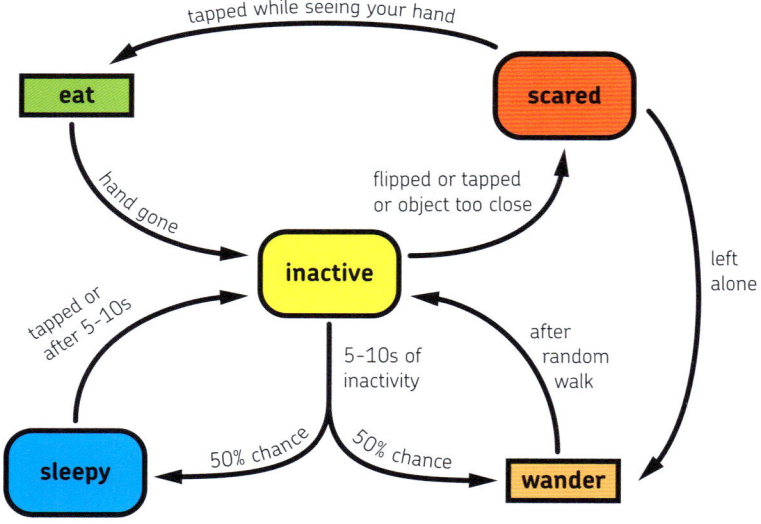

Figure 5-6: This diagram describes Shelly the Turtle's behavior. Each arrow represents a change in her mood caused by sensor or timer events.

want to store, and they have a name, which is a label you give to that data so you can find it again. You can also update the value of a variable at any time by storing a different piece of data under the variable's name. LEGO MINDSTORMS programs let you store two kinds of data in variables: numbers and strings (text).

NOTE You've encountered some blocks that have *logic* (or *Boolean*) inputs, but no such type exists for variables. If you need to store logic data, you can use the number 1 for true or 0 for false.

We'll create a variable called mood where we can store a value showing Shelly's current mood: INACTIVE, SLEEPY, or SCARED. Figure 5-7 shows how to do it step-by-step.

1. In the Variables palette, click the **Make a Variable** button.
2. Type in the name mood for your variable. Always choose a name that describes clearly what information the variable is going to contain.
3. Click **OK** to create the variable.
4. Three new blocks should appear in the Variables palette: a round variable block with the variable name on it, the set variable block, and change variable.

The round variable block reports the current value of the variable for use in the program. Whenever you create a variable, one of these blocks appears in the Variables palette with your variable's name on it. The set variable block sets the variable to the value you specify—either a string or a number. The change variable block changes the variable by the specified value. For example, if your variable contains the value 30, then change variable by 3 would set the value to 33, while change variable by -1 would set the value to 29. The Variables palette will always contain only one set and one change block, no matter how many variables you create. You can choose which variable to set or change with the drop-down menu on each block.

NOTE You can always change a variable's name by right-clicking its block in the Variables palette (or long-pressing on touchscreen devices) and selecting Rename Variable.

Now that you know how to manage variables in your program, you should be able to reproduce the advanced program for the Turtle by looking at Figures 5-8, 5-9, 5-10, and 5-11. You'll just have to create one more variable called randomTime.

understanding the advanced program

Let's start our look at the advanced Turtle program with the three additional custom blocks that, together with those already shown in Figure 5-3, control Shelly's movements. The new block definitions are shown in Figure 5-8.

The eat custom block, defined in stack A, moves both legs ❶ so that Shelly's head points down, then moves the head in

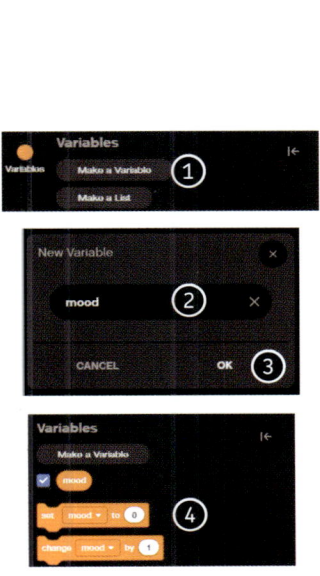

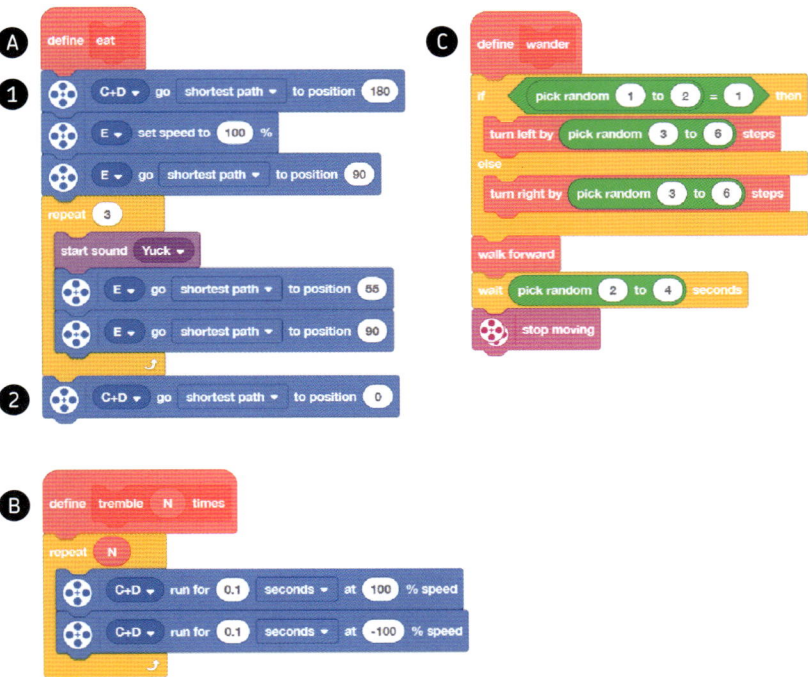

Figure 5-7: Follow these steps to create a variable.

Figure 5-8: More custom blocks defining Shelly's movements: eat, tremble N times, and wander

and out three times in a `repeat` loop while playing the Yuck sound (from Charlie's library) in sync with the movement of the head to simulate the Turtle's biting and gulping. Then the legs are moved to raise the head up ❷.

The `tremble N times` block, defined in stack B, makes the legs shake with the same Motor blocks we used in the `wait to be left alone` block definition in Figure 5-4. This custom block has an input controlling the number of times the loop should repeat.

The `wander` custom block, defined in stack C, makes Shelly turn randomly left or right by a random number of steps drawn from the range of 3 to 6. Then Shelly begins walking forward. After letting her walk for a random number of seconds between 2 and 4, we stop the motors.

Figure 5-9 shows the `inactive behavior` custom block stack. It defines Shelly's behavior when in the INACTIVE mood, during which Shelly doesn't appear to be doing anything (hence the name *inactive*!). In fact, Shelly uses two `if then` blocks, ❶ and ❸, to check whether certain conditions are true. The first `if then` block makes Shelly wait for a random number of seconds to pass. This random number is generated and stored into the `randomTime` variable every time Shelly's mood is changed to INACTIVE (see stacks E, F, and G in Figures 5-10 and 5-11). If the time returned by the `timer` block is greater than that random number of seconds, one of two things can happen based on a random draw between 1 and 2: Shelly's `mood` variable can change to SLEEPY, or Shelly can wander to stretch her legs and then remain in the INACTIVE mood. In either case, the timer is reset. If we set MOOD to SLEEPY, we also store a random number of seconds into the `randomTime` variable with block ❷. This will be the length of Shelly's nap (see stack E).

The other `if then` block ❸ makes Shelly react to disturbances: being flipped over, sensing an object too close, or being tapped on the shell. If one of these events occurs, Shelly hides her head, and the `mood` variable changes to SCARED.

The `sleepy behavior` custom block stack shown in Figure 5-10 defines Shelly's behavior when in the SLEEPY mood. First the Distance Sensor lights are turned off (to simulate closing the eyelids), and the head is slowly pulled into the shell. The timer is reset, and the Spinning animation from Blast's library is played on loop on the Hub display. The `wait until` block keeps Shelly asleep until one of three conditions becomes true: either Shelly is flipped over, the timer passes the random number of seconds stored in the `randomTime` variable (by block ❷ in stack D), or the shell is tapped.

To show that Shelly has woken up, we light up a custom pattern on the Hub display, and the head is brought out. Then the `mood` variable is changed to INACTIVE, the `randomTime` variable is updated with a new random number from 5 to 10, and the timer is reset so that it starts counting from 0 when the `inactive behavior` custom block is executed.

NOTE It might seem strange that flipping Shelly over while she's SLEEPY would change her mood to INACTIVE rather than SCARED. However, when the `inactive behavior` custom block executes, Shelly will immediately detect that she's flipped over, and her mood will be set to SCARED.

The `scared behavior` custom block, defined by stack F in Figure 5-11, produces Shelly's behavior in the SCARED mood. First she trembles six times. Then she keeps trembling until she's no longer upside-down. (If Shelly was frightened for some reason other than being flipped over, the `repeat until` block ❶ ends immediately, so Shelly trembles only six times.) The timer is then reset.

The next `repeat until` loop ❷ gives you 2 seconds after Shelly stops trembling to make her trust you and eat from your hand. You can make her eat by patting her back with one hand while your other hand is close to her head. If you meet these conditions within 2 seconds, the `eat` custom block will repeat until you move your feeding hand more than 20 cm (about 8 inches) away. If you fail to pat Shelly while showing your hand within the 2-second interval, she won't eat.

Next, Shelly checks if an object is very close ❸. If so, she wanders a bit. Then our custom shell pattern turns on, Shelly shows her head, and the `mood` variable is set to INACTIVE. We update the `randomTime` variable with a fresh random number, and the timer is reset.

Finally, stack G in Figure 5-11 is the main stack of blocks that executes when the program starts. We set up the motors, light up the custom shell pattern, and show Shelly's head. Next we set Shelly's initial mood to INACTIVE and store a random value into the `randomTime` variable for the first time ❹. This value will be used in the `inactive behavior` custom block. Then we use three `if then` blocks inside a `forever` loop to choose which custom behavior block to execute, based on the current value of the `mood` variable. For example, if `mood` is equal to INACTIVE, the `inactive behavior` custom block is executed.

EXERCISE 5-1

Try shaping a different personality for Shelly—for example, by tweaking the ranges for the random number generator blocks or by defining a new mood. Would you like her to be more adventurous? Or would you prefer her to be even more shy?

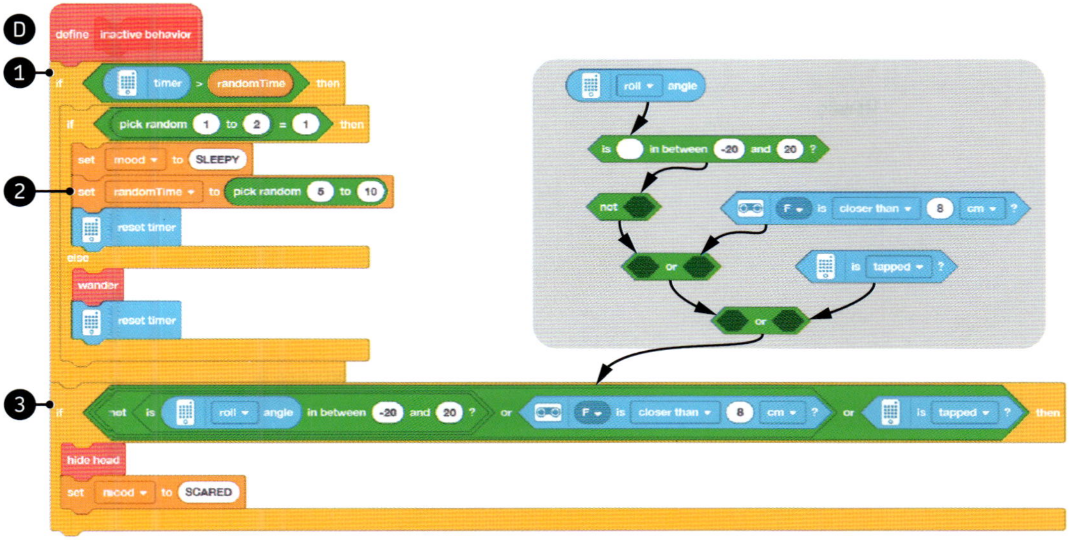

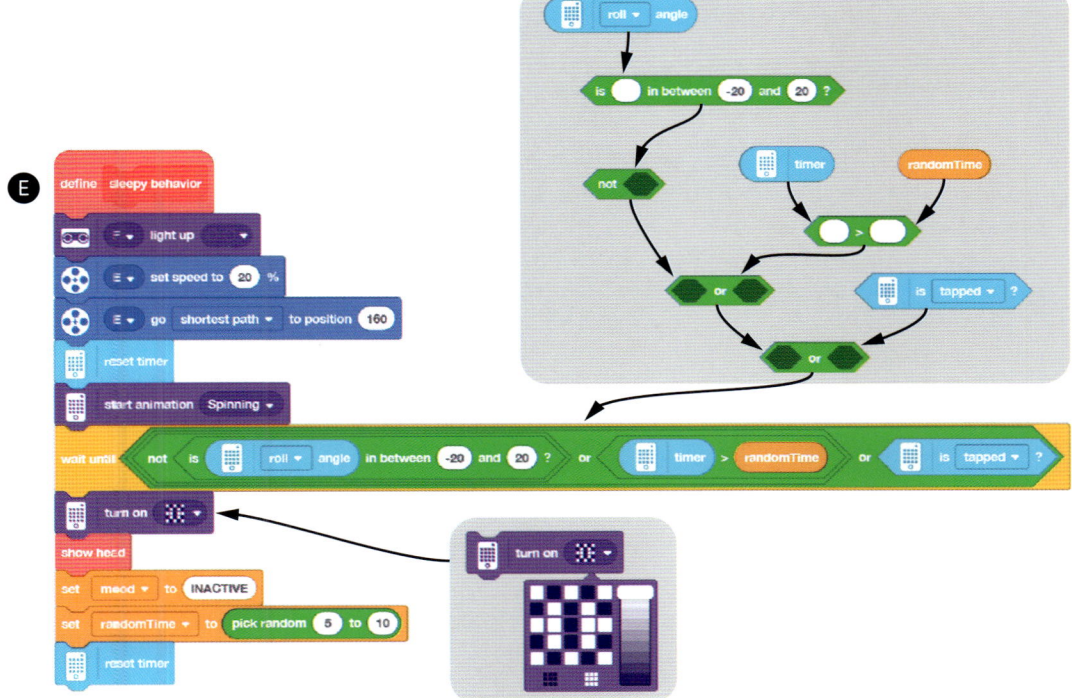

D define inactive behavior

1 if [timer > randomTime] then
    if [pick random 1 to 2 = 1] then
        set mood ▾ to SLEEPY
2   set randomTime ▾ to pick random 5 to 10
        reset timer
    else
        wander
        reset timer

3 if not is [roll ▾ angle in between -20 and 20 ?] or [F ▾ is closer than ▾ 8 cm ▾ ?] or [is tapped ▾ ?] then
    hide head
    set mood ▾ to SCARED

Figure 5-9: The custom block that implements Shelly's inactive behavior

E define sleepy behavior
    E ▾ light up ▾
    E ▾ set speed to 20 %
    E ▾ go shortest path ▾ to position 160
    reset timer
    start animation Spinning ▾
    wait until not is [roll ▾ angle in between -20 and 20 ?] or [timer > randomTime] or [is tapped ▾ ?]
    turn on ▦
    show head
    set mood ▾ to INACTIVE
    set randomTime ▾ to pick random 5 to 10
    reset timer

    turn on ▦

Figure 5-10: The custom block that implements Shelly's sleepy behavior

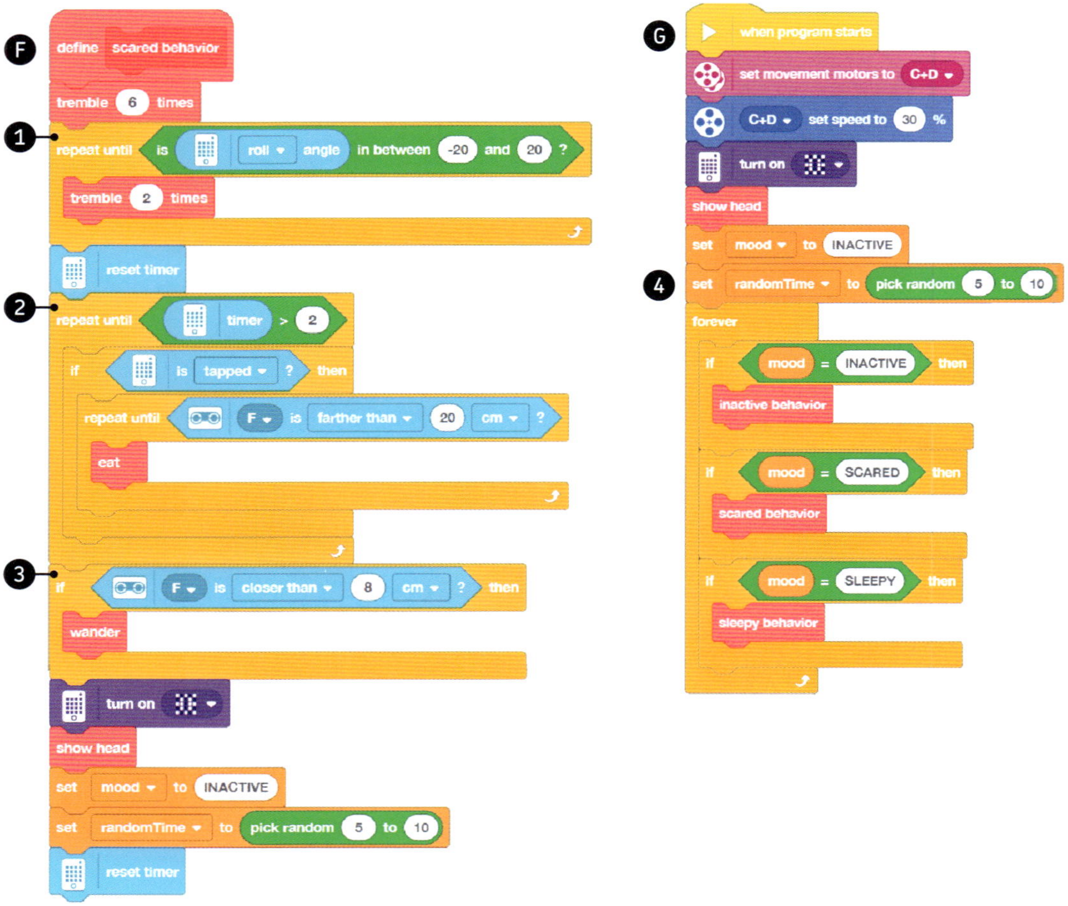

F define scared behavior
tremble `6` times
1 repeat until is [roll] angle in between `-20` and `20` ?
　tremble `2` times
reset timer
2 repeat until [timer] > `2`
　if [is tapped ?] then
　　repeat until [F is farther than `20` cm ?]
　　　eat
3 if [F is closer than `8` cm ?] then
　wander
turn on [:::]
show head
set mood to INACTIVE
set randomTime to pick random `5` to `10`
reset timer

G when program starts
set movement motors to C+D
C+D set speed to `30` %
turn on [:::]
show head
set mood to INACTIVE
4 set randomTime to pick random `5` to `10`
forever
　if mood = INACTIVE then
　　inactive behavior
　if mood = SCARED then
　　scared behavior
　if mood = SLEEPY then
　　sleepy behavior

Figure 5-11: The custom block that implements the scared behavior and the main stack of the program that manages Shelly's behaviors based on the value of mood

what you've learned

While building Shelly the Turtle, you learned another way to transform a motor's continuous rotation into an alternating motion to make a walking leg. You also transformed rotation into the linear motion of Shelly's head. You saw many examples of bracing, a technique to lock subassemblies together so a robot won't come apart while it's moving. While programming Shelly, you learned how to use variables to let your programs remember data and how to give your robots interesting behaviors by reacting to sensors and timer events.

In the next chapter, you'll build, program, and play with a version of the classic arcade game Whac-A-Mole, or Mole Buster. How good are your reflexes?

whac-a-mole!

The practice of repeatedly getting rid of something, only to have more of that thing appear.
—*The Online Slang Dictionary*

The LEGO MINDSTORMS set isn't just a toy. It's a toolbox that you can use creatively, even to make your own games. In this chapter, you'll build and program the Whac-A-Mole, a robot based on a popular arcade game (also called Mole Buster) invented by Kazuo Yamada in the 1970s.

As you build and program this game, you'll discover how to use the built-in rotation sensor of the LEGO motors to sense the moles being hit. You'll also learn how to use the Hub's timer to measure a player's reaction time and how to work with several variables within the same program. How fast are your reflexes?

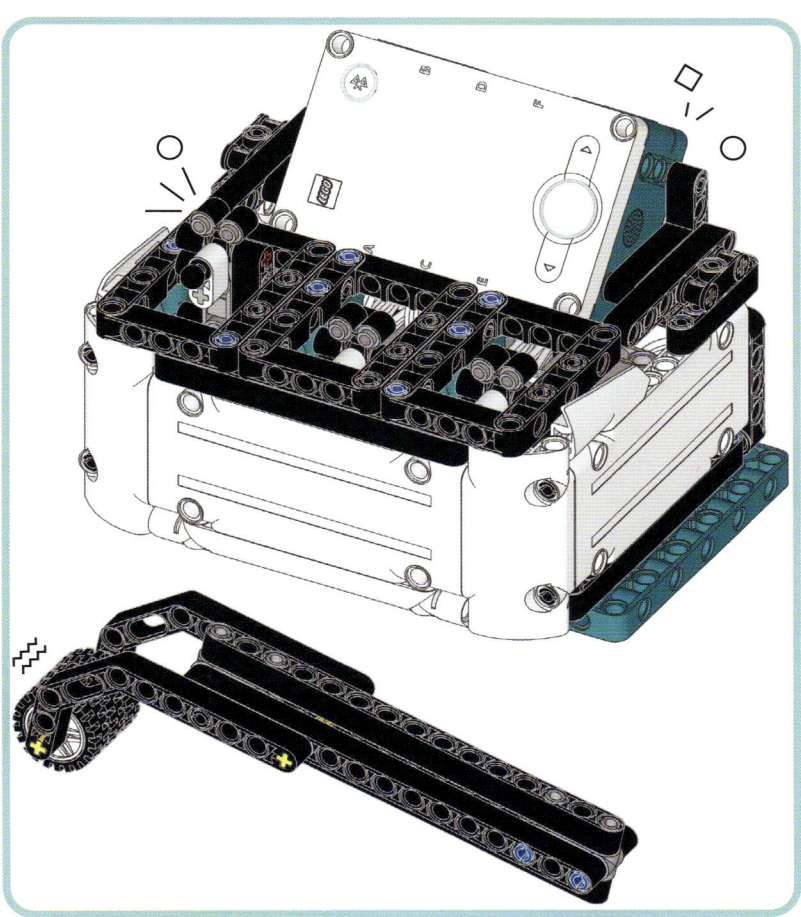

Figure 6-1: Whac-A-Mole is a frantic game that will challenge your reflexes.

building the whac-a-mole game

In this section, you will find step-by-step instructions to build the game. Throughout the instructions are notes describing the notable design choices and interesting techniques.

1

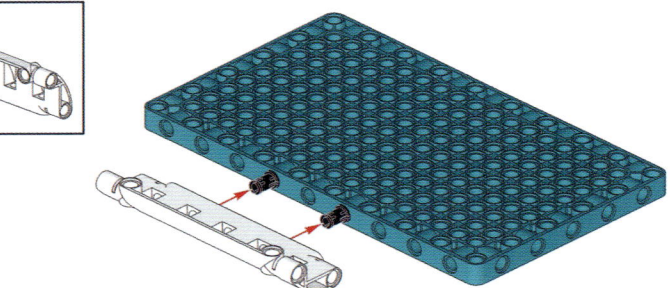

---

mole subassembly

2

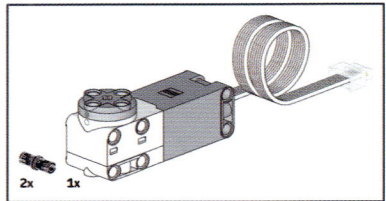

Build three copies of this submodel.

3

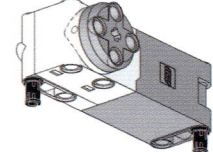

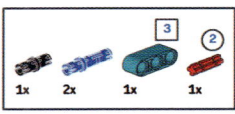

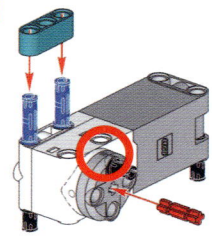

Insert the black pin in the marked hole of the shaft.

4

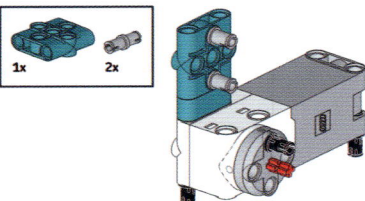

5

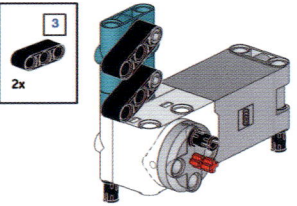

6

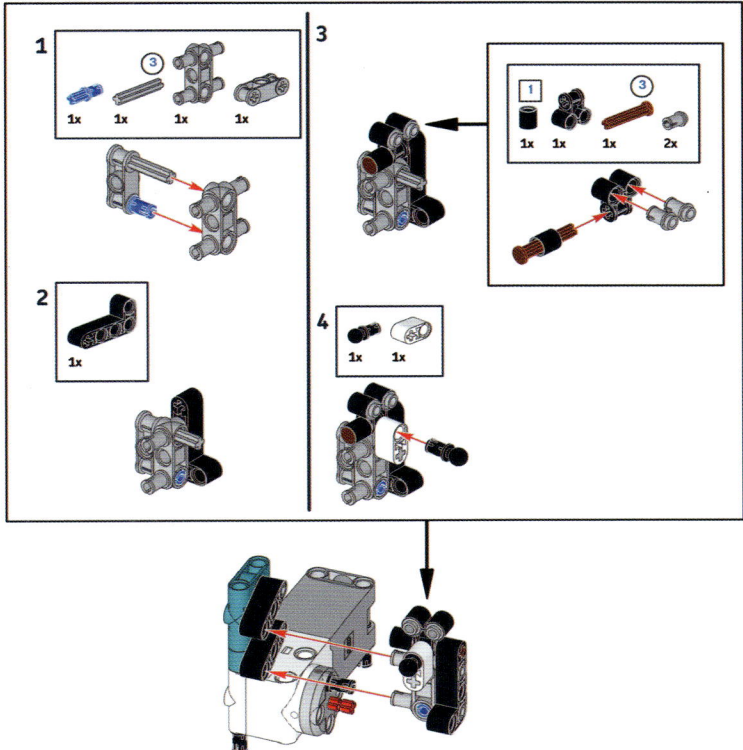

7

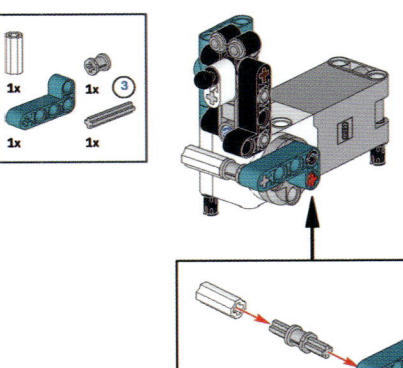

Because the black 3M beams are free to move, the mole will go up when pushed by the lever attached to the motor shaft and down by gravity when the lever goes down. This kind of assembly is called a *four-bar linkage*: a parallelogram that can change shape.

8

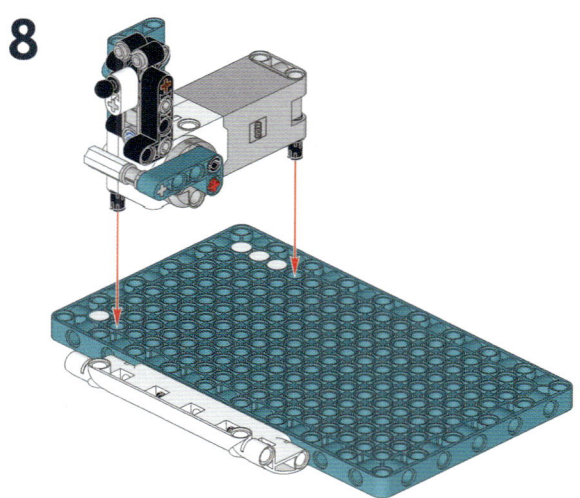

Use the white dots as a reference for adding the motor.

9

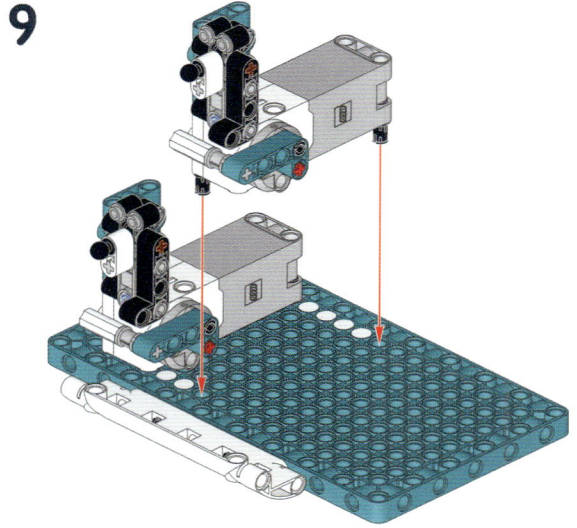

Use the white dots as a reference for adding the motor.

10

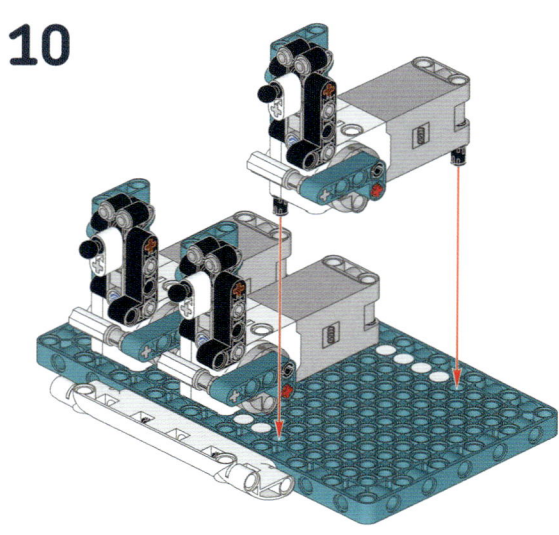

Use the white dots as a reference for adding the motor.

box subassembly

11

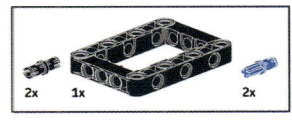

2x 1x 2x

12

2x 1x 2x

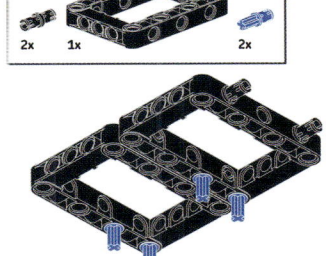

13

2x 1x 2x 2x

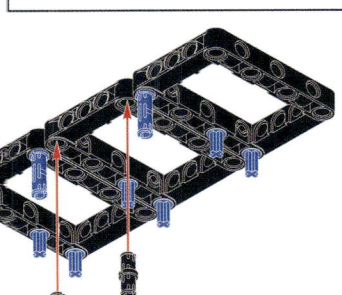

14

6x 3x [3]

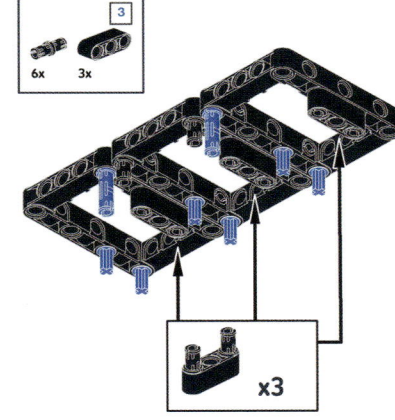

x3

15

1x [11]

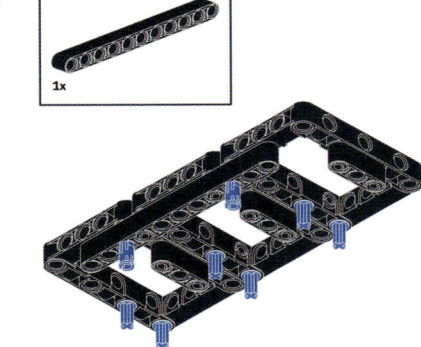

16

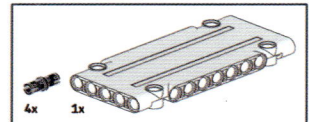

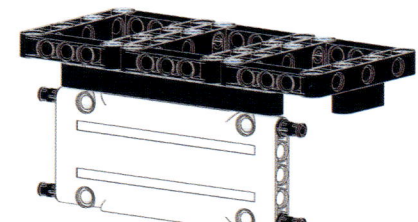

17

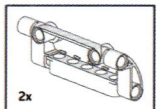

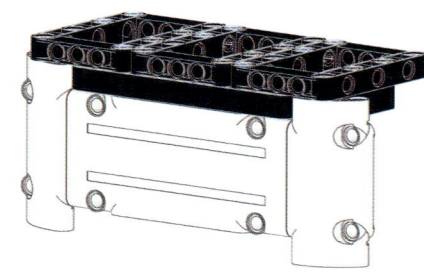

18

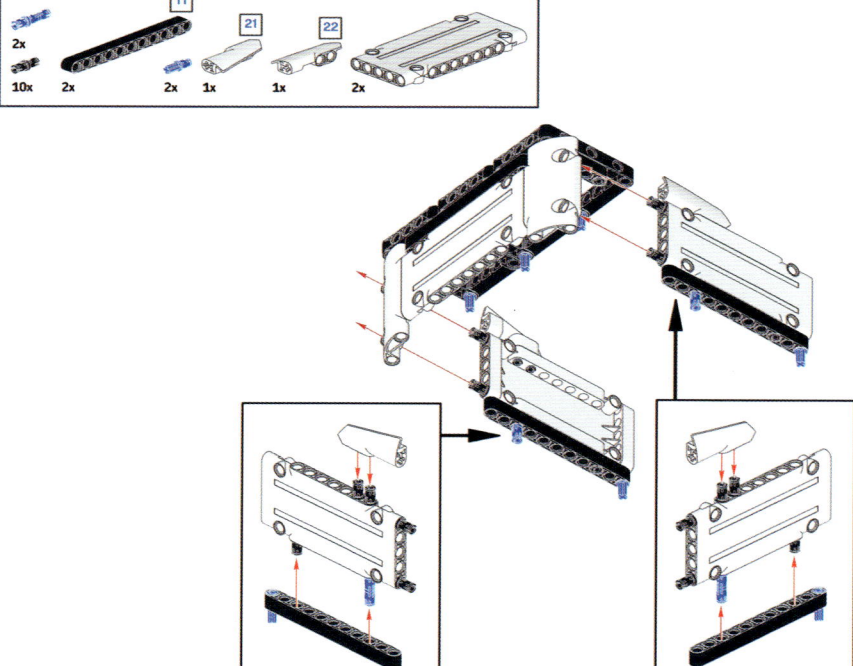

19

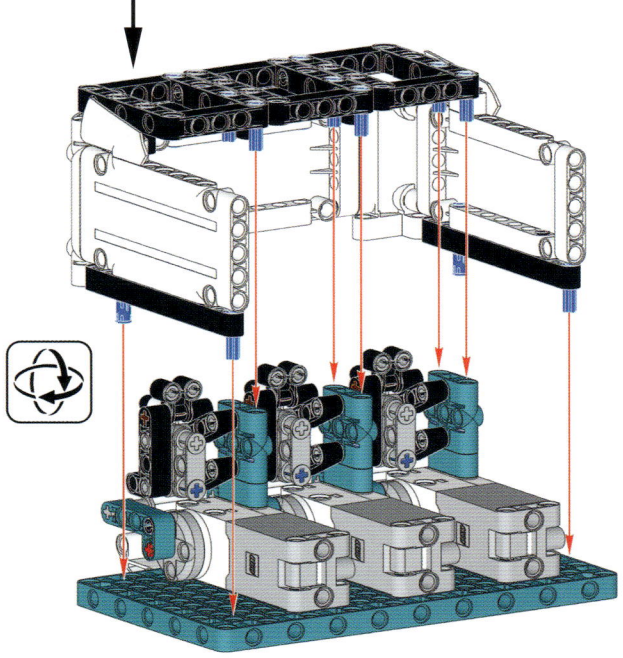

20

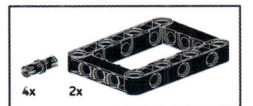

4x   2x

x2

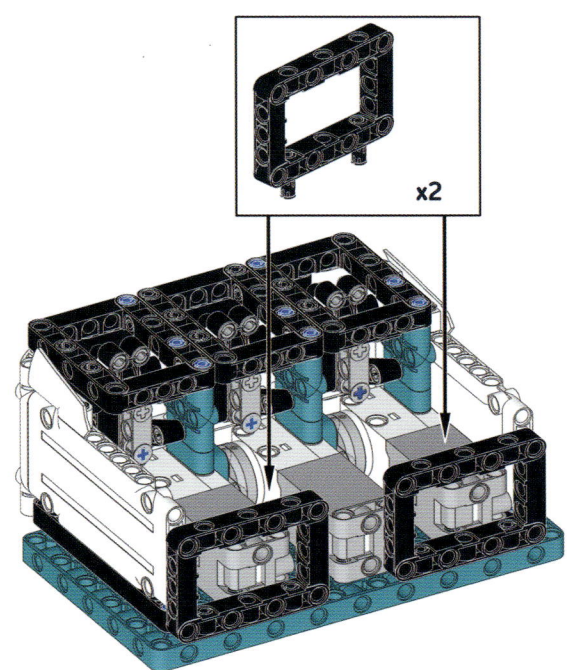

display subassembly

21

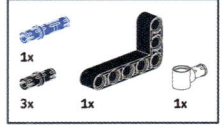

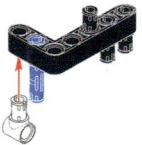

22

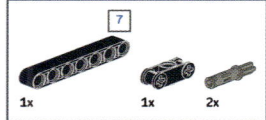

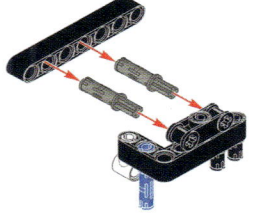

23

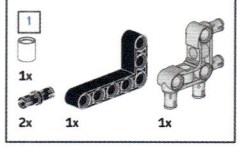

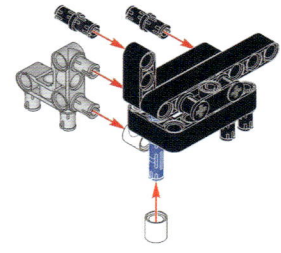

24

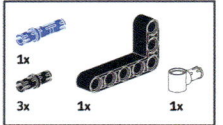

25

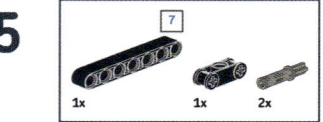

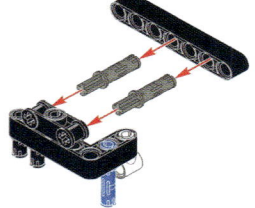

26

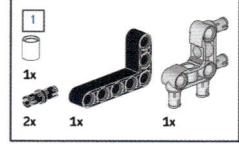

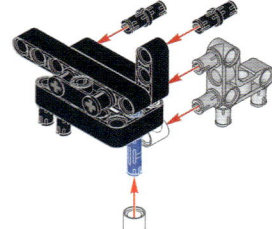

27

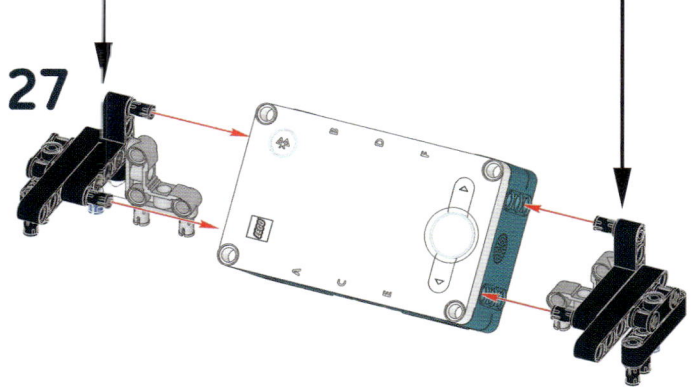

You can sturdily attach tilted assemblies to frames by matching their hole patterns.

28

Be careful not to pass any of the cables close to the moving parts.

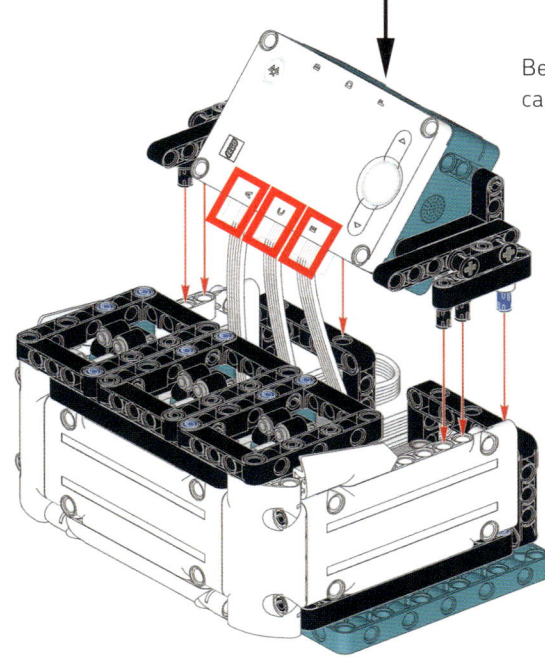

hammer subassembly

29

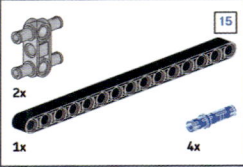

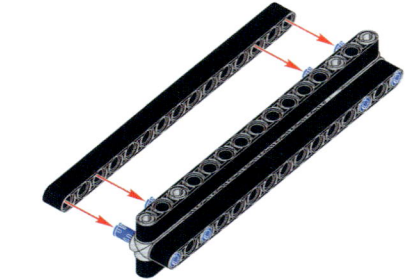

30

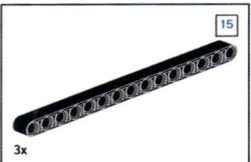

31

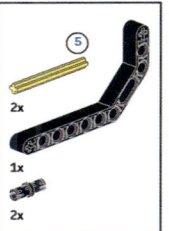

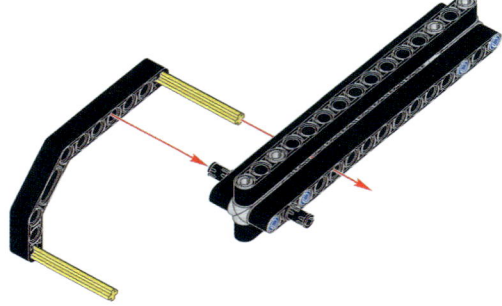

32

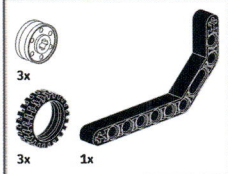

3x

3x   1x

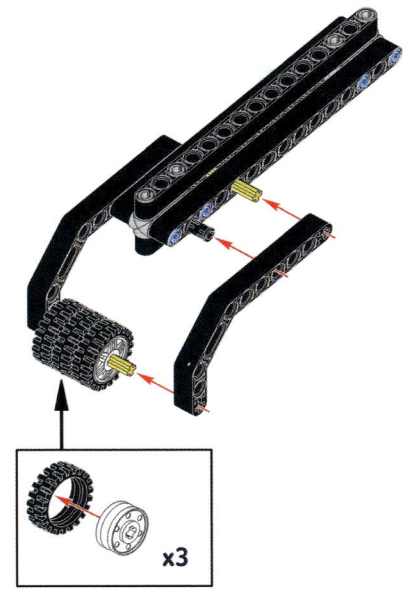

x3

33

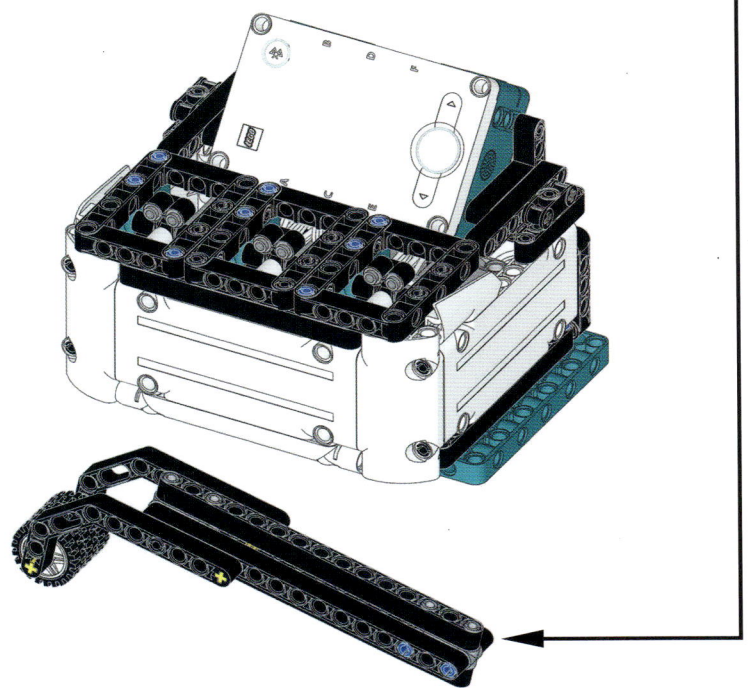

The Whac-A-Mole is complete!

USING MOTORS AS SENSORS

Inside each LEGO MINDSTORMS Medium Motor, there's an electric motor that drives the shaft by using a gear train, a series of gears that reduce the speed of the electric motor and increase its torque (its twisting force). The gears add backlash, or a bit of wiggle room, to the motor shaft because there are small gaps between the meshing teeth of the gears. Therefore, the shaft can move a little, even if the motor is still.

Previous LEGO MINDSTORMS motors had a rotation sensor at the beginning of the gear train, but the motors included in the Robot Inventor set have a rotation sensor at the end of the gear train, allowing the sensor to measure the actual position of the shaft, including when the shaft moves slightly while the electric motor is still. In the Whac-A-Mole program, we'll use this feature to sense when a mole has been hit. In Chapter 7, we'll also use this feature to know when the ball has hit the pinball machine's bumpers.

programming the whac-a-mole game

Create a new program and save it with the name *whacamole*. Then, go to the Variables palette and create these variables:

```
count
duration
hit1
hit2
hit3
pause1
pause2
pause3
pos1
pos2
pos3
score
time1
time2
time3
```

Reproduce stack A as shown in Figure 6-2. To set three motors in the Motor blocks, click the port drop-down menu, click the **MULTIPLE** button at the bottom of the box, and select the three ports **A**, **C**, and **E**. The gray box in the figure shows how to create the formula inside the write block, which calculates your final score. The round block, which rounds the input value to the nearest whole number, is in the Operators palette.

Next, create the custom blocks move up and move down, defined in Figure 6-3. These blocks move an individual mole up and down. You'll need to add the More Motors extension palette to access some of the motor blocks used in the stacks. To review how to add an extension palette, see "Adding an Extension Palette" on page 48 in Chapter 4.

The custom blocks have an input named port that should be dragged to the input of each Motor block in the stacks. The input lets you specify which motor to turn on and off when calling the custom blocks from other stacks. You'll type the motor's port letter in the input space of the custom block (for example, at ❶ in Figure 6-3). This way, you can use the same custom blocks to control each motor separately.

Now create the block stacks to control the moles, shown in Figure 6-4. Each mole is controlled by a separate stack, allowing them to operate independently. However, the stacks are nearly identical, so you can just create stack D and then make two copies of it. To copy the stack, right-click (long-press on smart devices) the first block in the stack and select **Duplicate**.

Once you've created two copies of the first stack, you must update the blocks with the correct motor ports and variable names. Any references to port A should be changed to port C in stack E and port E in stack F. Similarly, variables ending with the number 1 (for example time1, pause1, pos1) should be changed to variables ending with 2 (time2, pause2, pos2) in stack E and ending with 3 (time3, pause3, pos3) in stack F.

playing the game

Start the *whacamole* program by pressing the **Play** button on the Hub after selecting program slot **0** with the **Select Storage Position** button in the app. The moles will start popping up at random intervals. You can hit them with the brick-built hammer or with your fingers. There's no need to slam them hard, as the motors' built-in rotation sensors are very sensitive to little variations in the shaft position. If you don't hit a mole in time, it will hide back in its den.

The faster you hit the moles after they pop out, the higher your score will be. The maximum score you can get is 100, while you'll get a 0 if you never hit a mole. The game is programmed to last 30 seconds (block ❸ in Figure 6-2 sets the duration variable to 30), but you can change this value to make the game last longer if you want. Can you reach 100 points?

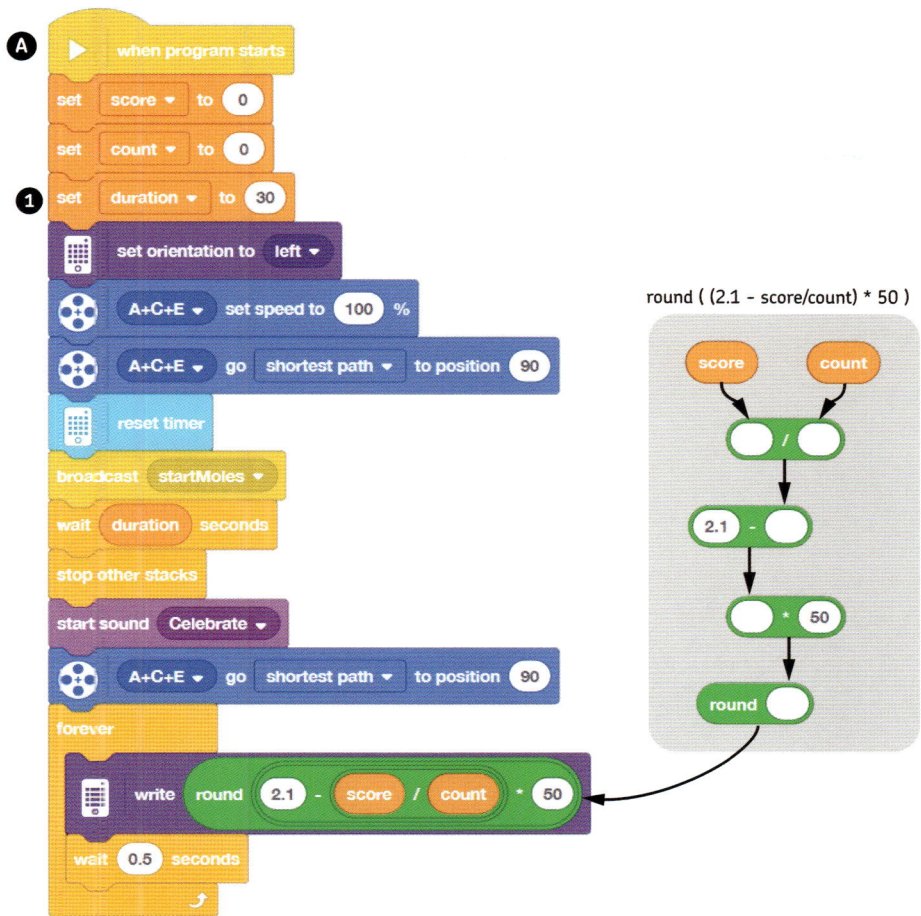

Figure 6-2: The main stack of the Whac-a-Mole program

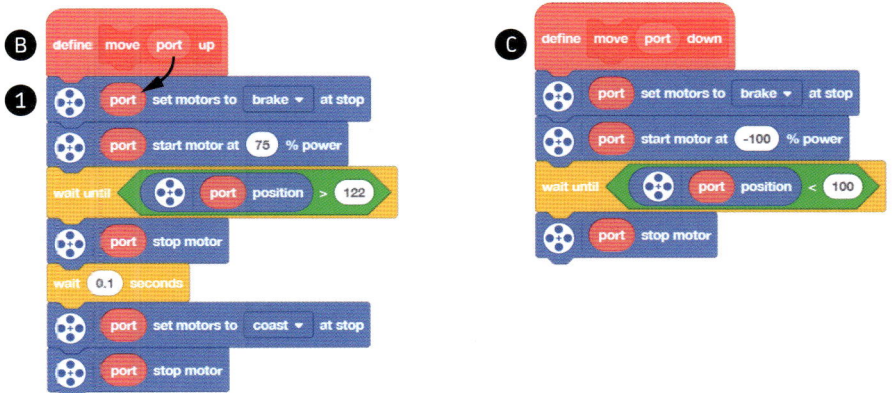

Figure 6-3: The custom blocks to move the moles

understanding the main stack

Now that you've played the game, let's see how it works. First take a look at stack A in Figure 6-2. When the program starts, we set the variables score and count to 0. We also set duration to 30, or however many seconds you want the game to last. Next the Hub orientation is set to left so your final score will display the right way up. Then we set the maximum speed for the motors and move them to position 90, which makes all the moles go down. We reset the timer to give you a full 30 seconds of gameplay.

Now the fun begins: we broadcast the message startMoles to run stacks D, E, and F, which control the events of the game (we'll discuss how these three stacks work soon). While you're busy whacking moles, the wait for seconds block in stack A simply waits for the number of seconds stored in the variable duration. After this set period of time, the stop other stacks block halts all other activity in the program, ending the game.

Once the game is over, we play the Celebrate sound (from Tricky's sound library) and move the motors to lower all the moles. Finally, the forever loop displays your score as scrolling text on the Hub. It will be there for everyone to see, forever . . . or at least until you press the Hub's center button to stop the program.

We compute the final score with this formula:

```
round ( ( 2.1 - score/count ) * 50 )
```

The formula calculates your average reaction time—that is, the average amount of time that passes from the instant a mole pops up until you hit it—and converts that value to a whole number between 0 and 100. Here's how it works:

1. The variable count holds the number of times a mole pops up, while score holds the sum of all your reaction times. If you miss a mole completely, your reaction time for that mole is recorded as 2.1 seconds, while a really fast reaction time is about 0.1 seconds.

2. We divide score by count to compute the average reaction time. If you miss every single mole, the average score/count will be 2.1, while in the best case of almost inhuman speed, the average will be 0.1.

3. We need to turn that best-case average of 0.1 into 100 points, while the worst-case average of 2.1 should become 0 points. To do this, we subtract the average from 2.1, then multiply the result by 50. As you can see:

```
(2.1 - 0.1) * 50 = 2 * 50 = 100
(2.1 - 2.1) * 50 = 0 * 50 = 0
```

Reaction-time averages between 0.1 and 2.1 will yield a score somewhere in the middle. For example, an average time of 0.8 yields a score of 65:

```
(2.1 - 0.8) * 50 = 1.3 * 50 = 65
```

4. Before displaying the result, we round the score to the nearest whole number to get rid of any decimals.

understanding the move up and move down blocks

Now take a look at the definitions for the custom blocks move up and move down, shown in Figure 6-3. Remember that both blocks have a port input, allowing you to control which motor to move. Beyond setting the motor port with an input, we control the motors a bit differently than we're used to. So far, when we've wanted to move a motor to a particular position, we've used the motor go to position block. However, the precision this block provides takes some extra time, which would make the game slower and less fun.

In the move up definition (stack B), instead of finely controlling the position of the motor, we just start the motor at 75% speed and wait until the position of the shaft is beyond a set angle. The angle 122 corresponds to the lever that lifts a mole almost fully up—but not completely raised, or the lever would jam against the mole assembly and resist your hit. We want the motor to sense the change in position from your hit and react immediately, to avoid damaging the motor.

Once the motor has reached the desired position, it takes a few different blocks to stop it. Since we already instructed the motor to use braking when it stops ❶, the first stop motor block stops the motor abruptly. This keeps the lever from bumping against the mole assembly and rotating back, which would cause the mole to fall. After waiting 0.1 seconds, we set the motor to coast and stop it again. This way, the motor can sense your hit without resistance, because it's been set to turn freely.

The move down block definition (stack C) is similar but simpler, as we don't need to coast the motor after it's been moved to an angle less than 100 degrees.

understanding the game logic

Let's now analyze the core game logic, executed by the three parallel stacks D, E, and F, shown in Figure 6-4. For each mole, we want to wait for a random amount of time, then raise the mole and start counting time. If you hit the mole within a certain interval, the program will store your reaction time to compute your score at the end of the game. If you don't hit the mole in time, it will go back down, and your reaction time will be assigned a penalty value of 2.1 seconds, lowering your final score. This process is repeated for the duration of the game, independently for each of the three moles. Since all three stacks work the same way, we'll just focus on stack D.

The stack starts when it receives the startMoles message, which is broadcast from stack A (see Figure 6-2). The rest of the stack is contained within a repeat until loop, which ends when the timer passes the value stored in the duration variable. (Remember, the timer was reset in stack A just before sending the startMoles message.) Before raising a mole, the program waits for a random time ranging from 1 to 2 seconds in steps of 0.1 seconds. To get a random value with a decimal place, we pick a number between 10 and 20 and divide it by 10. Thus, the random number 13 becomes 1.3 seconds, for example.

After the pause, the move A up custom block lifts the mole, accompanied by the squeaky Hi sound from Charlie's sound library. At this point, we set the values of four variables:

✳ The variable time1 stores the current value of the timer. We'll use this value to see how much time has passed since the instant the mole popped up. (Unfortunately, we can't

simply reset the timer when the mole pops up because other stacks need the timer, too.)

✳ The variable pos1 stores the current position of the motor shaft. We'll check whether that value changes to sense the mole being hit.

✳ The variable hit1 is set to 0. We'll change it to 1 when the mole is hit or too much time has passed. This kind of variable is called a *flag*: it's read later in the program to see if a certain event has happened.

✳ The variable pause1 is set to a random value ranging from 1 to 2 seconds in steps of 0.1 seconds, just as we set a time for the wait for seconds block previously. This is the maximum time you have to hit the mole before it goes back down into its den.

Next we start a repeat until loop that ends when variable hit1 changes from 0 to 1. This will happen either at ❸ when you hit the mole or at ❻ if the time stored in pause1 runs out. In the loop, an if then else block checks whether the motor shaft position has changed from the position saved into variable pos1 ❷. Specifically, we check whether the current position is more than 1 degree less than the stored position,

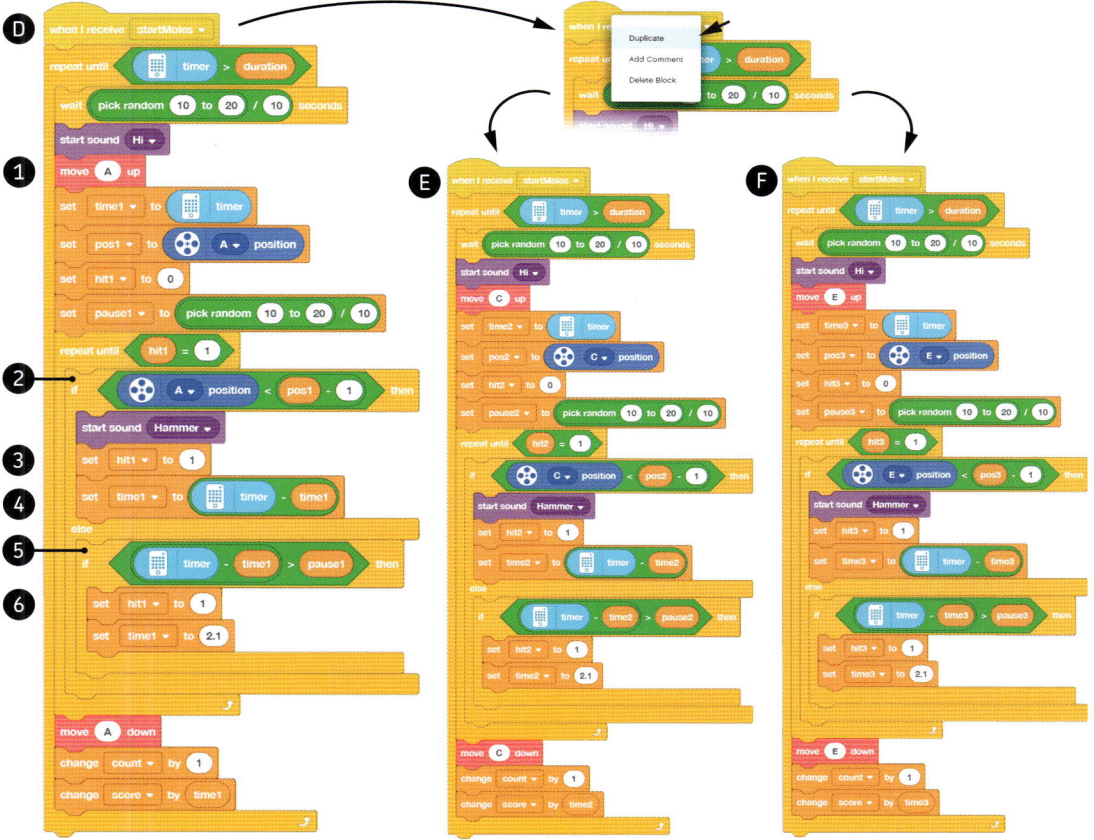

Figure 6-4: The three parallel stacks that manage the moles

indicating that the mole has dropped slightly from being hit. If that happens, we play the sound Hammer from Blast's sound library, the variable `hit1` is set to `1`, and your reaction time is calculated by subtracting the time stored in variable `time1` from the current reading on the timer. The result is stored back into the `time1` variable, whose old value is replaced, or *overwritten*, by the new one ❹.

If the motor position hasn't changed, we jump to the `else` space of the `if then else` block. There we check whether the time passed since the mole popped up (`timer - time1`) is greater than the value of `pause1` ❺. When that happens, it means you were too slow to hit the mole. The variable `hit1` is set to `1`, and the `time1` variable is set to a penalty value of `2.1` seconds.

In either case (hit or too slow), the variable `hit1` becomes `1`, so the `repeat until hit1 = 1` loop ends. The mole is rapidly lowered by the `move A down` custom block. Then the `count` variable is increased by `1` to take into account that a mole has gone through its up-down cycle (we don't care whether it was hit or not), and the `score` variable is increased by the value of the `time1` variable. Remember, we're accumulating the player's reaction times into the variable `score`, and the number of moles that have popped up into the variable `count`, allowing us to compute the average reaction time and final score in stack A at the end of the game, as explained previously.

EXERCISE 6-1

1. A game is more fun when played with friends! Try modifying the Whac-A-Mole program so up to three people can play together. Each player should focus on hitting a single mole, and the player who is fastest at whacking moles (on average) wins. You'll have to update the program to change how the score is calculated. You should use variables to separately keep track of the average reaction time of each player; then compare these values to see which is the lowest. Rather than displaying the results on the Hub, try lifting just the winning player's mole at the end to announce the winner. Try updating the code yourself. Then check out the complete program code for the multiplayer version of the game at the companion website *https://nostarch.com/lego-mindstorms-robot-inventor-activity-book/*.

2. Try adding a fourth mole to the game. Use the spare LEGO motor and adapt the program (either the single-player version or the multi-player version, if you've created it) to manage the extra mole.

what you've learned

While building the Whac-A-Mole, you learned how to move the moles up and down using a four-bar linkage. You also saw how to transform the rotation of the motor into the linear motion of the moles. While programming the Whac-A-Mole, you learned how to use the motors' rotation sensors to detect when moles are whacked. You also got more comfortable working with variables. For example, you measured time by storing the current timer value into a variable, and you plugged variables into math formulas to compute the final score.

In the next chapter, you'll have a blast with another classic arcade game, a pinball machine!

7

pinball

They wanted me to build them a bomb, so I took their plutonium and in turn I gave them a shiny bomb casing full of used pinball machine parts!
—Doc Emmett Brown, *Back to the Future*

When I started designing a MINDSTORMS pinball machine, I realized there weren't enough parts in the Robot Inventor set to build a large, smooth base for the ball to roll on. I had to think outside the box, and ended up . . . well, using the box! The base for Pinball is actually the Robot Inventor box lid itself.

In this chapter, you'll build Pinball, a fully functional pinball game complete with bumpers that use the LEGO motors' built-in rotation sensors to sense even the slightest contact with the ball. As you program Pinball, you'll learn how to use lists, which are a special kind of variable. You'll also learn how to display two-digit numbers on the Hub LED matrix display and how to change the intensity of the Distance Sensor lights by using text strings.

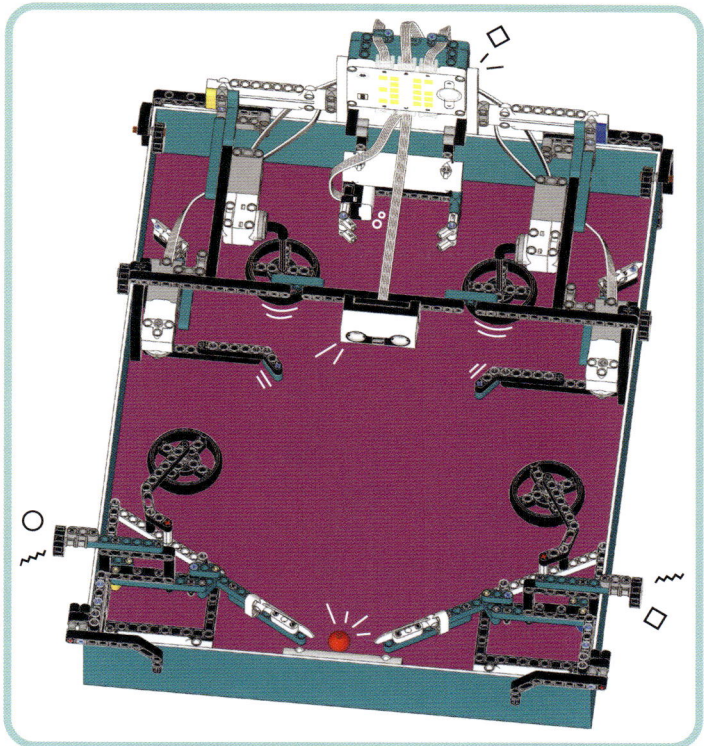

Figure 7-1: This pinball machine is a super fun and engaging game, built inside the Robot Inventor box lid.

building pinball (part 1)

In this section, you'll find step-by-step instructions to start building Pinball. For now, we'll leave off the bumpers; we'll add those later in the chapter. To save space in the layout, sometimes you'll see cross sections of the box lid, but you shouldn't actually cut anything, of course.

In the unfortunate event that you've thrown the original cardboard box away, you can try to find a different box lid or create one yourself. It should be 36.5 centimeters wide, 46.5 centimeters tall, and 6 centimeters deep (inner dimensions). The walls should be about 3 millimeters thick.

scoreboard subassembly

1

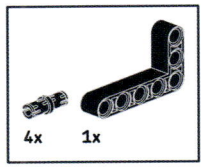

4x   1x

2

1x

1x   1x

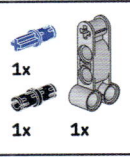

3

7

3x   1x

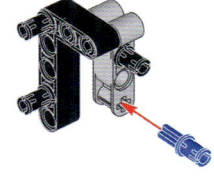

4

13

2x   1x

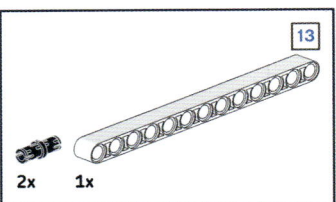

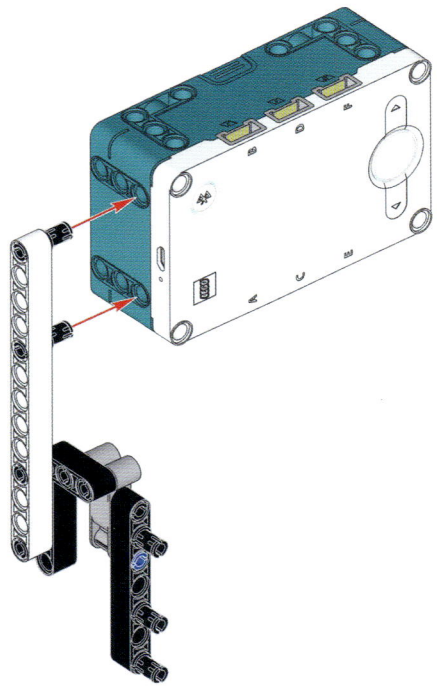

5

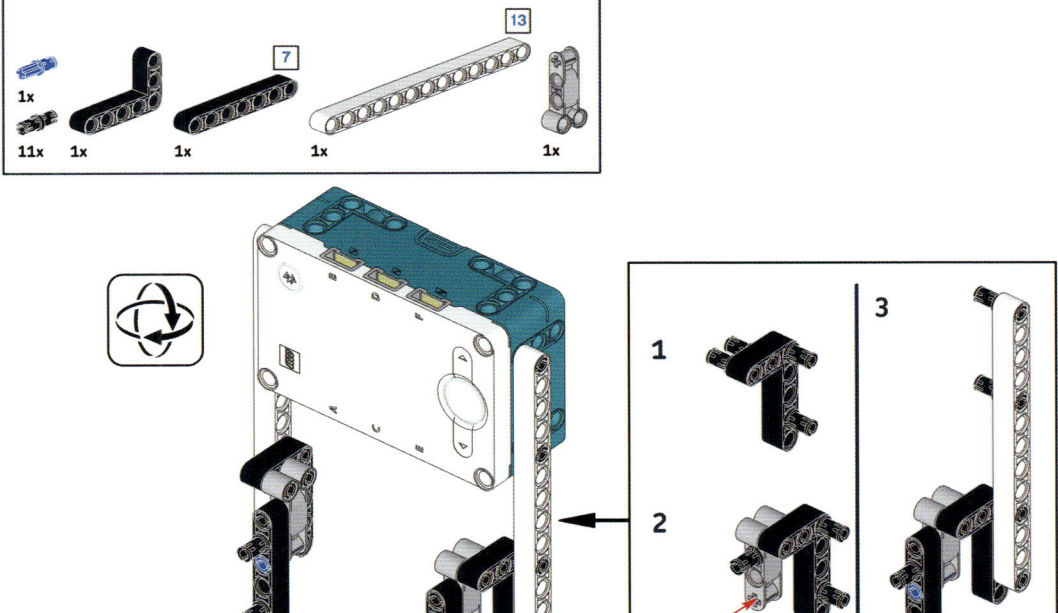

6

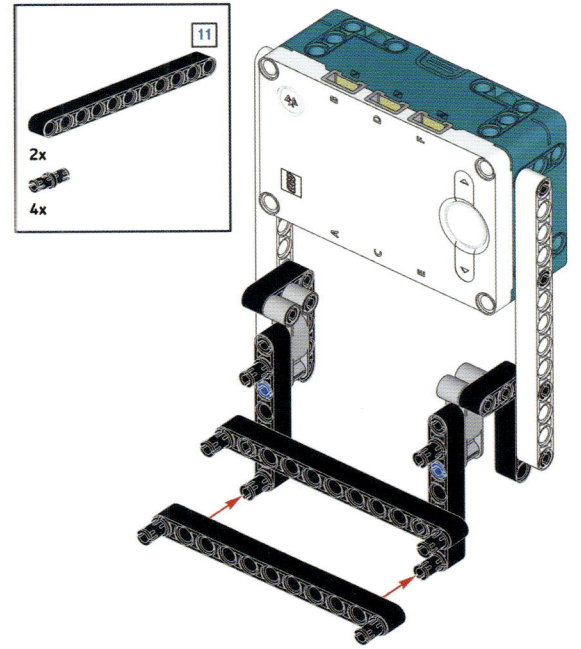

7

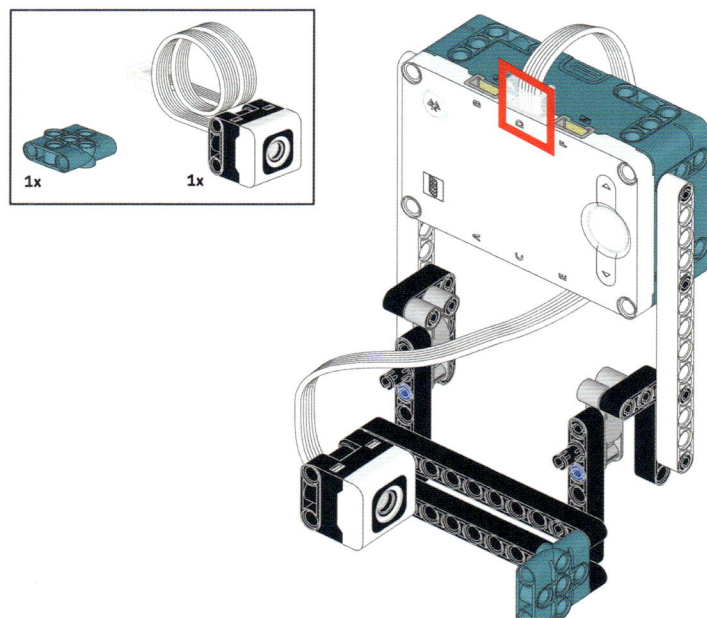

8

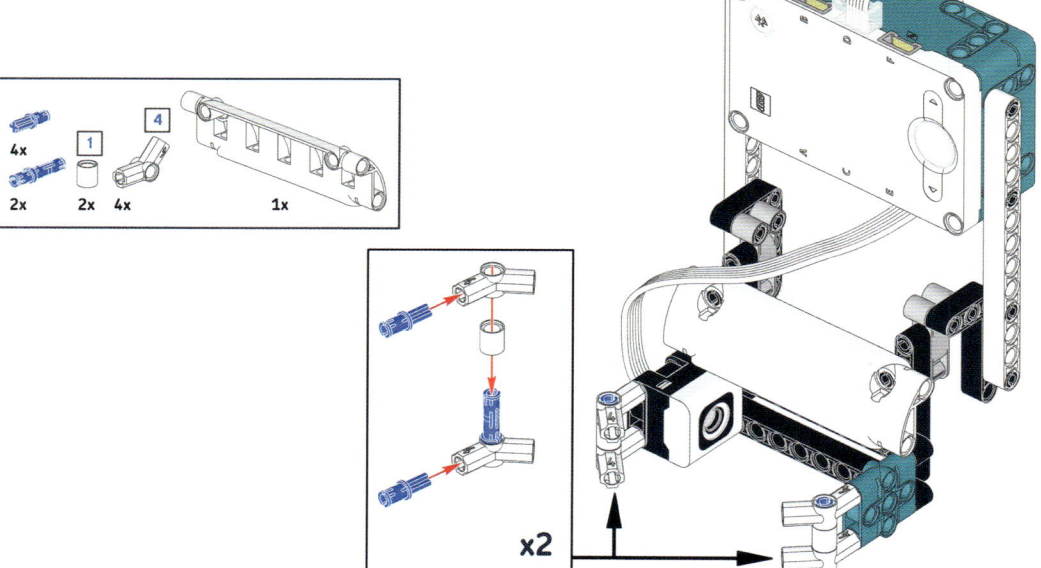

x2

9

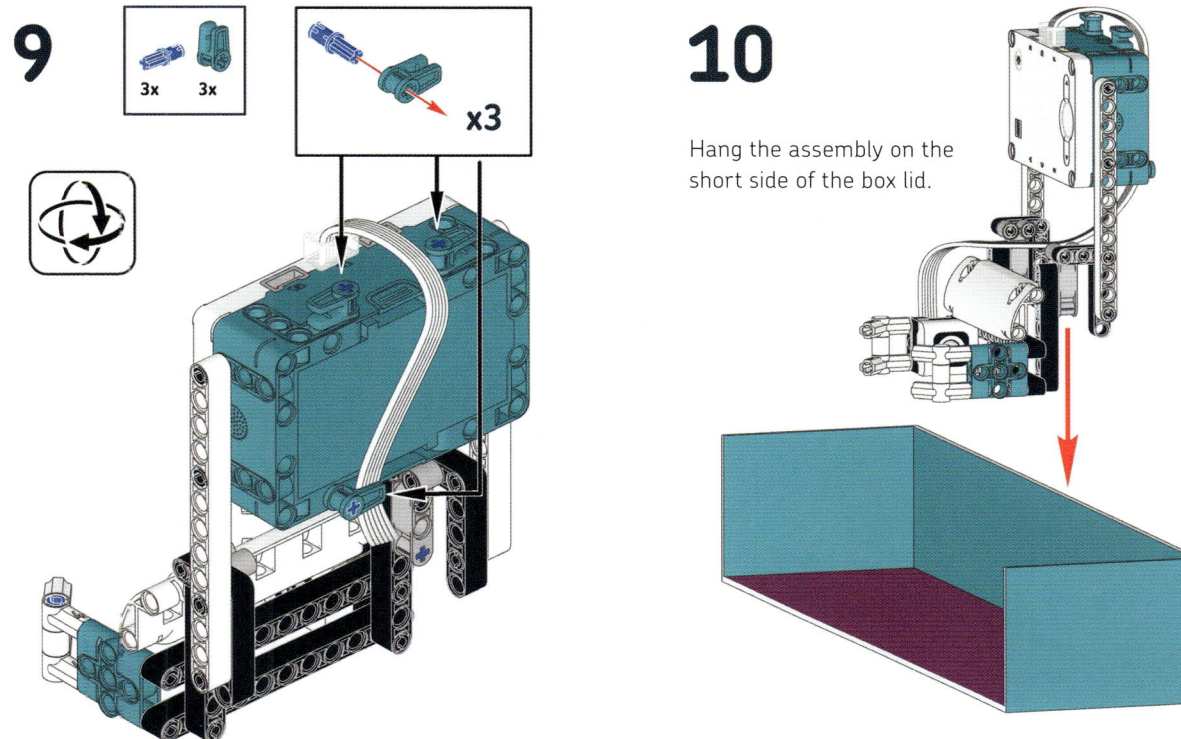

3x **3x**

x3

10

Hang the assembly on the short side of the box lid.

---

left foot subassembly

11

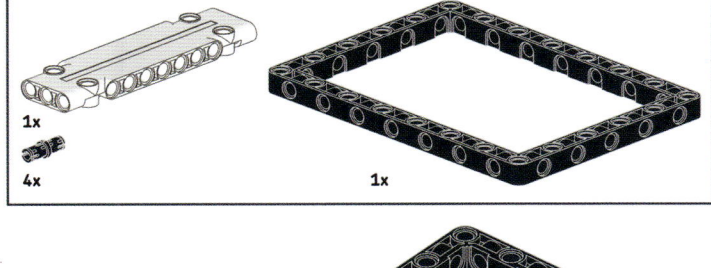

1x

4x

1x

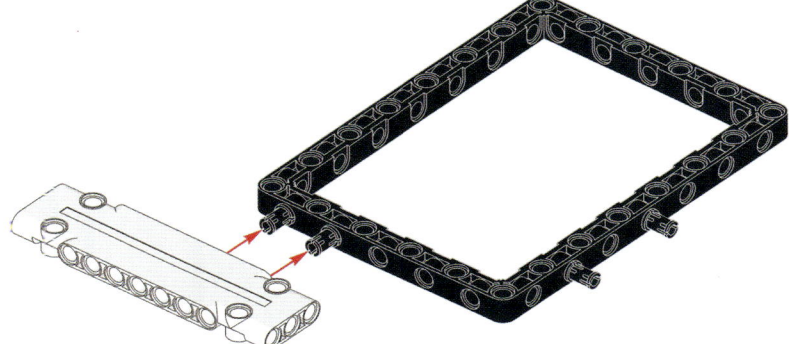

12

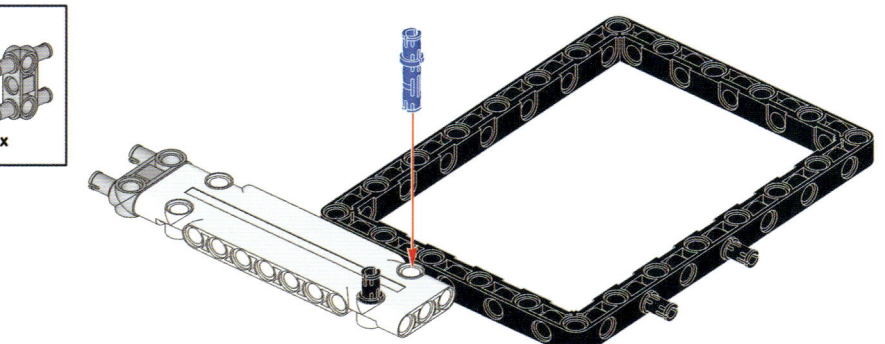

13

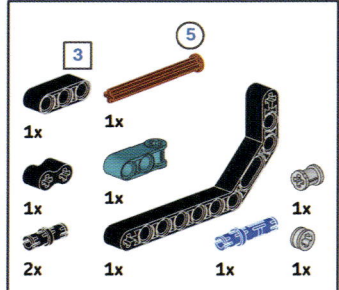

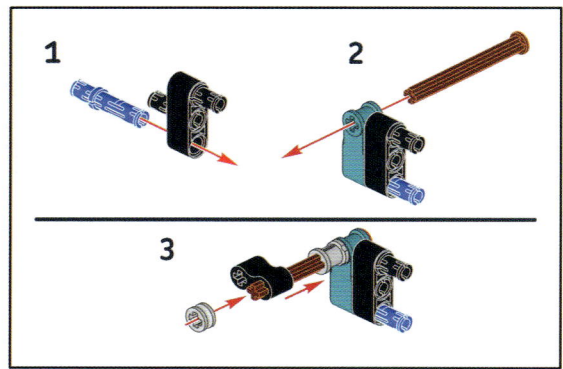

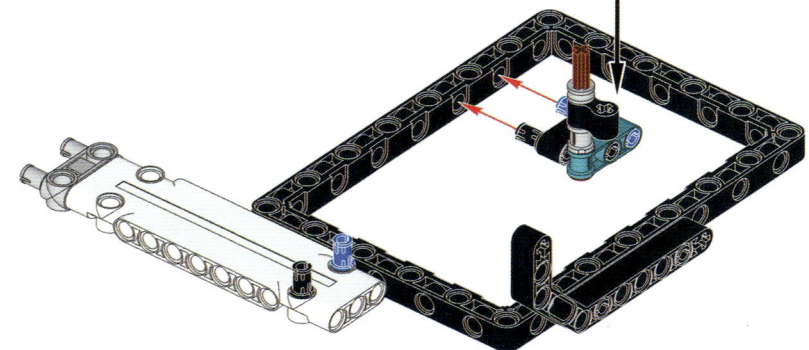

right foot subassembly

14

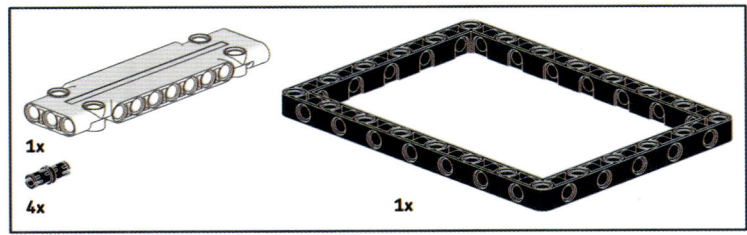

1x
4x
1x

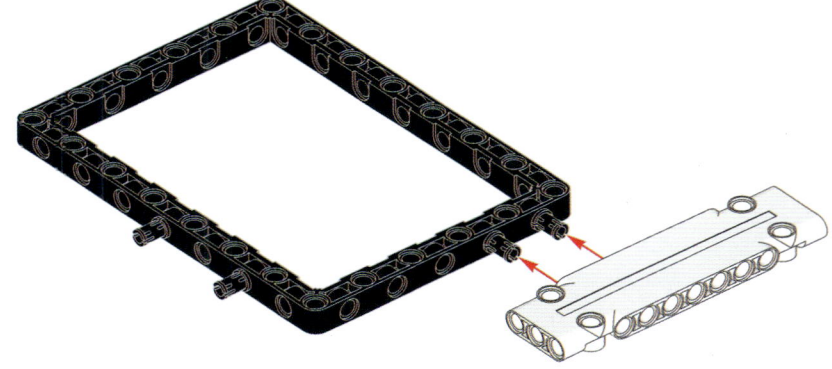

15

1x
1x
1x

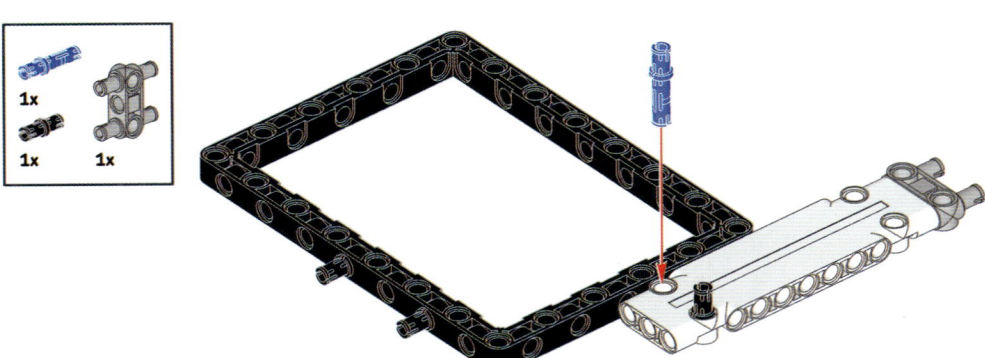

16

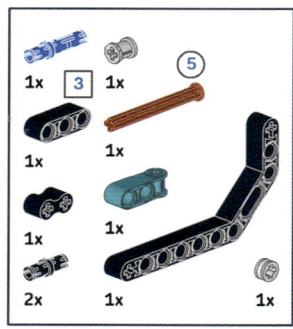

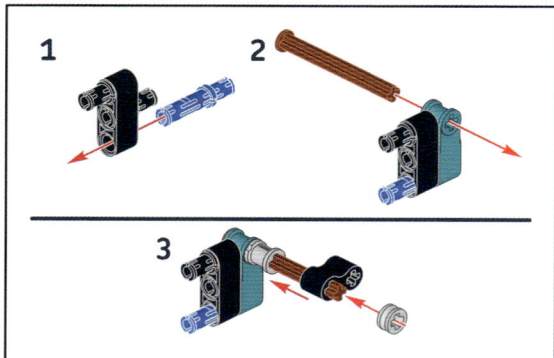

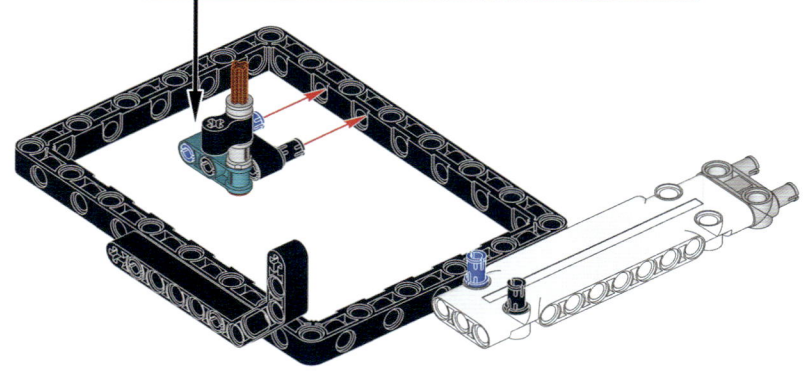

17

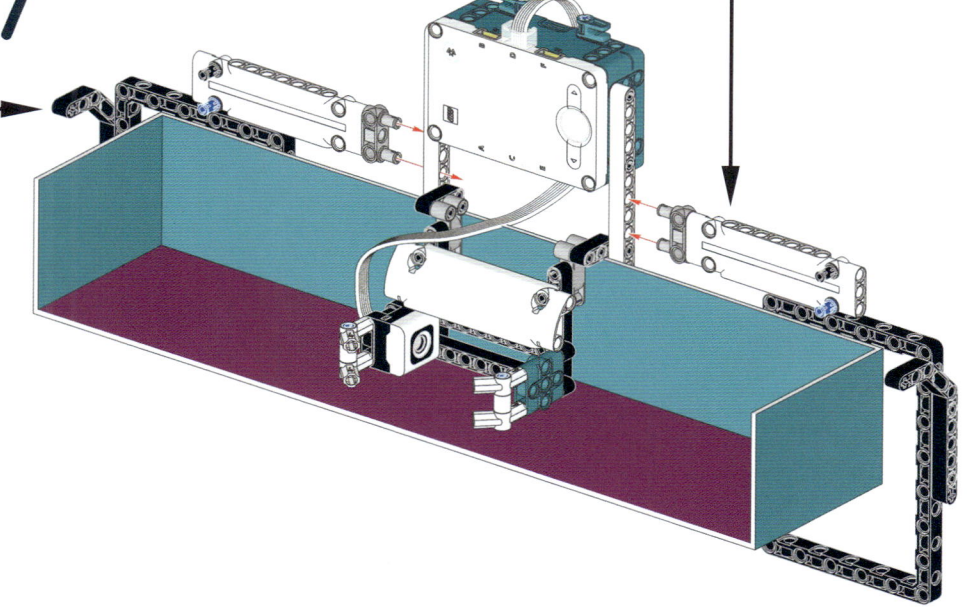

18

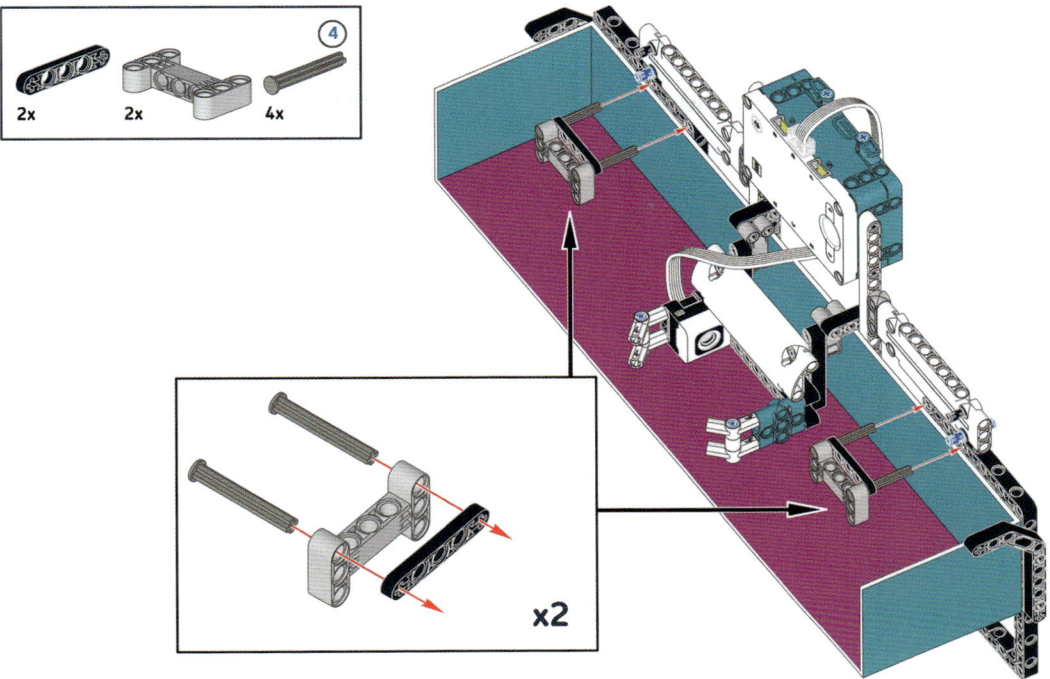

x2

19

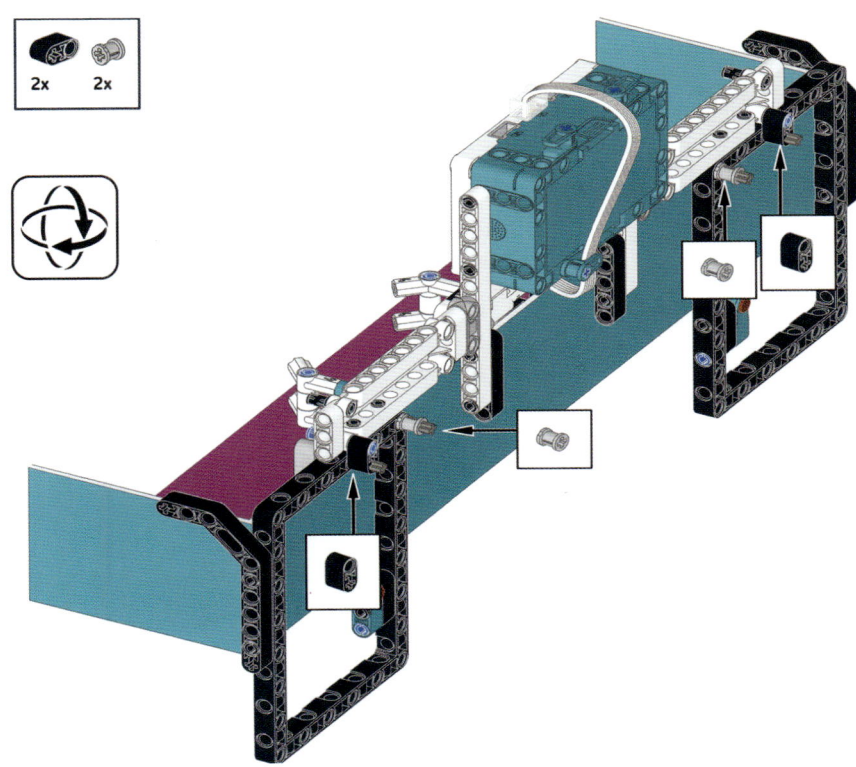

20

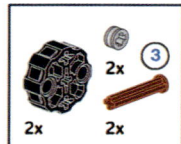

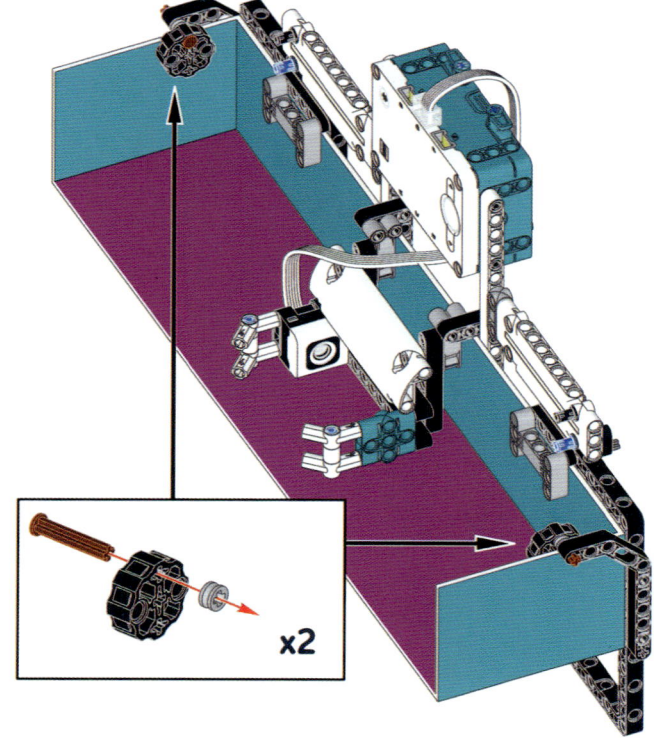

x2

---

right flipper subassembly

21

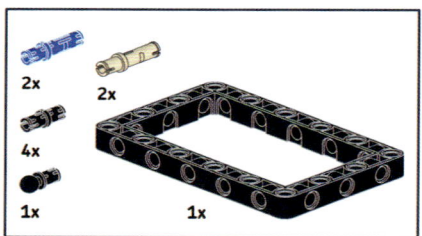

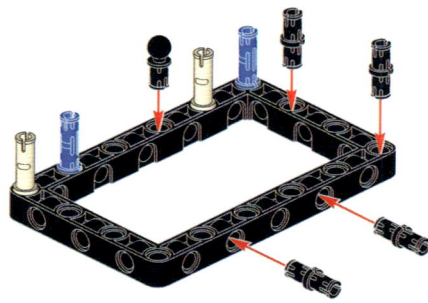

22

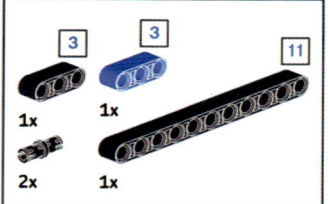

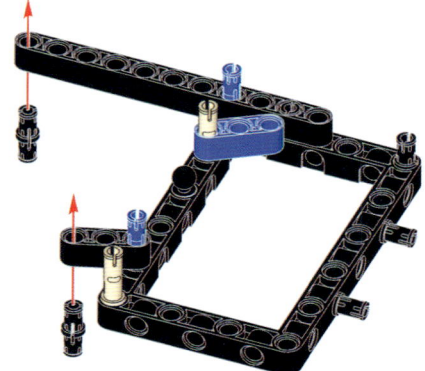

23

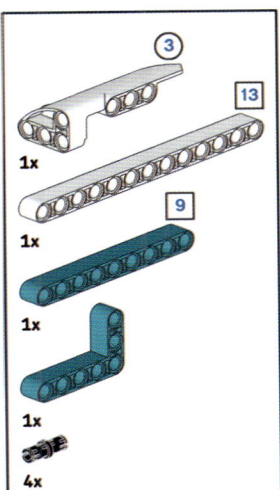

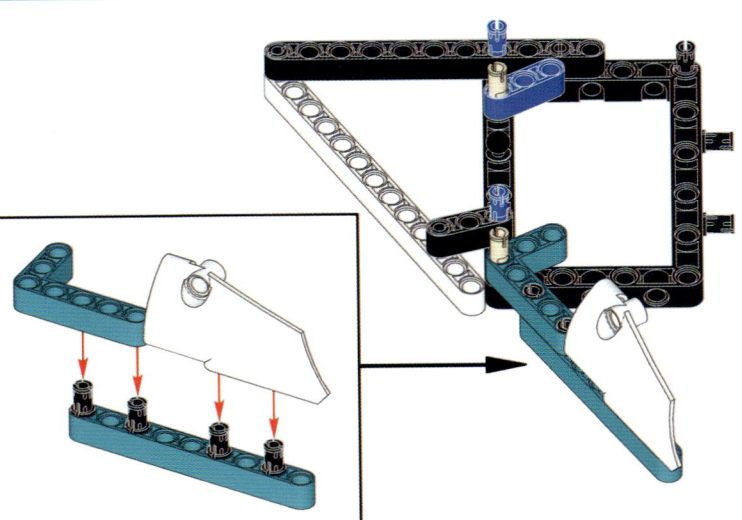

24

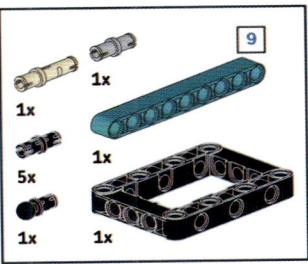

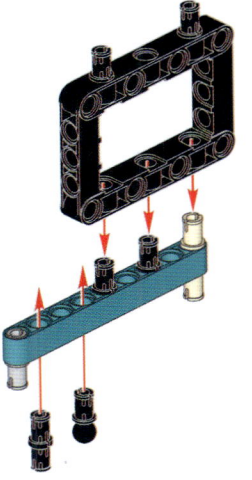

25

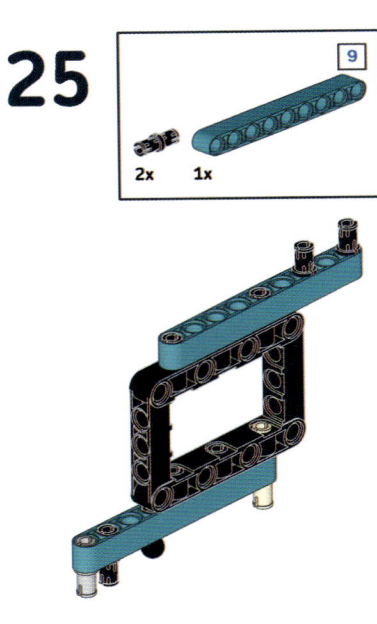

26

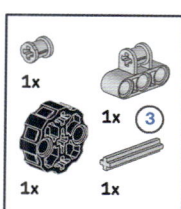
1x 1x ③
1x 1x

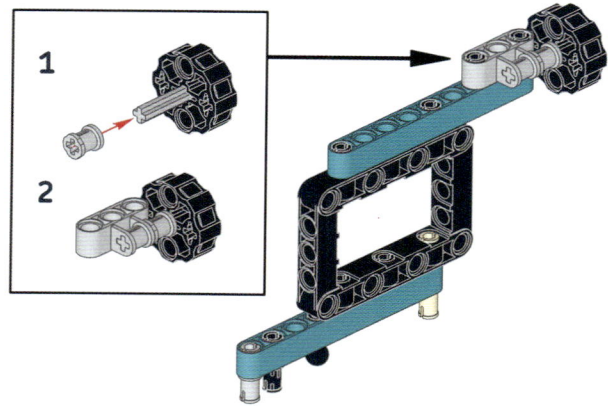
1
2

27

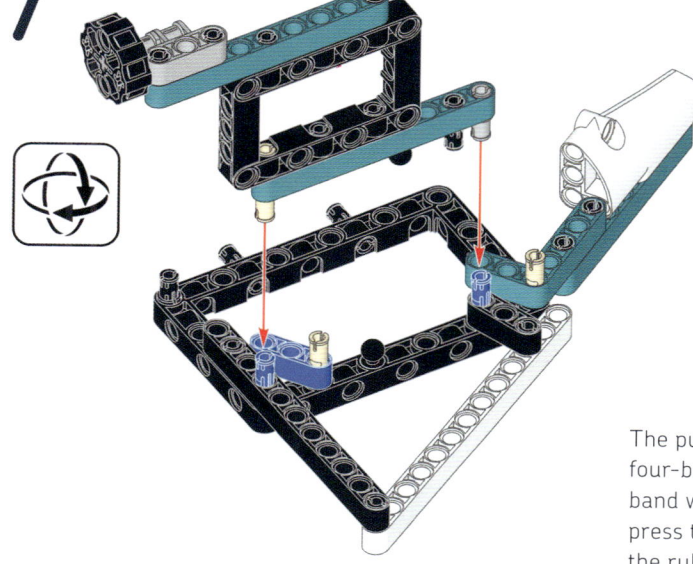

The push button drives the flipper through a four-bar linkage (parallelogram). The rubber band works as a spring that stretches when you press the button. When you release the button, the rubber band makes the flipper return to its resting position.

28

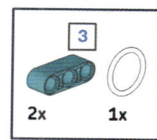
③
2x 1x

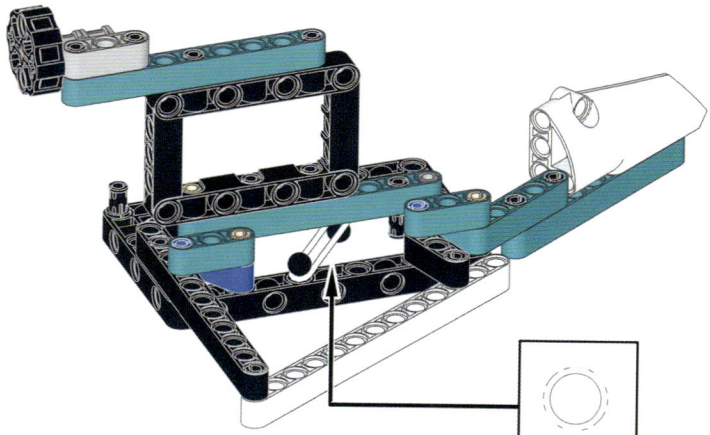

29

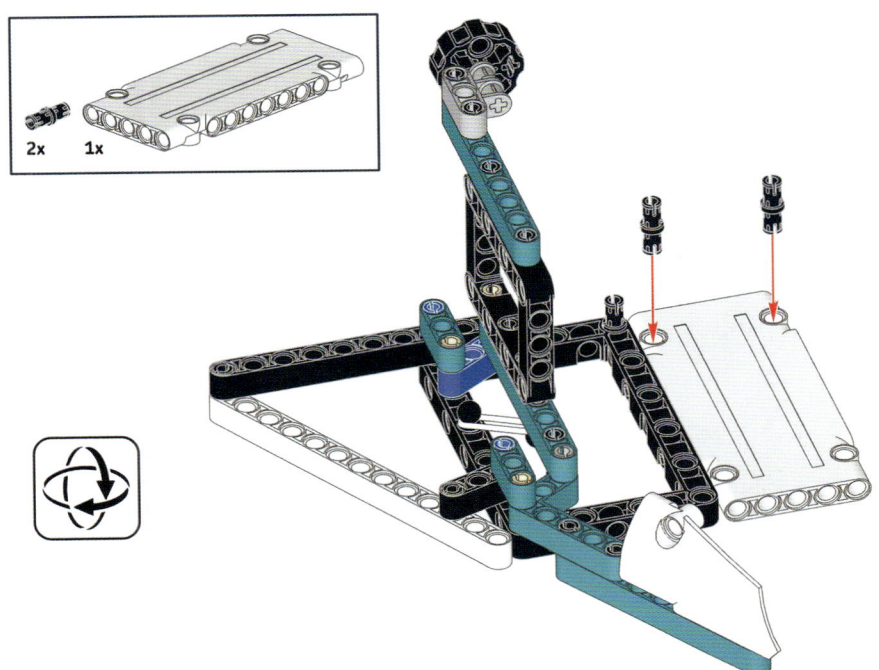

30

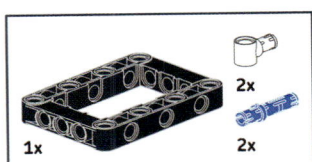

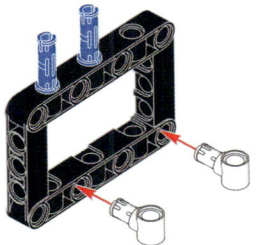

31

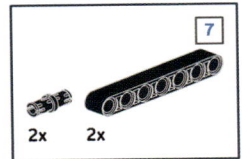

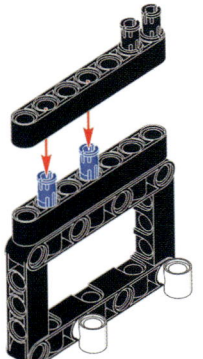

32

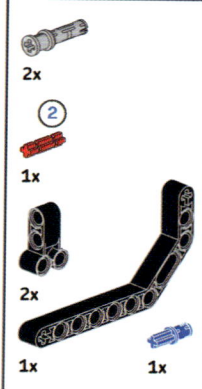

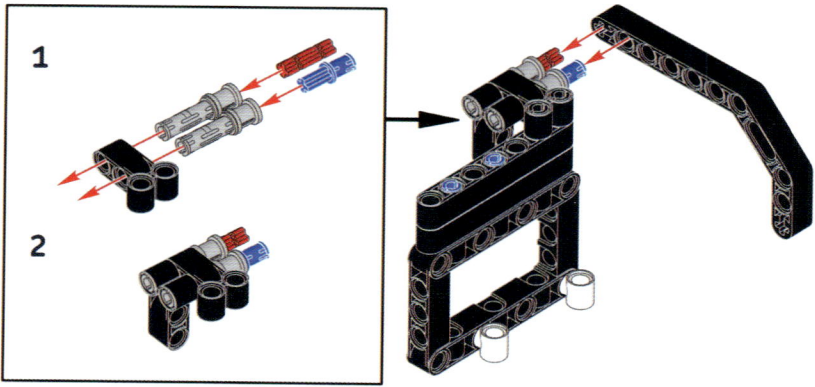

33

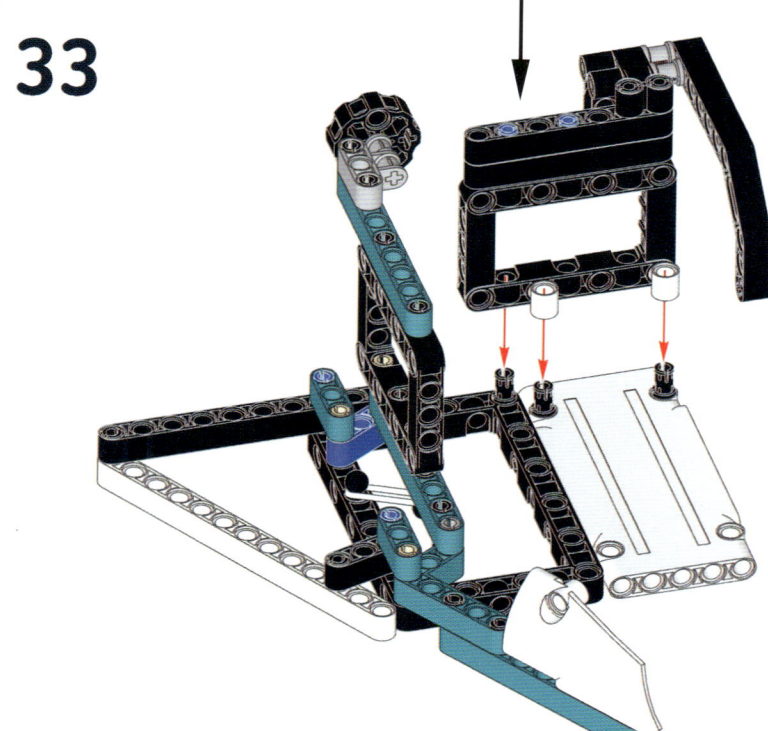

34

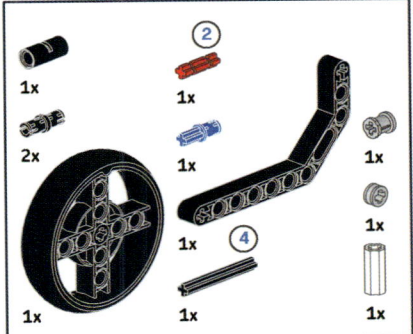

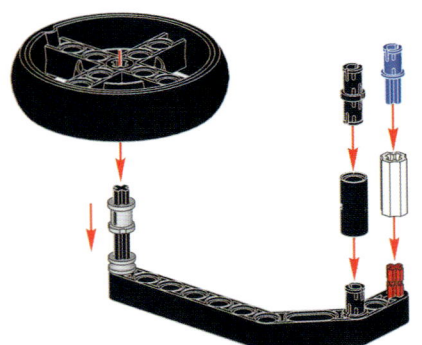

35

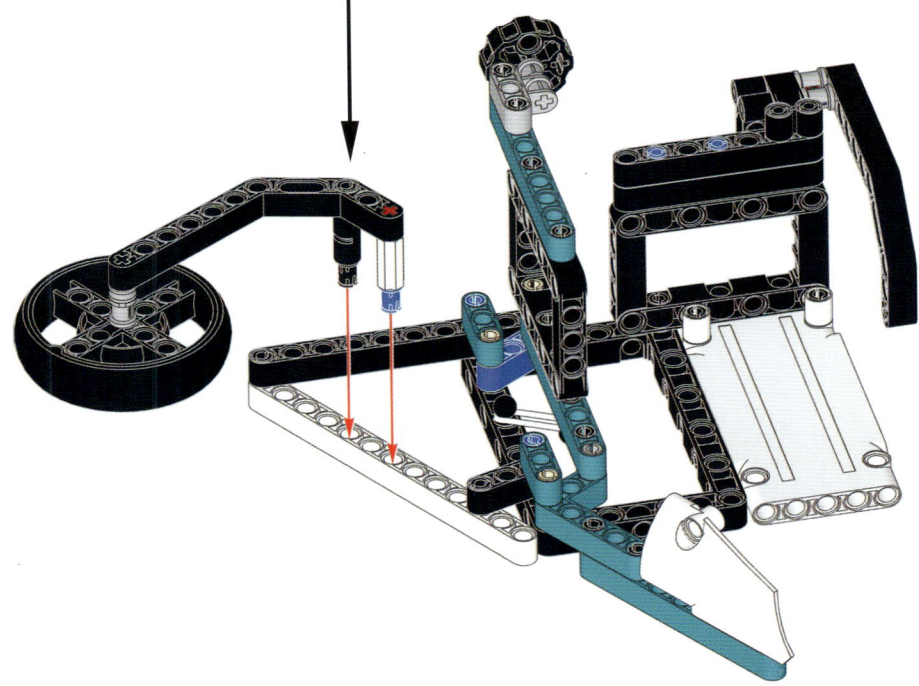

bottom bar subassembly

36

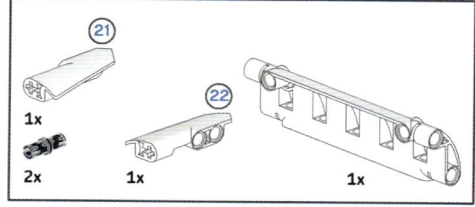

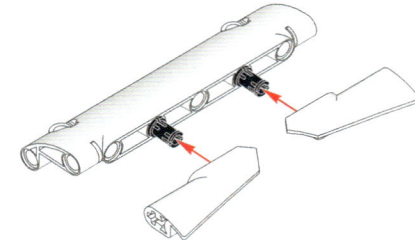

37

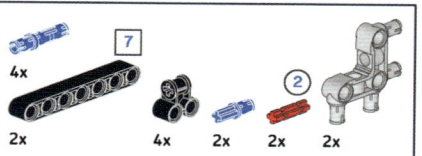

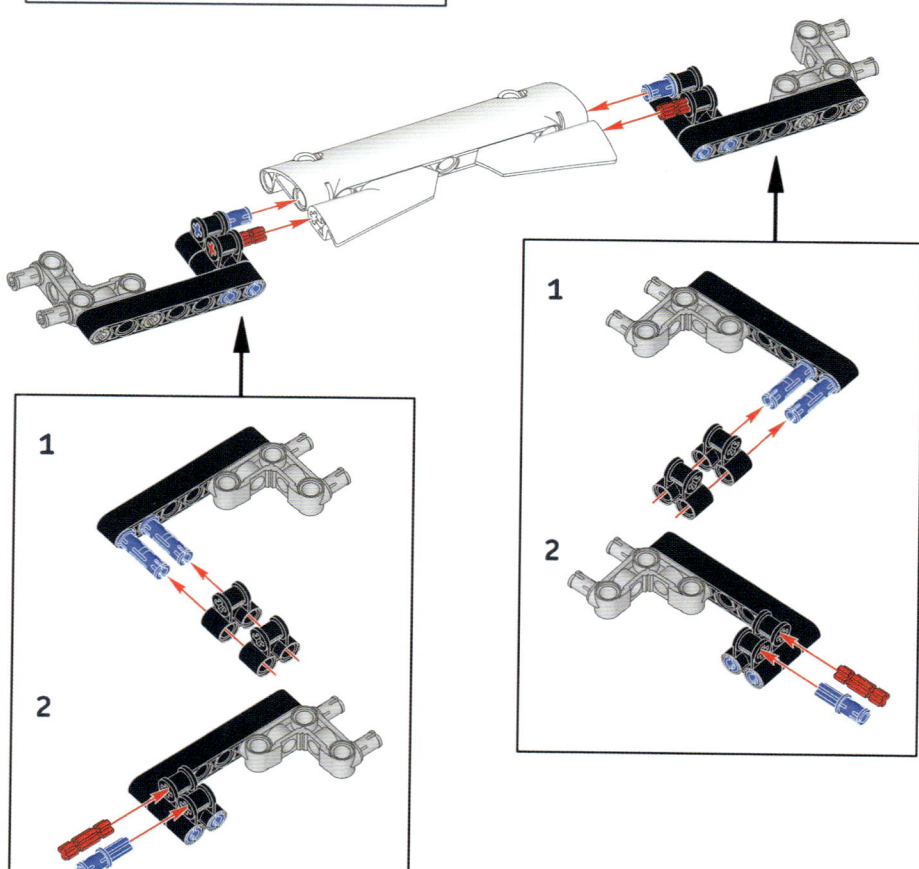

38

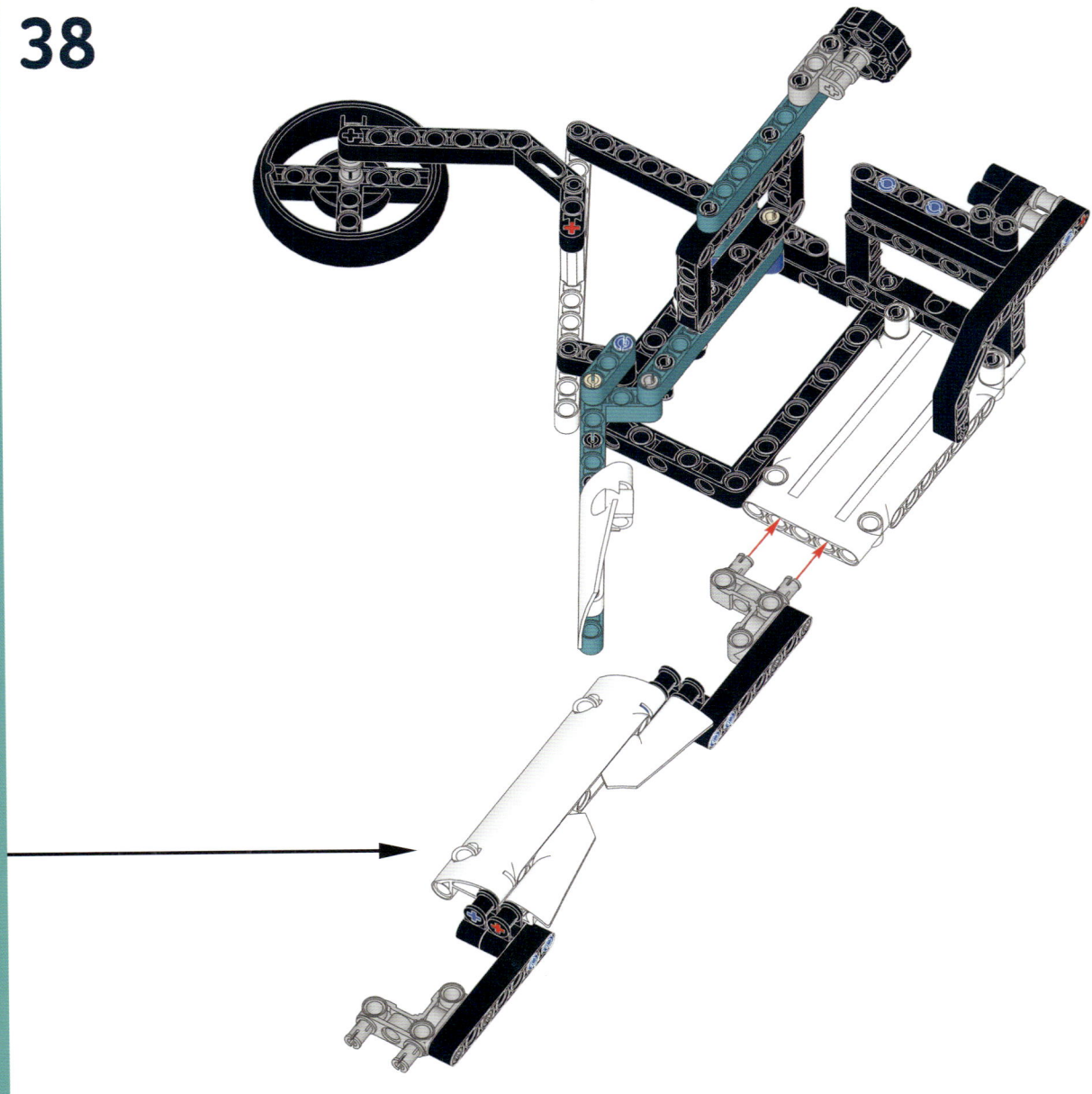

left flipper subassembly

39

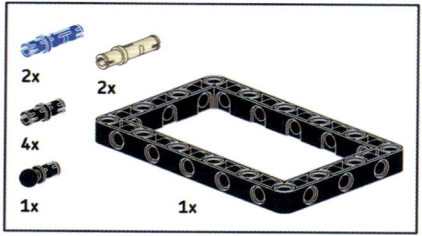

2x 2x 4x 1x 1x

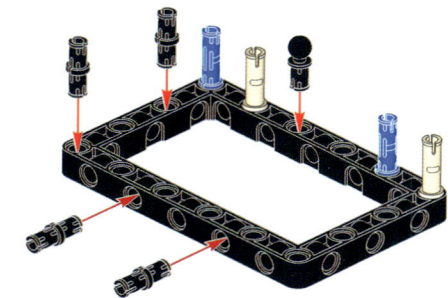

40

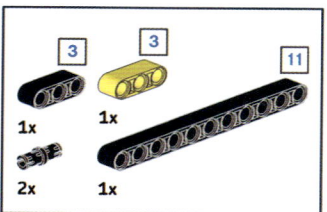

3 3 11
1x 1x
2x 1x

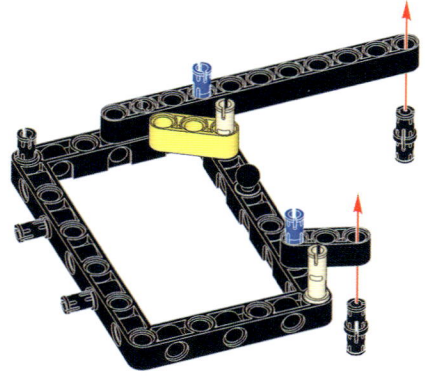

41

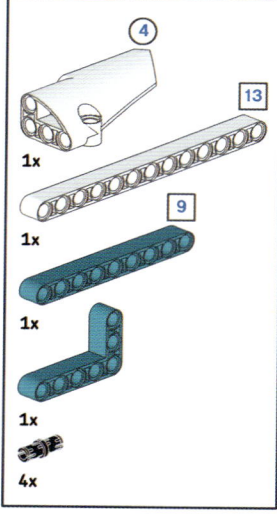

4 13
1x
9
1x
1x
1x
4x

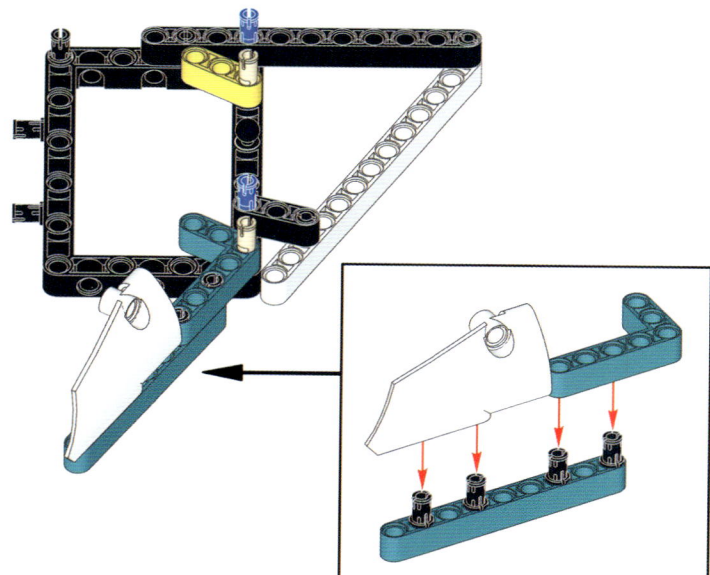

42

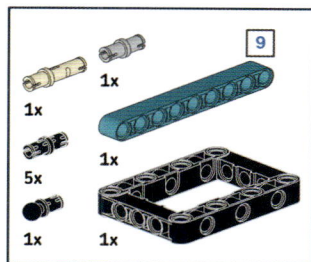

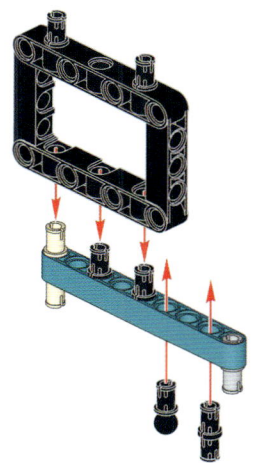

43

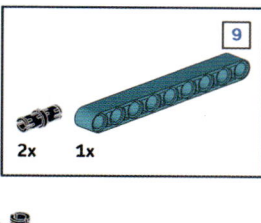

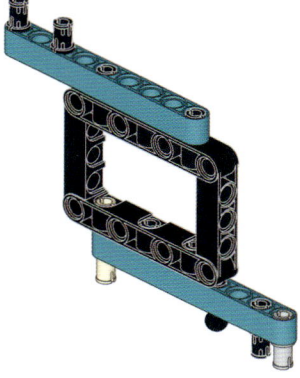

44

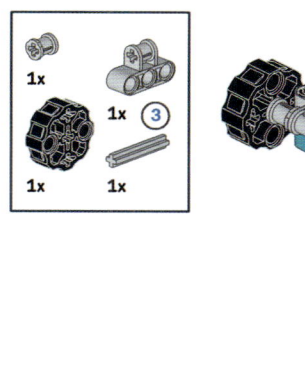

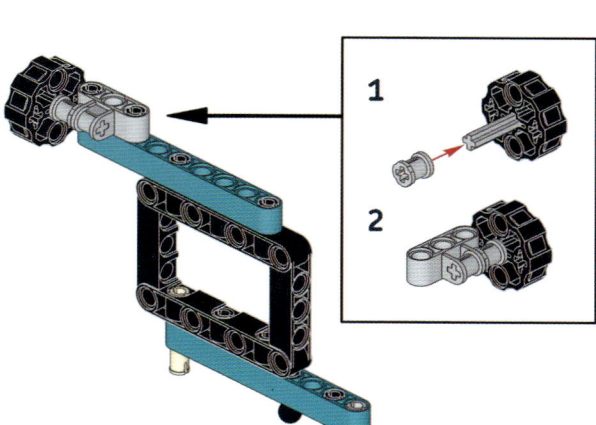

45

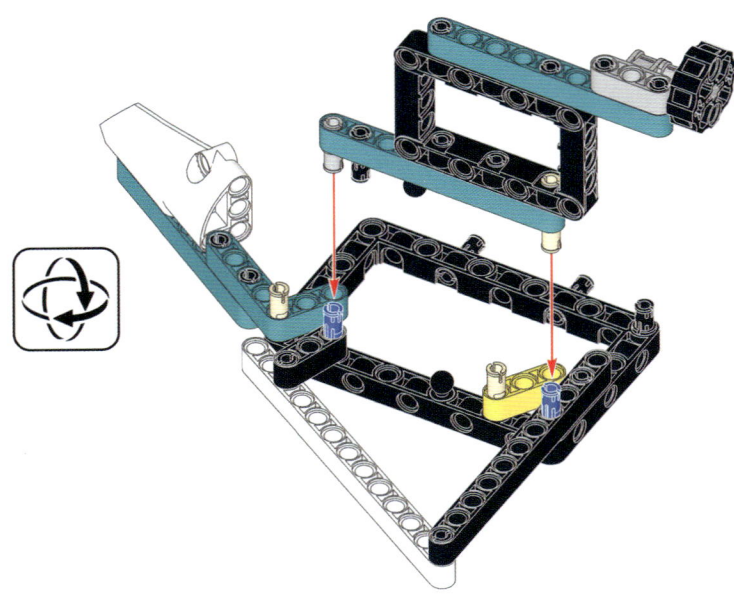

46

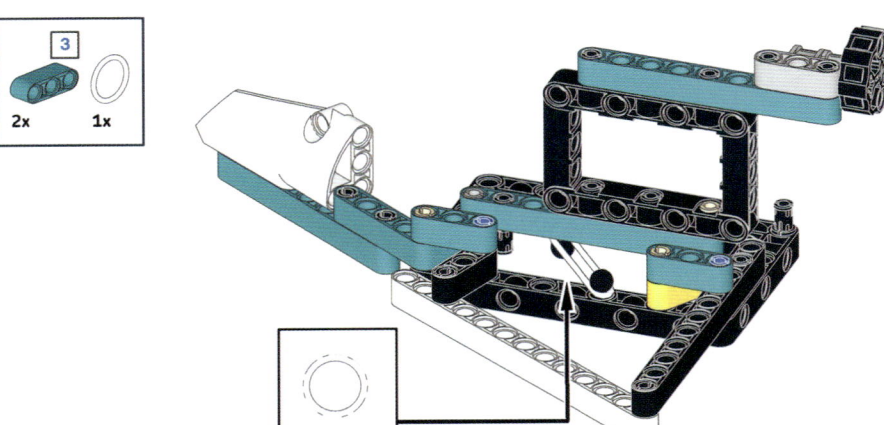

47

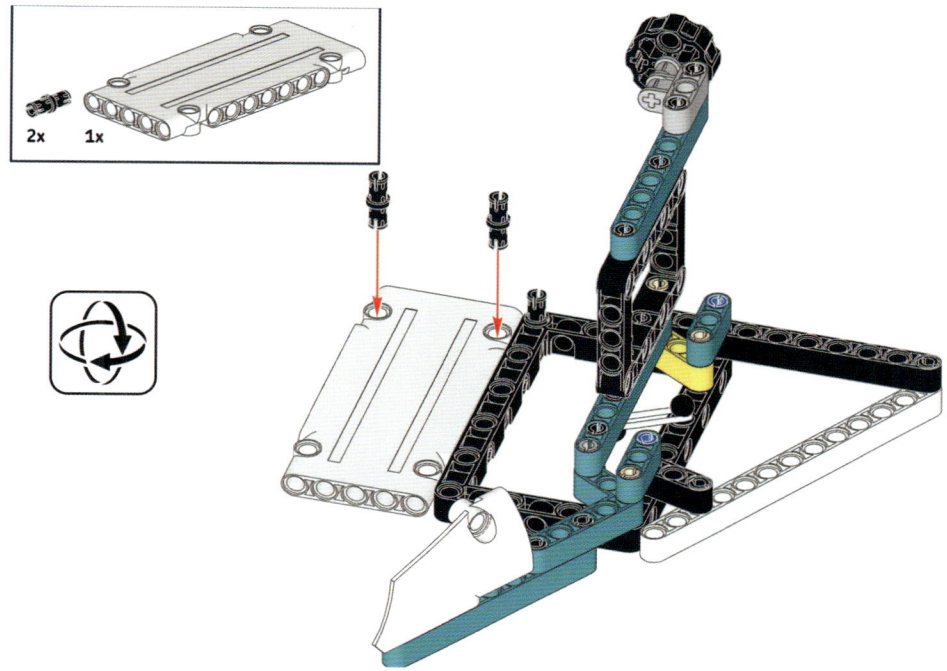

2x 1x

48

1x 2x 2x

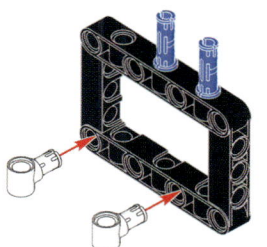

49

2x 2x

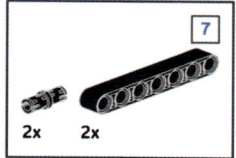

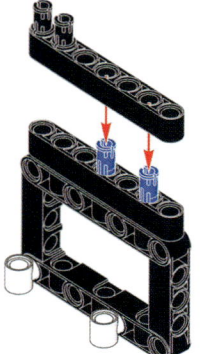

50

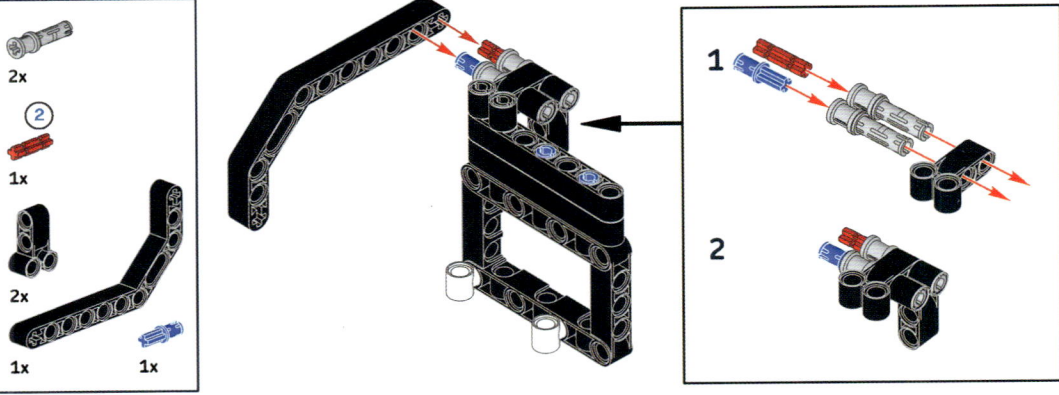

1

2

51

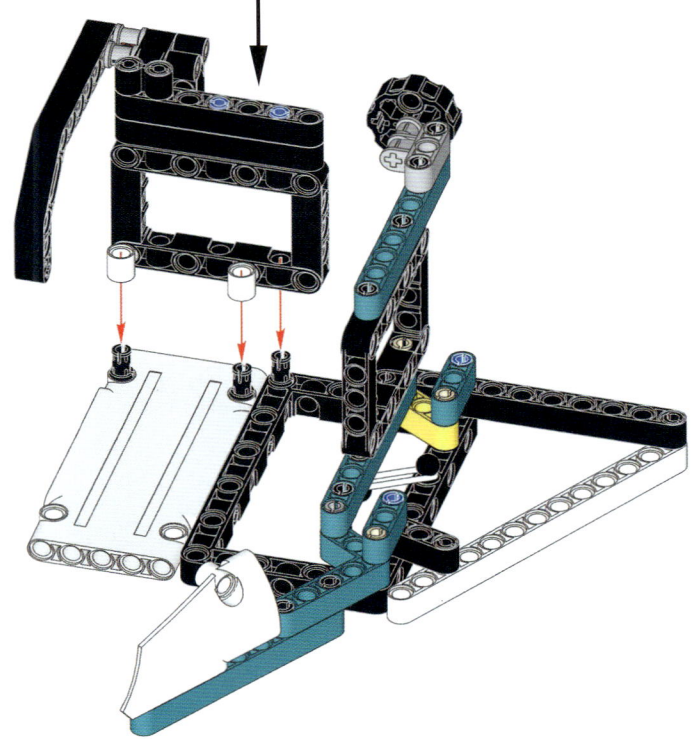

52

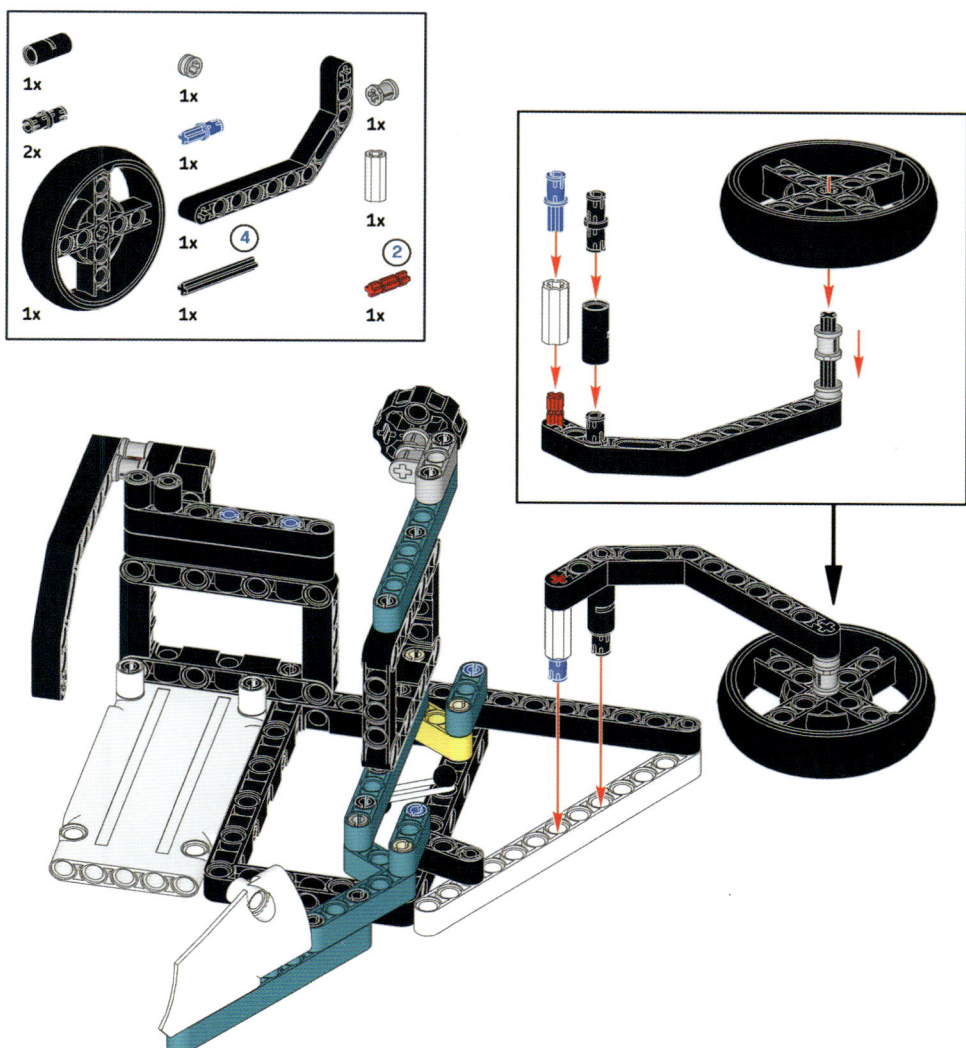

53

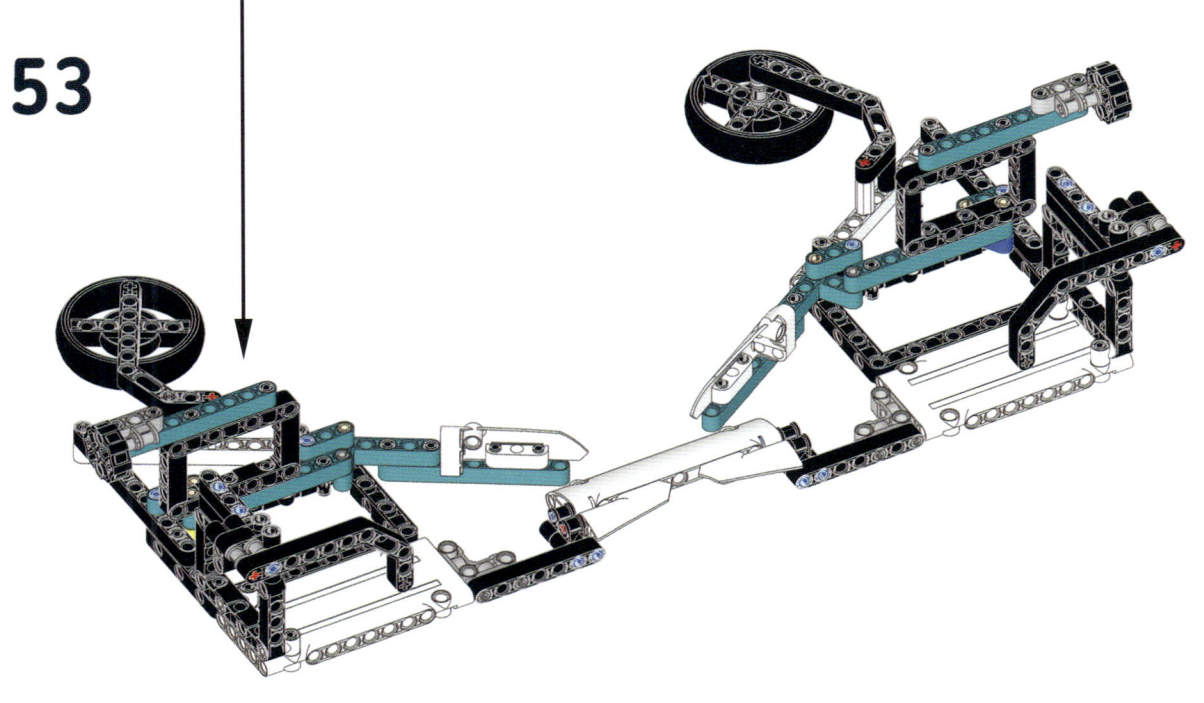

54

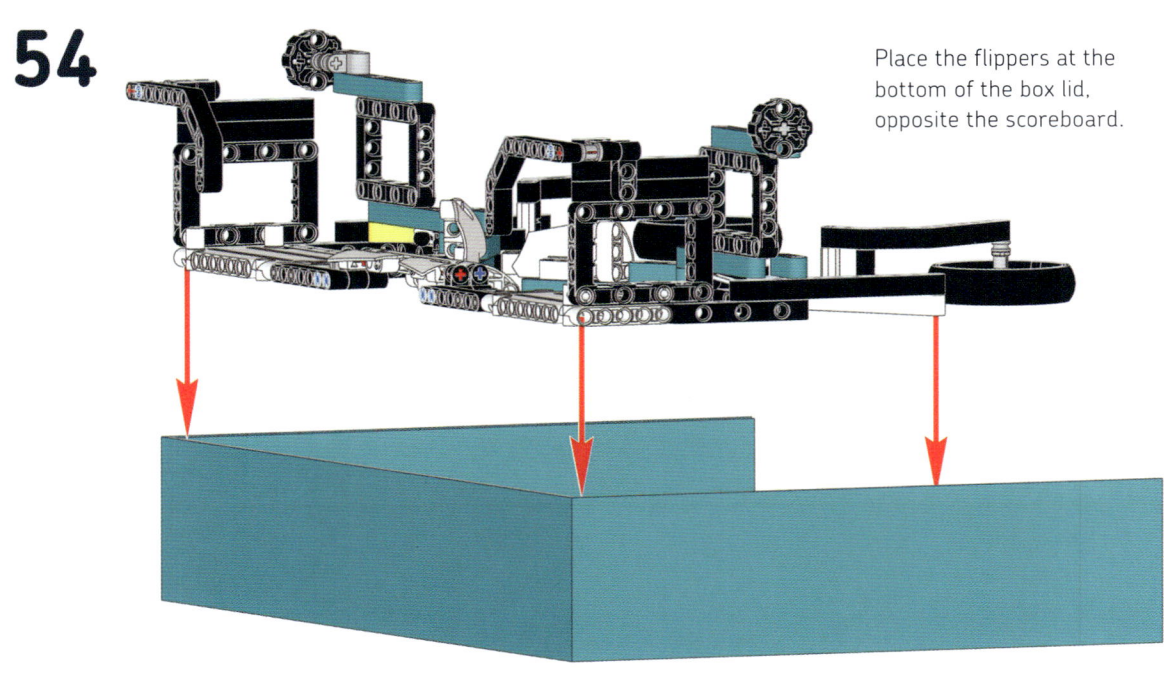

Place the flippers at the bottom of the box lid, opposite the scoreboard.

55

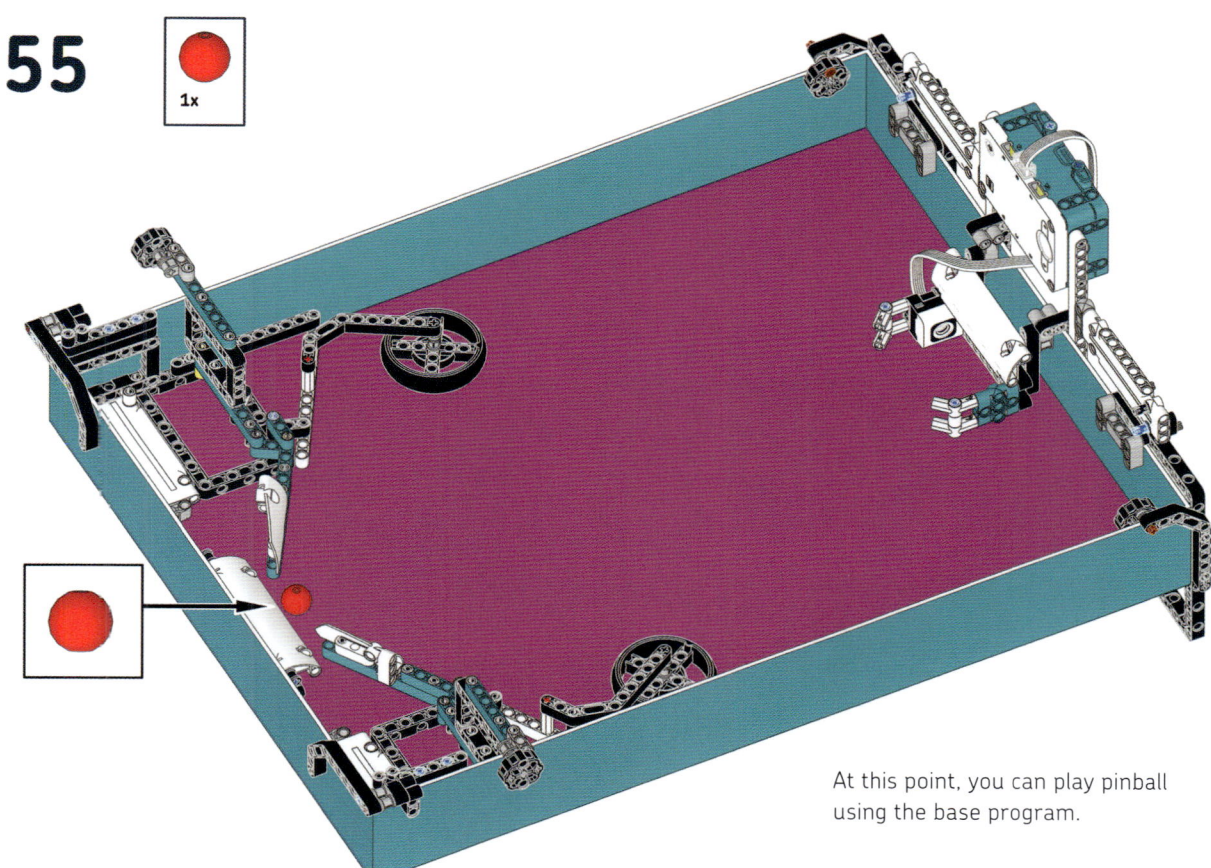

At this point, you can play pinball using the base program.

using lists

A *list* (sometimes also called an *array* in some programming languages) is a collection of items grouped under the same name. It's like a variable that can store several values at once. You can think of a list as a box divided into lots of little compartments. The box can store multiple pieces of data: one piece of data in each compartment.

Each item in a list has a number, called an *index*, that corresponds to the item's position in the list. The first item has an index of 1, the second item has an index of 2, and so on. You can access an item in a list by using its index.

In our pinball programs, we'll use a list to store the data needed to light up your score on the Hub display. Each item in the list will have the information for lighting up a single digit, from 0 to 9. Placing all that information in one list will make the programs more readable.

Follow these steps to create a list in the LEGO MINDSTORMS App. Use Figure 7-2 as a reference.

1. In the Variables palette, click the **Make a List** button.

2. Type in the name for your list. Choose a name that describes clearly what the list is going to contain. For example, our list will store information about how the pixels on the Hub display should be lit up to show different digits, so we'll call it **digitsPixels**.

3. Click **OK** to create the list.

Once you create a list, several new blocks will appear in the Variables palette, as you can see in Figure 7-2. We'll focus on some of the most commonly used ones. For each list you create, there will be a round `list` block with the name of that list. This block reports the items contained in the list as a single text string, with the different values separated by blank spaces. For example, if a list includes the items [1, 2, 3], this block returns the string "1 2 3".

The rest of the blocks for working with lists are shared among all the lists in your program. Each block has a drop-down menu where you choose the list you want the block to affect. The `add item to list` block adds the specified item to the end of the specified list, while the `delete item cf list` block deletes the item with the specified index. For example, `delete 1 of list` deletes the first item from the list. Once an

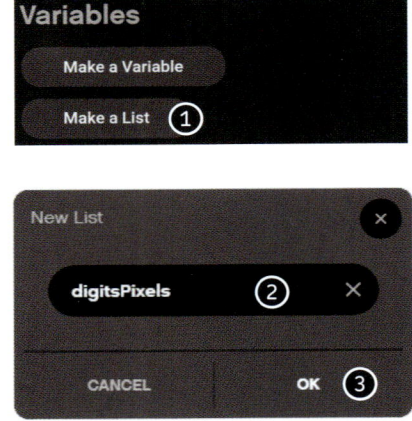

Figure 7-2: Creating a list (left) and the Variables palette blocks that manage lists (right)

item is deleted, any items that come after it shift to fill in the gap. In this example, once the first item of the list is deleted, what used to be the second item will now have an index of 1.

The `delete all items in list` block clears every item from the specified list. Until a list is cleared, the MINDSTORMS App remembers that list's contents, even if you stop the program or close the project. Therefore, it's important to use this block at the start of your program, to make sure the list is empty before filling it with the needed values by using the `add item to list` block. Otherwise, you'd just keep adding more and more elements to the end of the existing list.

Perhaps most importantly, the round `item at index in list` block returns the value stored at the specified index of the specified list. For example, `item 1 of digitsPixels` will return the value of the first item in the `digitsPixels` list. We'll use this block in our pinball programs to look up the information needed to light up specific digits on the Hub display.

controlling the hub display with text strings

So far, you've used blocks from the Light palette to show a letter (Chapter 3), an image (Chapter 5), and a scrolling number (Chapter 6) on the Hub LED matrix display. Remember that the `write` block can show only a single character at a time, so numbers with more than one digit must be shown by scrolling them across the display from right to left. For Pinball, however, we want to display two digits at once so we can write numbers from 0 to 99 without scrolling. We can do this by using customized versions of each digit that take up only two columns of the display, as shown in Figure 7-3.

NOTE I would like to acknowledge Frank Ellerkamp (phobricks.ch) for the two-digit number idea, as posted on the Facebook SPIKE Prime community page. His idea served as the basis for the shape of the numbers here.

We could draw these digit designs directly into `turn on light matrix` blocks (as we drew the turtle shell pattern in Chapter 5), but we'd need a separate block to display each number from 0 to 99. That would be a lot of blocks!

Instead, our program will just use one `turn on light matrix` block with a text string as its input. The text string tells the Hub which pixels of the matrix display to light up and how bright they should be. Each character in the string

corresponds to one pixel in the display. A **0** indicates a pixel should be off, a **9** indicates a pixel should be fully lit up, and the numbers **1** through **8** indicate different brightness levels in between.

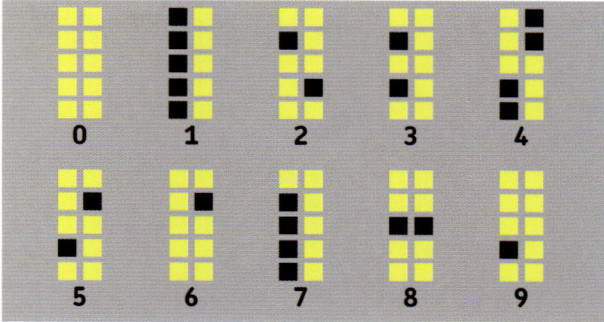

Figure 7-3: The digits 0 through 9 rendered using two 5-pixel columns of the Hub's LED matrix display

Figure 7-4 shows how each character of the string corresponds to each light on the display, using the number 24 as an example. By setting the display orientation to upright, even though the display itself is lying on its left side, we're able to use each group of five characters in the string to control one column of the display. The first group of five characters controls the five pixels in the leftmost column, from bottom to top, and so on. As you can see, the string begins with **99909**, which lights up all the pixels in the left column except the second pixel from the top.

Notice in Figure 7-4 that each group of five digits in the display string must be separated from the next group by some other character. I've used a colon for easy readability, but any character (such as a space) would work. Notice also that we've lit up some of the pixels in the middle column of the display at a lower intensity of **6** out of **9**. In our pinball programs, we'll use this middle column to show how many balls you have left in the game. In this case, two lit pixels means two balls left.

NOTE If you want to correctly display two-digit numbers on the Hub when it's standing upright, you should set the display orientation to right.

Figure 7-5 shows two ways of feeding a text string into the input of a `turn on light matrix` block. In (A), we store the string into the variable `display`, then place that variable into the `turn on` block's input. This isn't much better than drawing out the pixels by hand, since we'd need to create a separate string for each number from 0 to 99, as well as for the pixels that indicate the balls left.

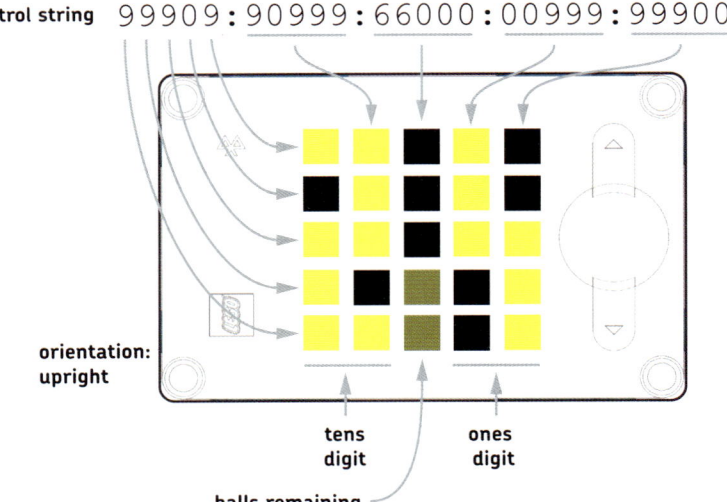

control string   99909:90999:66000:00999:99900

orientation:
upright

tens
digit

ones
digit

balls remaining

Figure 7-4: The correspondence between pixels on the Hub display and the digits in the string for controlling the display

Alternative (B) shows a preview of what we'll do in our pinball programs. Inside the `turn on light matrix` block, we'll use data from two lists. The `digitsPixels` list stores 10 different strings corresponding to the patterns to light up each of the 10 digits (0 through 9). Similarly, the `ballsPixels` list stores strings to light up the middle column with the number of balls remaining in the game. We retrieve two items from `digitsPixels`, one for the tens place of a two-digit number and one for the ones place, and we retrieve a pattern from the `ballsPixels` list in between. We bring them all together with two `join strings` blocks. Each `join strings` block merges two strings into one longer string. As a result, the Hub should display the number 24, with two pixels lit up in the middle column to indicate two balls remaining, as shown in Figure 7-4.

NOTE The strings to display digits are stored in `digitsPixels` in order from 0 to 9, but their index numbers within the list are 1 through 10. This means the index number will always be 1 greater than the digit we want to display. That's why we use `item 3 of digitsPixels` in Figure 7-5 to display the digit 2, for example.

Figure 7-5: Two ways to display the number 24 on the Hub

making the base pinball program

Let's make the base program for Pinball. It will let you play for a set amount of time, adding points when the ball rolls into the area in front of the Color Sensor. Later in the chapter, we'll add more features to the program.

To start, create a new program and save it with the name *pinball*. Then, go to the Variables palette and create the following variables:

```
ballDuration
balls
ones
score
tens
timeOld
```

Next, create two lists:

```
ballsPixels
digitsPixels
```

Now create the `initialize display` custom block (A), the `update display` custom block (B), and the stack for when the program starts (C), using Figure 7-6 as a reference.

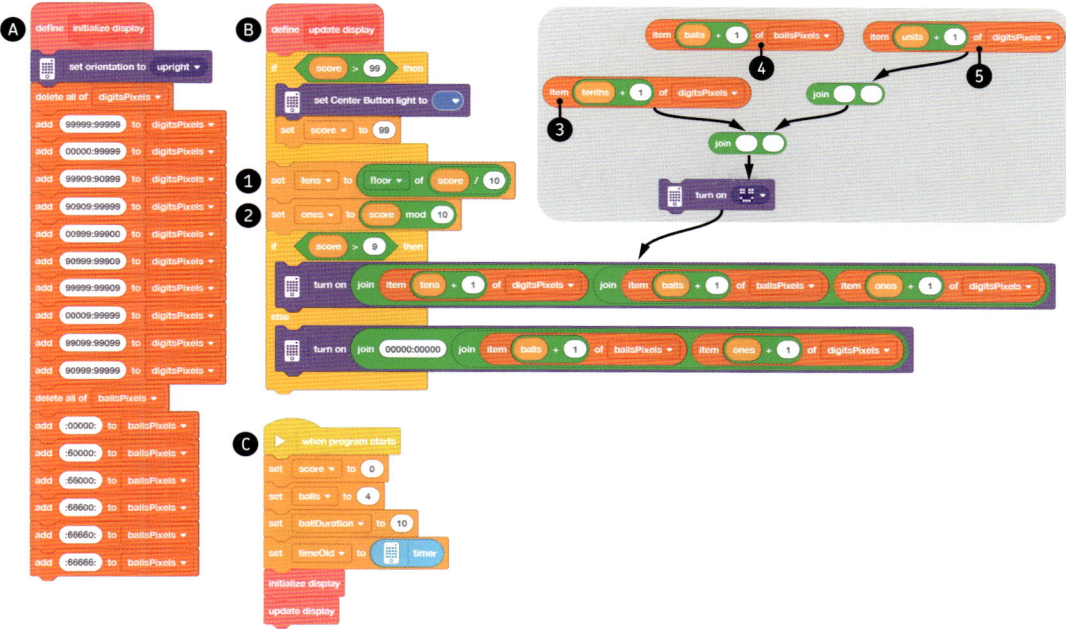

Figure 7-6: Stack C of the pinball base program uses the custom blocks defined in stacks A and B, which help display the score.

Refer to Table 7-1 to fill the digitsPixels and ballsPixels lists.

Table 7-1: digitsPixels and ballsPixels List Items

digitsPixels	ballsPixels
99999:99999	:00000:
00000:99999	:60000:
99909:90999	:66000:
90909:99999	:66600:
00999:99900	:66660:
90999:99909	:66666:
99999:99909	
00009:99999	
99099:99099	
90999:99999	

Then, looking at Figure 7-7, create stack D, which makes Pinball react to the ball's passing in front of the Color Sensor by playing a sound and changing the score.

If you run the program as it is now, you can play forever, because the balls variable is set to 4 in stack C and isn't changed anywhere else in the program. To decrease the balls variable and let the game end after a while, create stack E, shown in Figure 7-8.

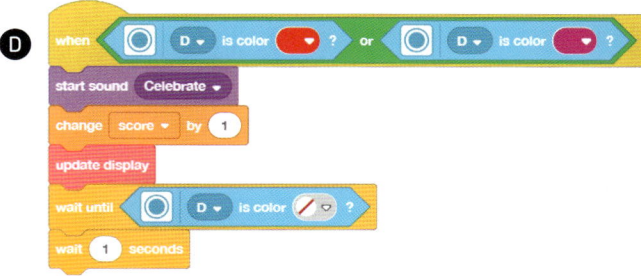

Figure 7-7: This stack makes the game react to the passage of the red ball in front of the Color Sensor.

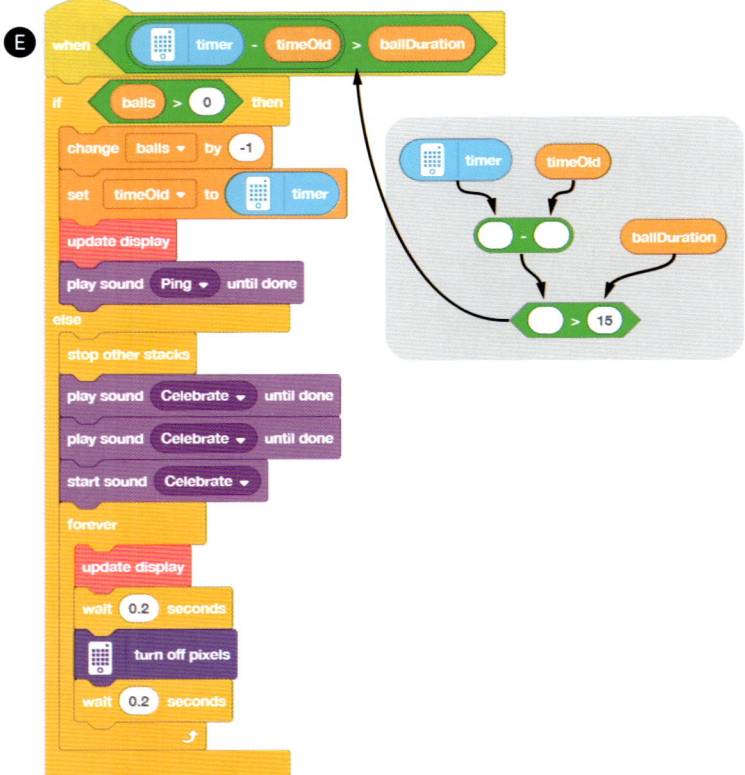

Figure 7-8: This stack decreases the value of the balls variable until the game ends, then shows the final score.

playing the game

Since there isn't a plunger (a spring-loaded rod with a handle) to launch the ball into the playing field, you should just throw the ball by hand into the playing field when you start the program. Push the side buttons at the bottom to move the flippers and hit the ball up, trying to reach the Color Sensor. When you succeed, you should hear a sound and see the score increase by 1. Unlike in real pinball, the red ball will never escape from the playing field. Instead, every 10 seconds the value of the balls variable decreases. When the ball count reaches 0, the game ends, and your final score flashes on the display.

To play another game, simply stop and restart the program. The value of the balls variable is set to 4 at the beginning of the program, and the ballDuration variable is set to 10; the game thus lasts 4 × 10 = 40 seconds. To make the game last longer, you can increase the value of ballDuration, or you can set the value of balls to 5. Setting balls higher than 5 would cause errors in the display of the ball count on the Hub, since we're using a single column of five pixels to show the number of balls.

understanding the base program

Now let's take a closer look at the program stacks shown in Figures 7-6, 7-7, and 7-8 to see how the program works.

initialize display custom block

The initialize display custom block, defined in stack A, sets up the lists needed for displaying the score and ball count on the Hub. It starts by setting the orientation of the display to upright, even though the Hub is lying on its left side. This allows us to show the score by using strings controlling the columns of the Hub display, as described earlier in the chapter. Next, we clear the digitsPixels list and fill it with 10 string items for displaying the digits 0 through 9. Remember that you should always clear a list at the start of a program before filling it up again.

We then go through a similar setup process for the ballsPixels list, first clearing it and then filling it with strings for showing the number of available balls in the center column of the display. The string 00000 at index 1 indicates no balls left, while the string 66666 at index 6 indicates five balls left (at a lower brightness to avoid interfering with the display of the score). Each element of the ballsPixels list starts and ends with a separator character (colon). This way, when a ballsPixels item is joined with two digitsPixels items, each group of five digits, representing one column in the display, will be separated from the next by a colon.

update display custom block

The update display custom block, defined in stack B, reads the value of the score and balls variables and controls the Hub display to show the two-digit number for the score and the column of lights for the number of balls left. Since we won't be able to show a three-digit number, we first use an if then block to check whether the score is greater than 99. If so, we turn the Hub's button light to blue and set the score back to 99 so it won't go any higher. That's the highest score you'll be able to get—sorry.

To display the score, we need to break down its value into a digit for the tens place and a digit for the ones place. To get the value to store into the tens variable, we divide score by 10, then round the answer down to the nearest whole number with floor, which is an option on the math functions block, the last block in the Operators palette ❶. For example, if the score were 24, the result would be floor(24 / 10) = floor(2.4) = 2.

To compute the value to store into the **ones** variable, we "divide" **score** by **10** by using the **mod** block (short for *modulo*) instead of the **divide** block ❷. The **mod** block divides one value by another and reports the remainder. For example, if the score were 24, the result would be **24 mod 10 = 4**, because 4 is the remainder of 24 divided by 10.

The values of the **tens** and **ones** variables, increased by 1, will be used to select the required items of the **digitsPixels** list with the value of **item at index in list** blocks ❸ and ❺. (Remember, the index of each digit's display information is 1 greater than the digit itself.) For example, if the score were 24, the **tens** value would be 2, so we would retrieve the third item of the **digitsPixels** list, which encodes the digit 2. Similarly, we add 1 to the value of the **balls** variable to retrieve the correct item from the **ballsPixels** list ❹. We use two **join strings** blocks to join the strings encoding the tens digit, the number of balls, and the ones digit. The result is one long string with all the required display information, which we can feed into a **turn on light matrix** block.

But what if the score is a one-digit number? We address this situation with an **if then else** block at the end of stack B, which checks whether **score** is greater than 9. If so, the score is displayed as described in the previous paragraph. Otherwise, the two columns of the Hub display corresponding to the tens digit are turned off by the **00000:00000** in the first input of the **join strings** block, and we look up display information only for the number of balls and the ones place of the score.

initializing the game

When the program starts, stack C begins by setting up four variables: **score**, **balls**, **ballDuration**, and **timeOld**. As explained in "Playing the Game," on page 168, you can make the game last longer by changing the value of the **balls** and **ballDuration** variables in this stack. We'll use **timeOld** to know when the time has run out for each ball in the game. To start, it gets a value of 0 from the timer. After setting the variables, we execute the **initialize display** and **update display** custom blocks.

reacting to the red ball

Stack D starts when the Color Sensor sees red or magenta. I chose to use two colors as valid conditions because sometimes the Color Sensor mistakes the red ball for magenta when the ball is moving very fast. When the ball is detected, we play a cheerful sound, increase the score by 1, and execute

the **update display** block to show the new score on the Hub display.

The **when** block shouldn't retrigger the stack if the condition remains true when the stack ends. During testing, however, I found that the ball's movements were causing the Color Sensor's readings to change rapidly, meaning the same hit of the ball could increase the score multiple times. To avoid this, the **wait until** block halts the stack until the ball isn't detected anymore; then we wait for one more second just to be safe.

ending the game

Stack E is executed when the difference between the current time measured by the **timer** block and the time stored in the **timeOld** variable is greater than the value of the **ballDuration** variable. In other words, since the **ballDuration** variable is set to **10** and since we set **timeOld** to an initial timer value of **0** in stack C, stack E is first executed 10 seconds after the program starts.

Once the stack starts, we use an **if then else** block to check whether the **balls** variable is greater than 0. If so, we decrease the value of **balls** by 1 and store the current time in the **timeOld** variable. Updating **timeOld** ensures that stack E will be triggered again in another 10 seconds (or after however many seconds you store in the **ballDuration** variable). Additionally, we update the display to show the decreased number of balls left and play the Ping sound to mark the change in the ball count.

When the value of **balls** reaches 0, the blocks in the **else** space are executed instead. All the other stacks are stopped, the Celebrate sound plays three times, and we use a **forever loop** to repeatedly flash the final score on and off. Within the loop, to achieve the flashing effect, we show the score with the **update display** custom block, wait for 0.2 seconds, turn off the display, wait for another 0.2 seconds, and then repeat until the player stops the program.

finishing pinball

Now let's finish building Pinball by adding four sensitive bumpers. Hitting a bumper will be another way to score points. We'll also add the Distance Sensor, although we won't actually use it to measure anything. Instead, we'll use its lights to show when the ball hits one of the bumpers.

56

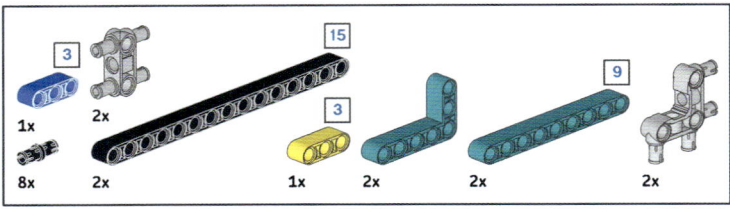

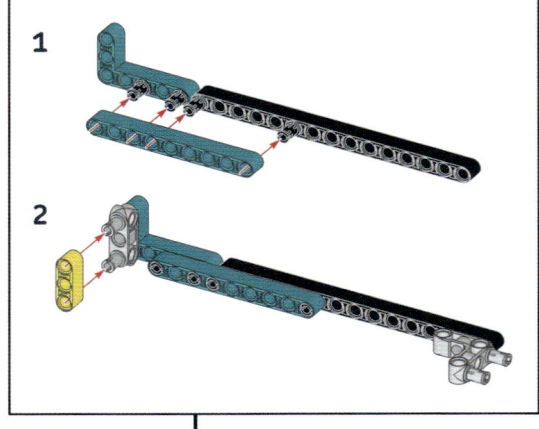

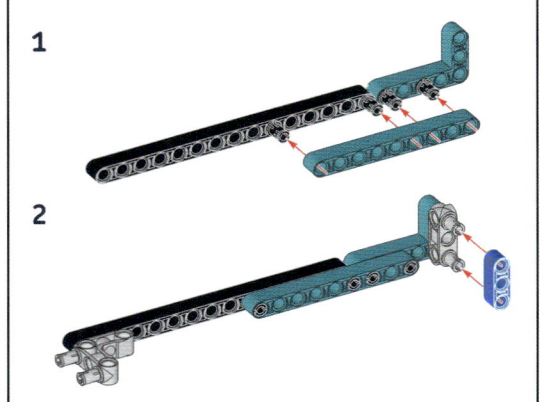

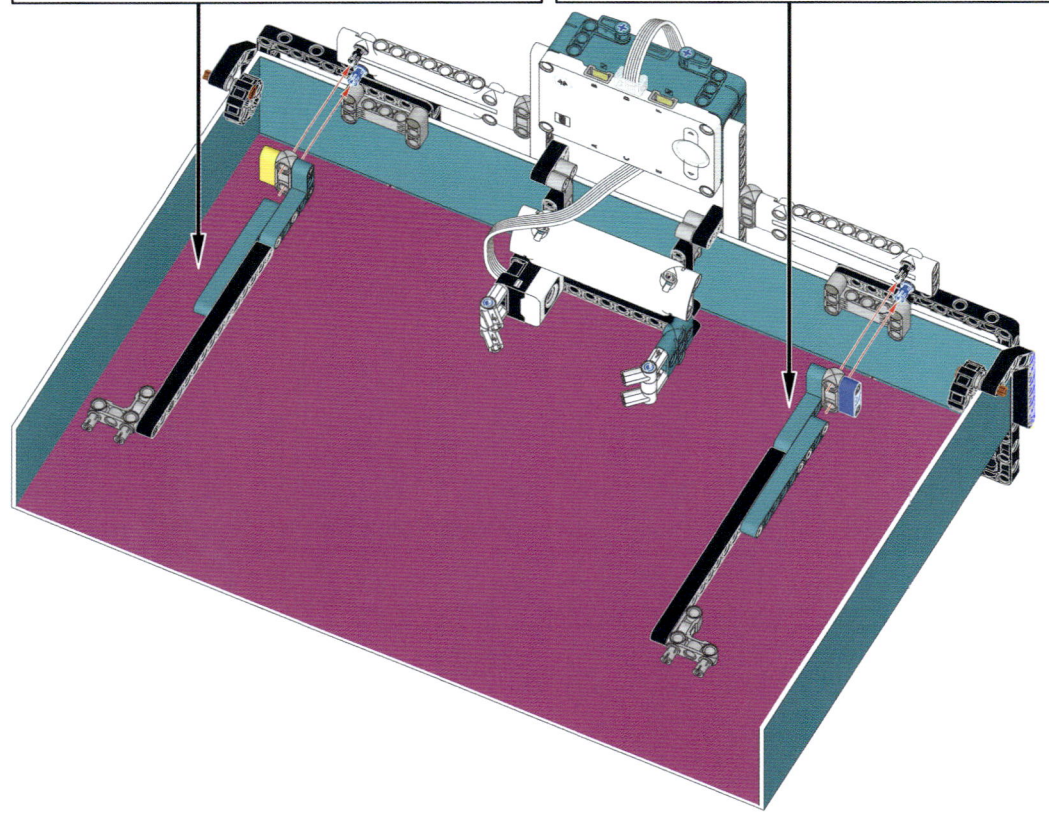

57

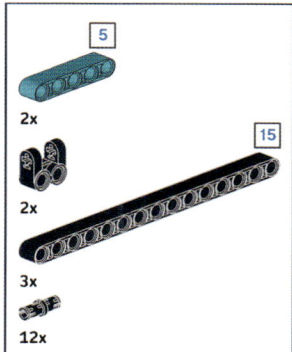

2x
2x
3x
12x

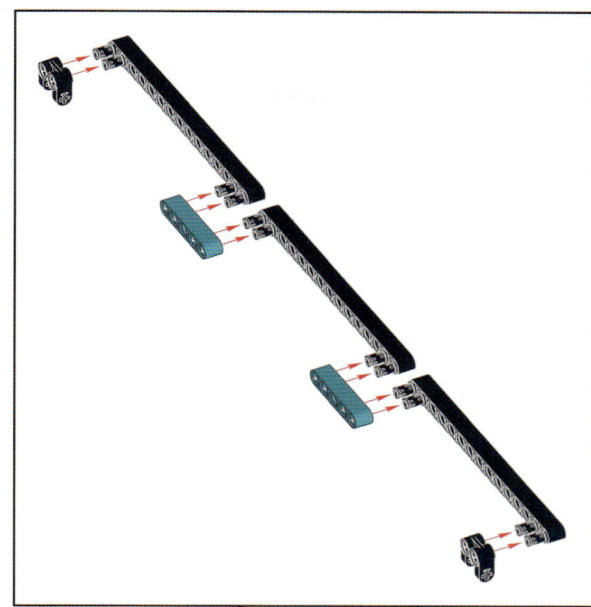

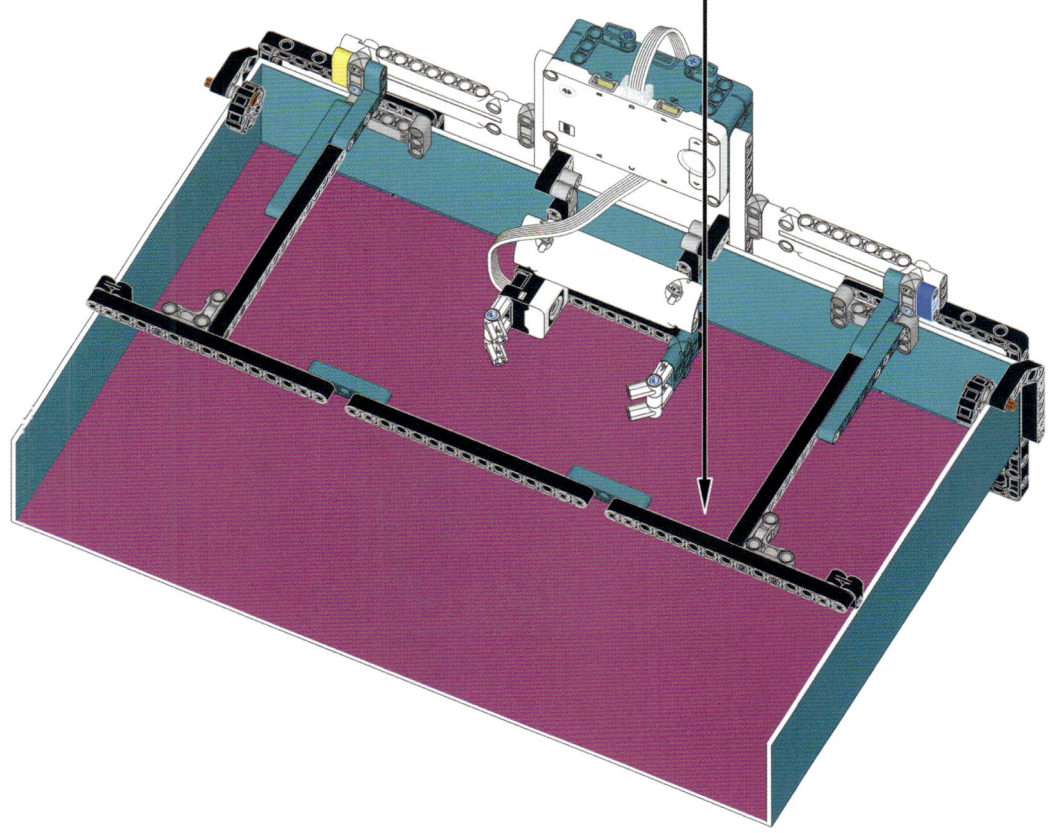

bottom-right bumper subassembly

58

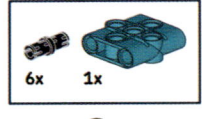

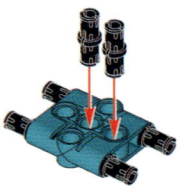

59

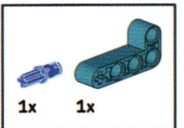

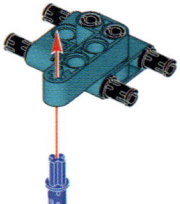

60

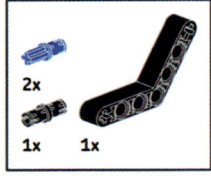

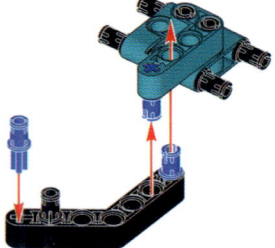

61

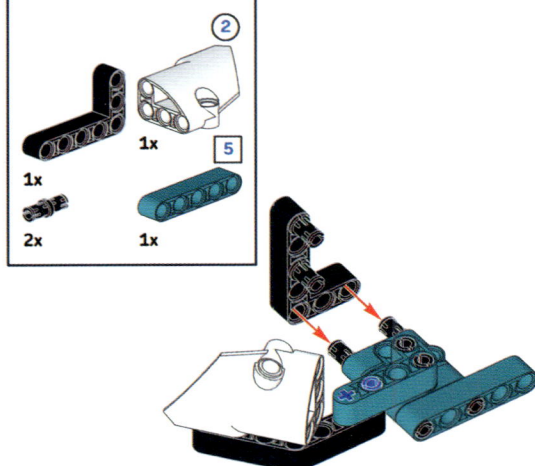

62

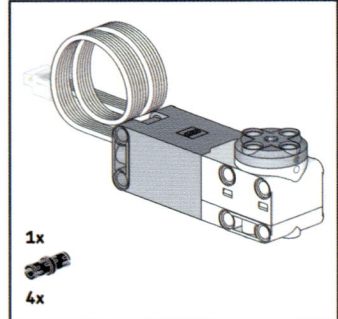

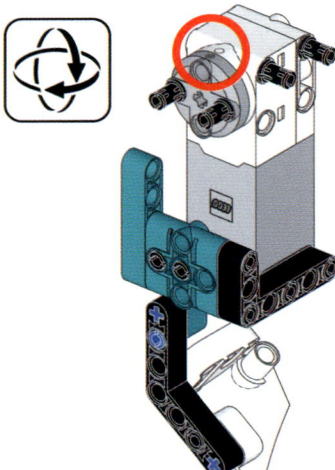

63

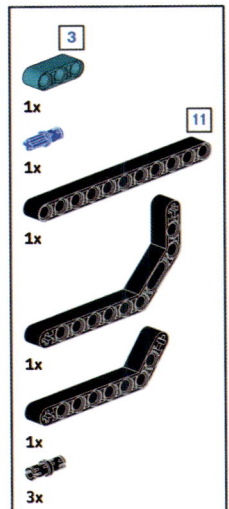

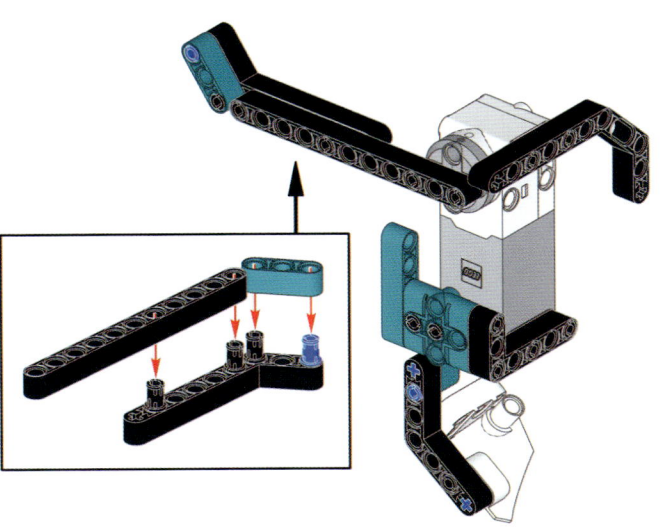

64

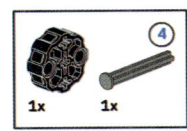

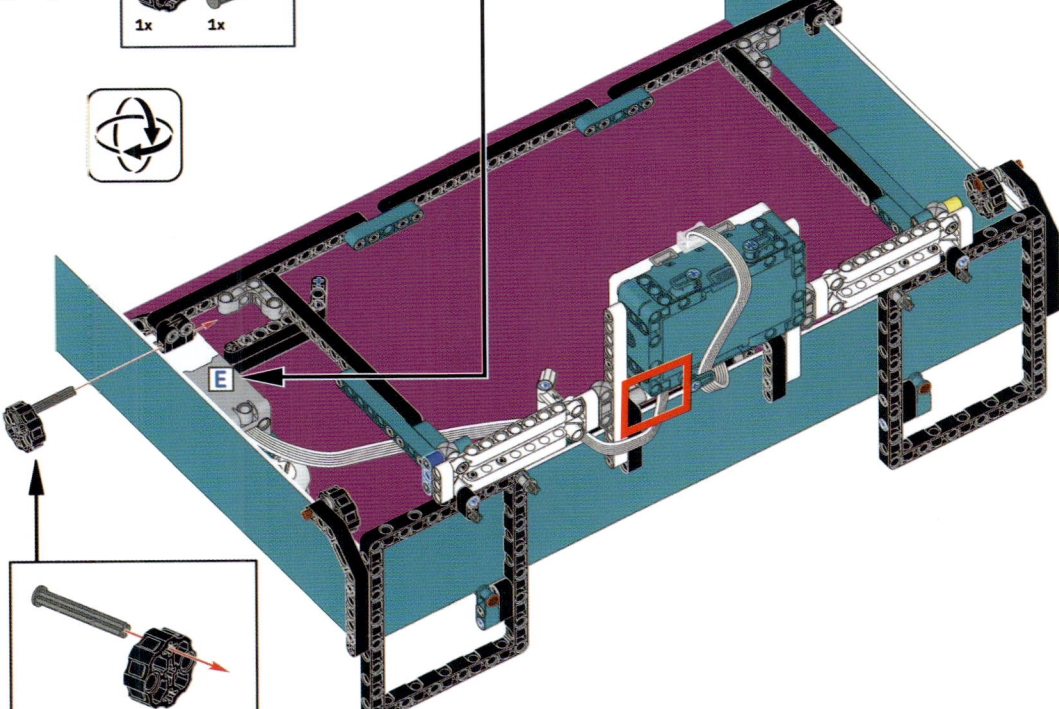

bottom-left bumper subassembly

65

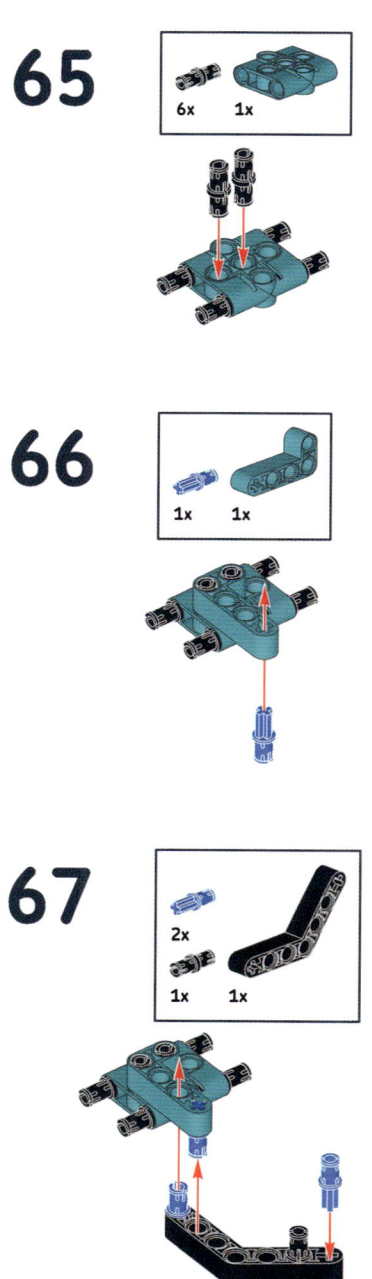

66

67

68

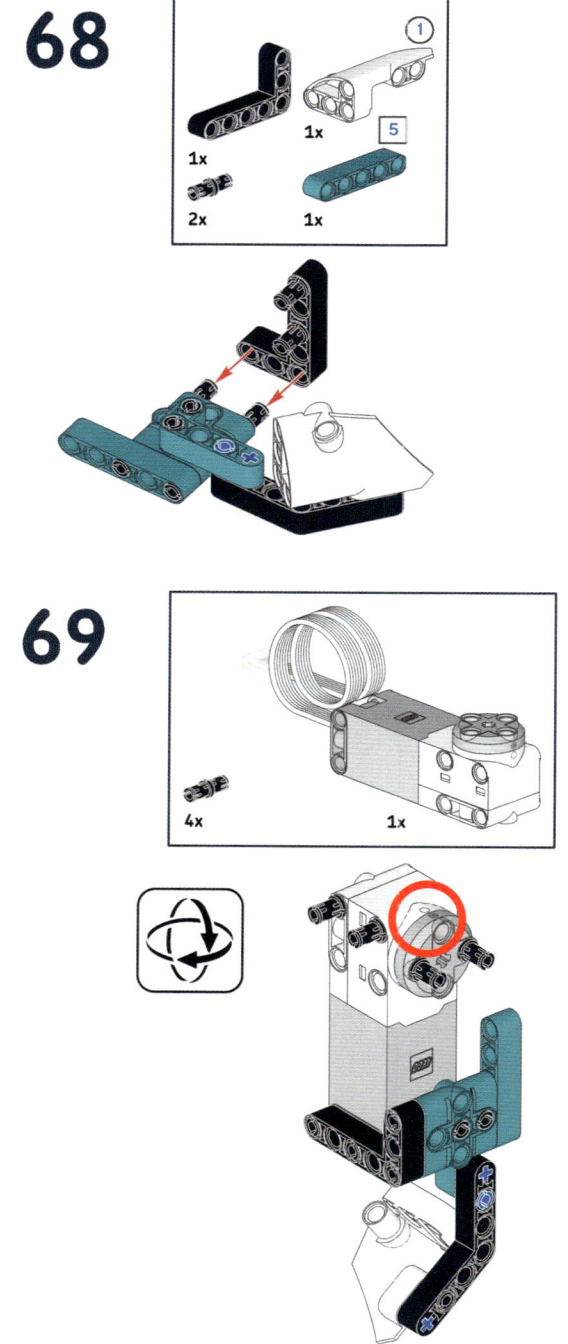

69

70

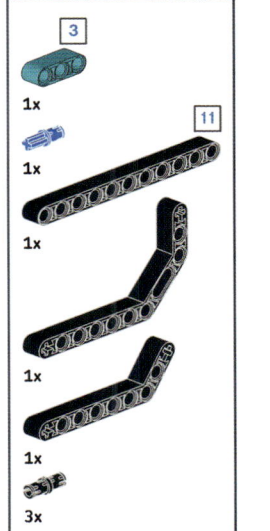

3
1x

11
1x

1x

1x

1x

3x

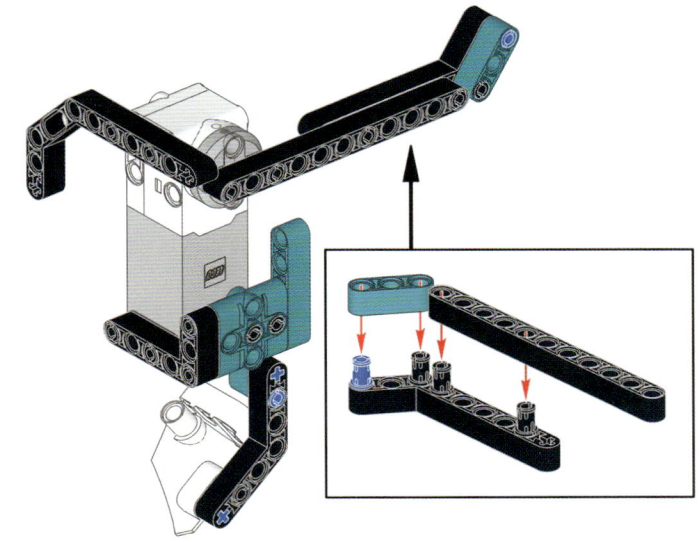

71

4
1x   1x

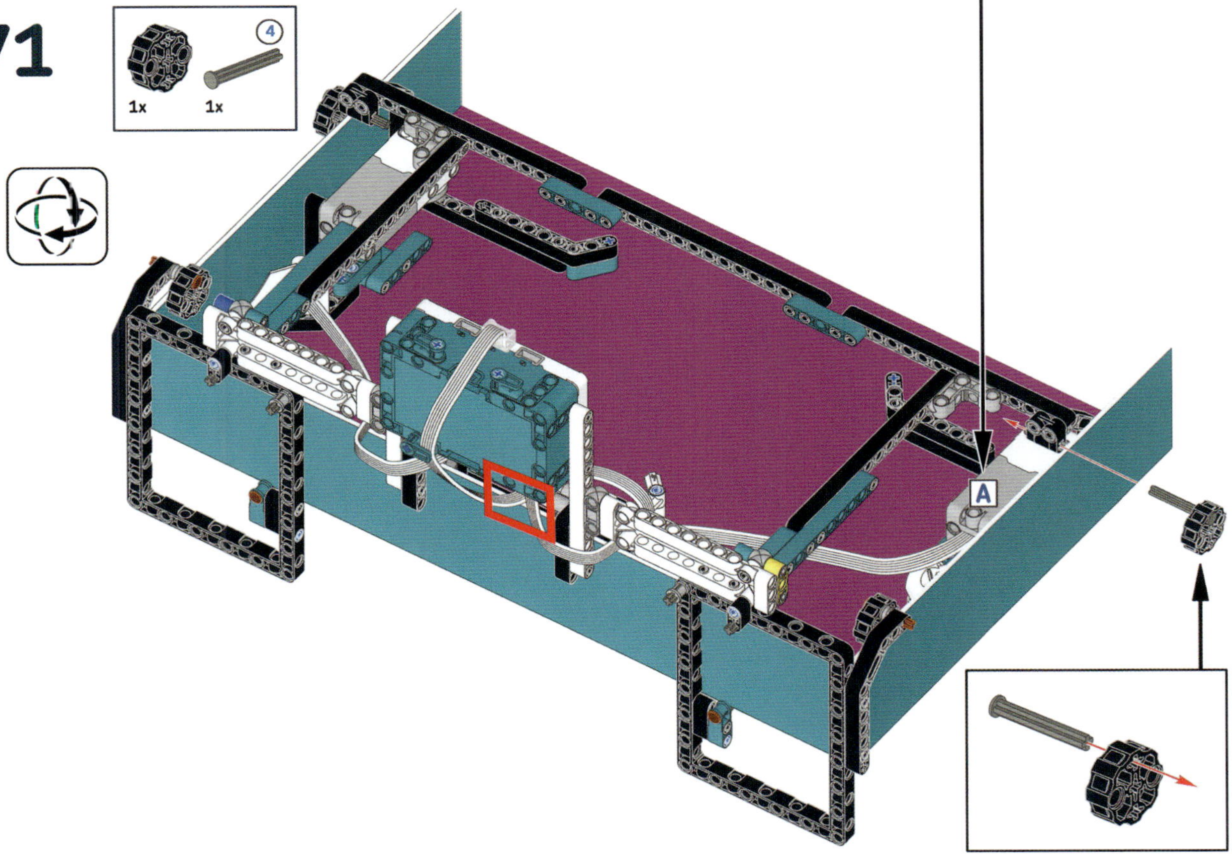

72

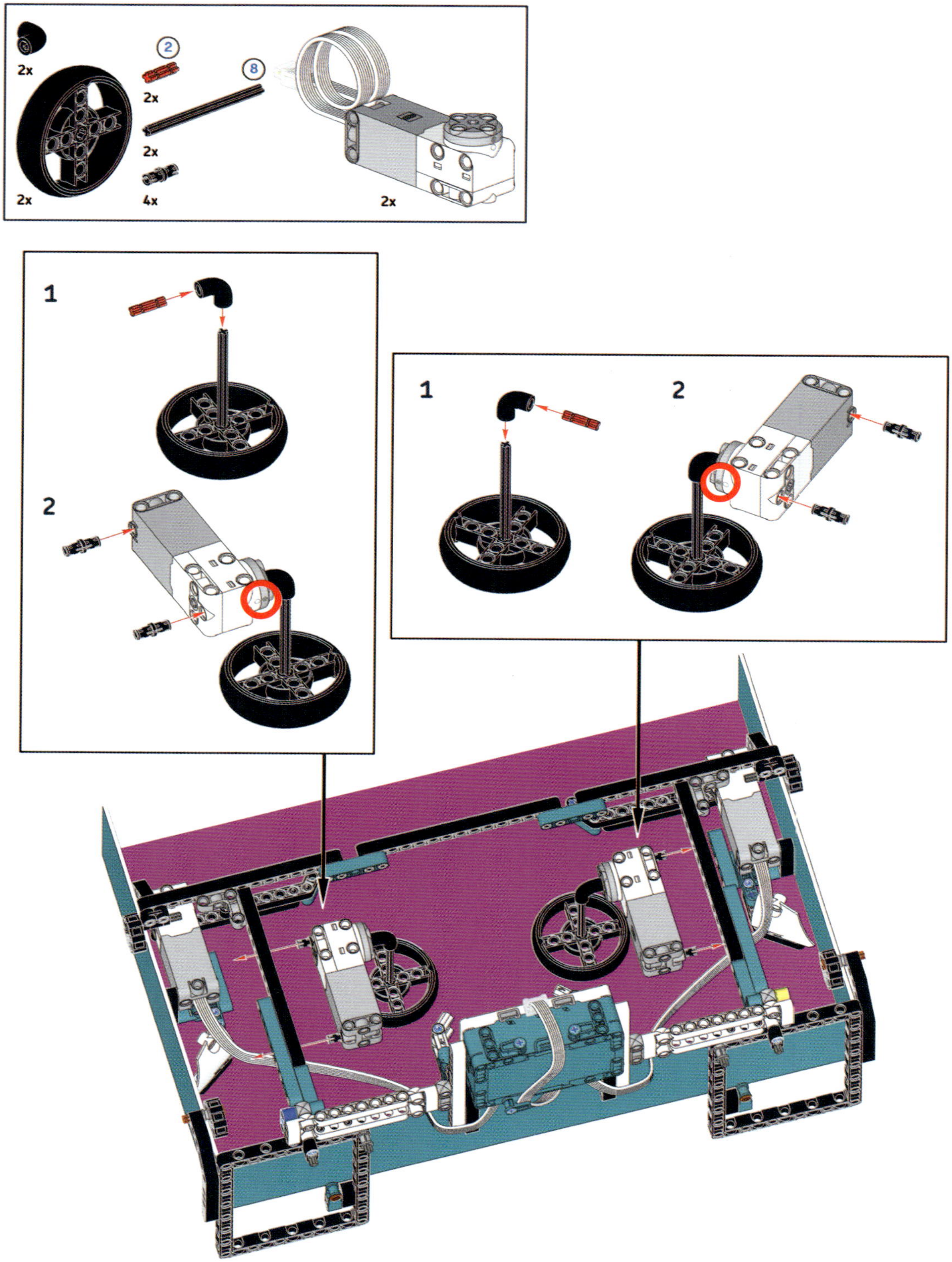

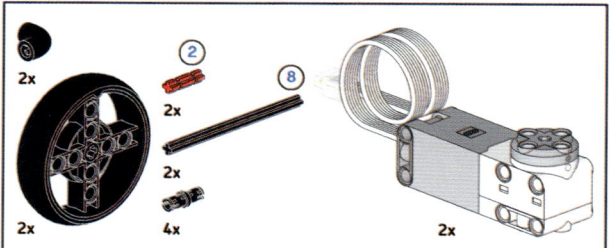

2x 2x 8 2x
2x 2x
2x 4x 2x

1

2

1 **2**

73

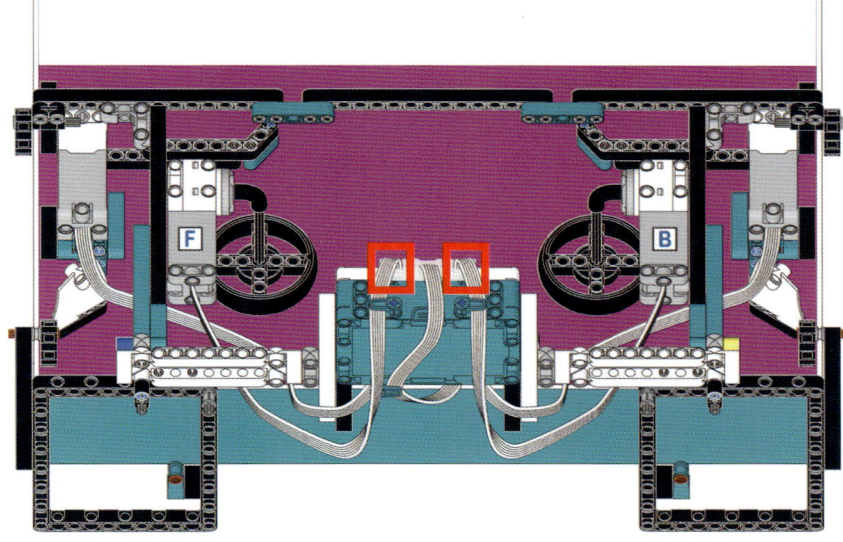

74

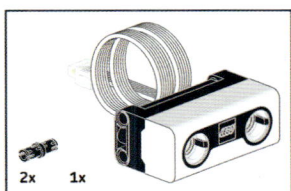

2x   1x

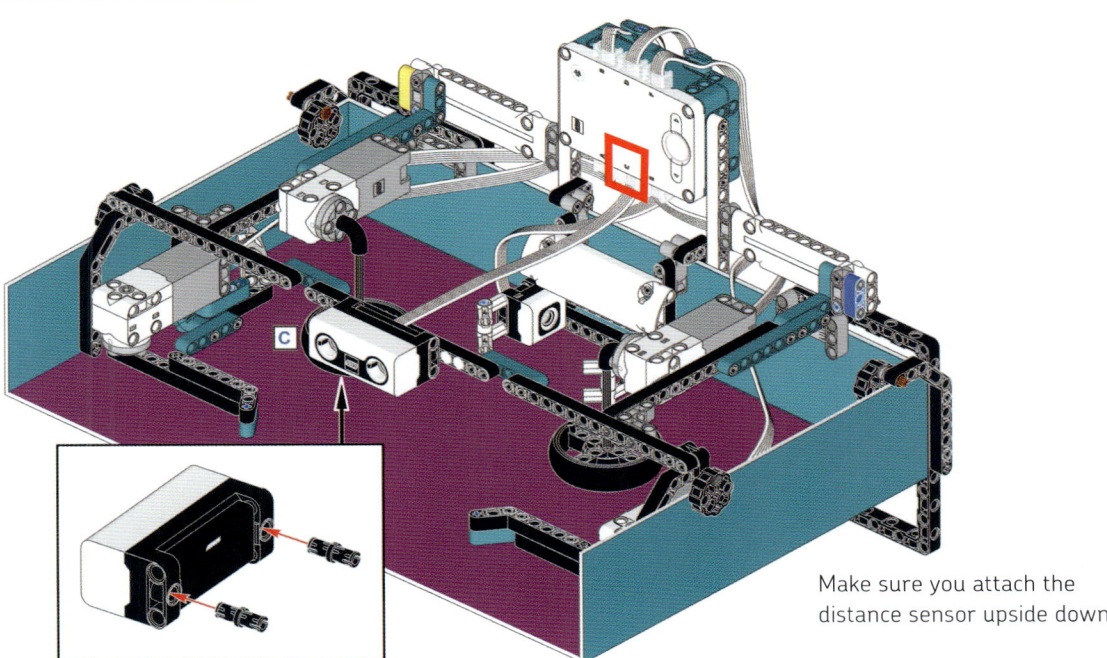

Make sure you attach the
distance sensor upside down.

75

Pinball is complete!

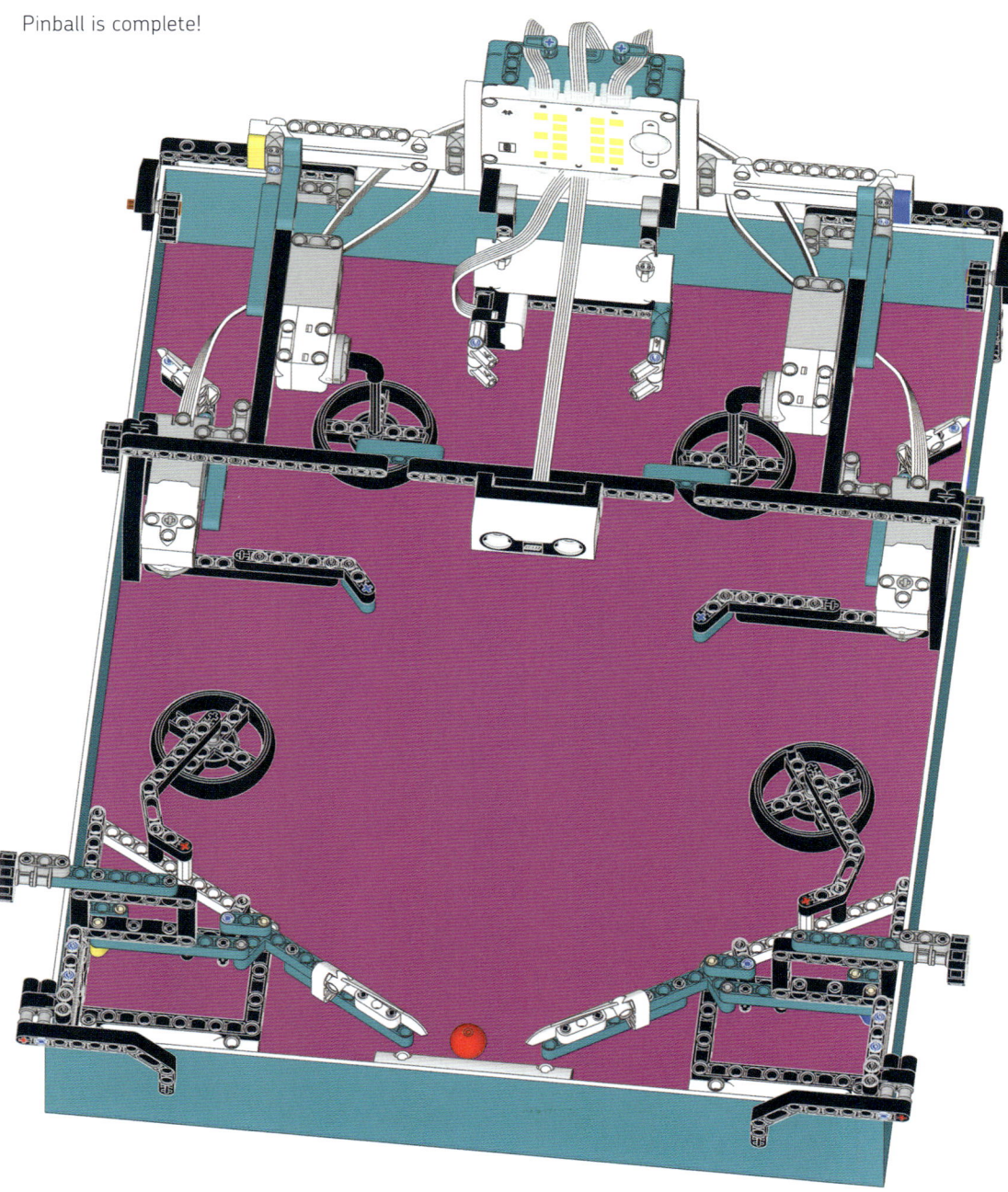

controlling the distance sensor lights with text strings

Since the Distance Sensor has four lights around its "eyes," we can use these lights to show which bumper was hit by the ball. Just as we did with the Hub's display, we can control the Distance Sensor's lights with a text string placed at the input of the `light up distance sensor` block. The string should include four numbers ranging from 0 to 100, separated by spaces (no other separator character will work). The higher the number, the brighter the light.

As shown in Figure 7-9, the first number controls the top-left light, the second number controls the top-right, the third number controls the bottom-left, and the last number controls the bottom-right. So, for example, the string `0 100 0 0` will light up only the top-right light at maximum brightness. We store that string in the variable LED, then feed the variable into the input of the `light up distance sensor` block.

NOTE You could also build the string by having the `join strings` operator read the control values for the lights from separate variables, as we did in Figure 7-6.

enhancing the pinball program

In this section, we'll update the pinball program with the blocks and stacks needed to react to bumpers, earn extra balls to make the game last longer, and prevent cheating.

programming the bumpers

To make Pinball react to the ball hitting the bumpers, we'll create a custom block with a rather long name: `increase score by (amount) and light up (LED) when motor (port) speed is greater than (speed)`. The words in parentheses are the inputs. The block, defined by stack F in Figure 7-10, detects when a bumper has been hit by checking whether the speed of the motor connected to the specified port is greater than the number specified as the speed input. This is similar to how we used the motors to detect hits in the Whac-A-Mole game, except this time we're checking the motors' speed rather than the position of the shaft.

This works because the `motor speed` block returns a value different from 0 when the motor shaft is moved—even only a tiny bit—by the ball hitting the bumper. We feed the `motor speed` block into the `abs` operator (short for *absolute value*, one of the options on the `math functions` block at the

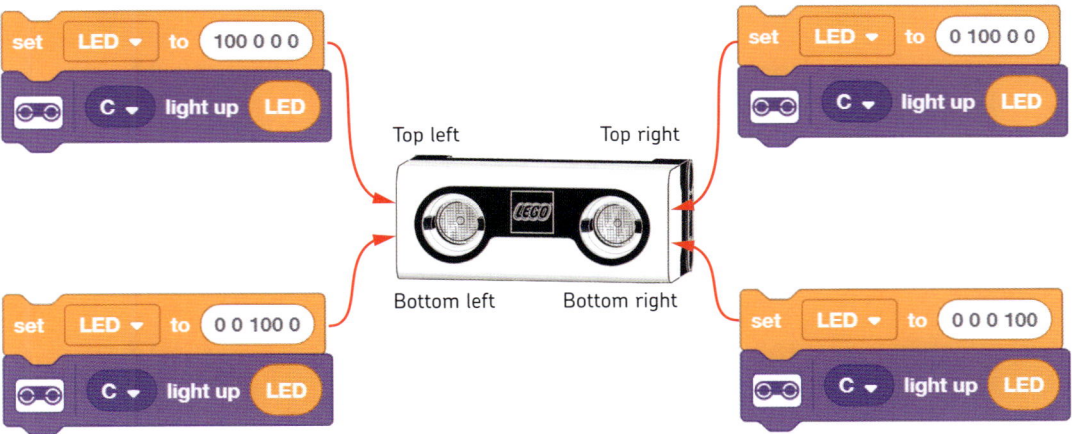

Figure 7-9: How to control the four lights of the Distance Sensor with text strings

bottom of the Operators palette), since the motor's speed might be positive or negative, depending on the direction the shaft is moved. Taking the absolute value of the speed converts any negative numbers to positive numbers.

If a bumper hit is detected, the Distance Sensor lights are turned on using a string fed into the LED input (taking into account that the distance sensor is upside down), the Laser sound is played, and the score variable is changed by the specified amount. Then we use the update display custom block to show the new score, and after a short pause, the Distance Sensor lights are turned off again.

Once you've made the custom block, create the variables extra and extraTimer, which you'll need later, and update stack C by adding several more blocks to the end, as shown in Figure 7-10. In the updated sequence of blocks, we tell the motors to hold their position when stopped, then run the motors to move the bumpers into place. We have the motors hold their position when stopped so the bumpers will act as springs: they'll resist the hits of the ball by continuously trying to return to their original position. Next, we turn off the Distance Sensor lights and wait for 2 seconds to allow all motors to settle back into place without counting that motion as another hit.

Lastly, we enter a forever loop, where we continuously call the stack F custom block four times, once for each of the four bumpers. With each call of the custom block, we specify that the score should be increased by 1 if the bumper is hit, we

provide the string to light up the corresponding Distance Sensor light, and we indicate which motor to check. With the final input of the block, we set what speed the motor needs to be greater than in order to register a hit. By changing this speed input, you can change the sensitivity of the bumpers. The value 0 sets the maximum sensitivity, while higher speeds mean the ball must hit the bumper harder to earn a point. I wouldn't recommend values greater than 2, as the bumpers would become unreactive.

Try playing pinball with this enhanced program. With four bumpers in addition to the Color Sensor added previously, the game is much more fun!

earning extra balls

Real pinball games allow the player to earn extra balls by hitting the same target multiple times within a limited amount of time. Let's add this feature: if you hit the ball into the Color Sensor detection range three times within 10 seconds, you'll earn 10 points and an extra ball (though the maximum number of balls you can have is still five).

To add this feature to the game, add stacks G and H to the program and change stack D as shown in Figure 7-11. To keep track of whether you're eligible for an extra ball, we use the variables extra and extraTimer. We initially set extra to 0 when we updated stack C (see Figure 7-10). In the updated stack D', we've added an if then block ❶ that checks whether

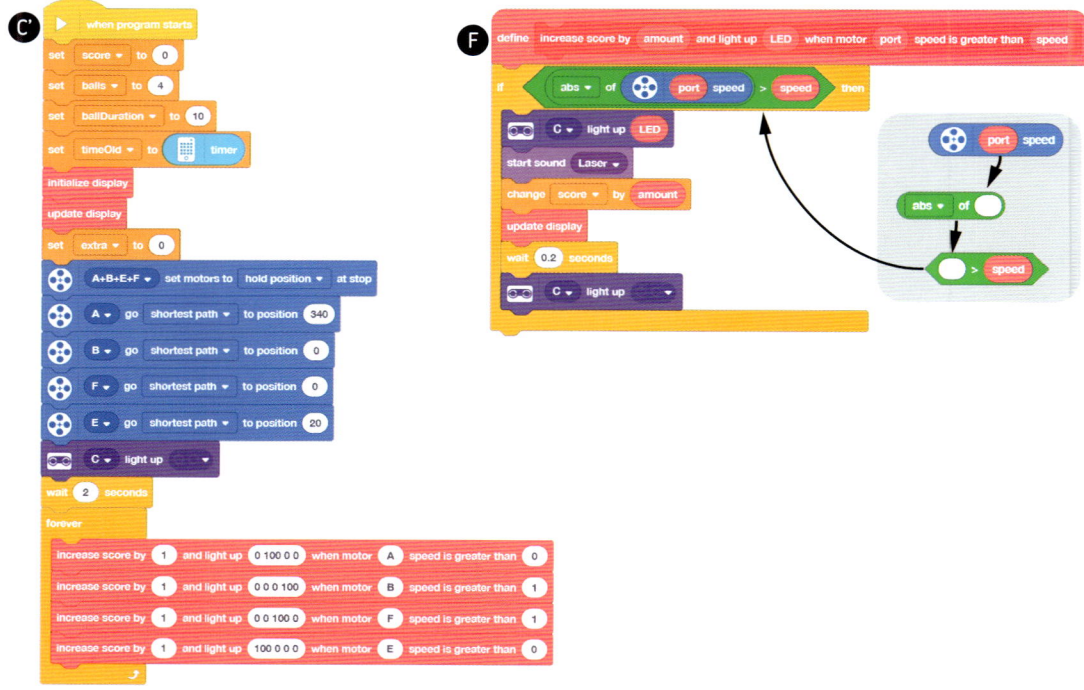

Figure 7-10: The updated stack C and the custom block for reacting to bumper hits

extra is equal to 0. If so, we set the **extraTimer** variable to the current value of the Timer (starting a 10-second count) and change the Hub's center button light to red, to show the player that the race for the extra ball has started. We've also added block ❷ to stack D. It adds 1 to **extra** once the Color Sensor has detected a ball. This way, the next time this stack is triggered, the condition of the **if then** block will be false, preventing the **extraTimer** variable from being overwritten.

Stack G is executed when the **extra** variable is greater than 2 (in other words, equal to 3). This happens when the player has hit the ball into the Color Sensor three times within 10 seconds. If the player achieves this feat, the Explosion sound plays, the Hub's center button light changes back to white, the **extra** variable resets to 0, the score increases by 10, and the **balls** variable increases by 1. Then we use an **if then** block to prevent **balls** from exceeding the value of 5. Finally, the display updates to show the new score and ball count.

Stack H will execute when the difference between the current timer reading and the value of **extraTimer** is greater than 10. Therefore, stack H will execute before stack G if the player can't hit the ball into the Color Sensor detection range three times within 10 seconds. Within stack H, we reset the **extra** variable to 0 and change the Hub's center button light back

to white, indicating the player has failed to earn an extra ball. However, the next time they hit the ball at the Color Sensor and trigger stack D, they'll have another chance.

Try playing the pinball game now. If your aim is good enough, you can earn lots of extra balls and keep playing until the Hub's battery runs dry!

adding an anti-tilt system

Real pinball games have an anti-tilt mechanism, so we should give our LEGO pinball machine tilt detection as well. If you try to cheat by tilting the whole machine up, you'll get a penalty. Stack K in Figure 7-12 adds this feature. The stack is executed when the Hub's roll angle is greater than –90 degrees, an angle that would indicate the player has tilted the machine up. The cheating attempt is punished by halving the score. Block ❶ overwrites the **score** variable by dividing the old **score** value by 2 and rounding down to the nearest whole number. For example, if the score were 31, after the tilt, the score would be **floor (31 / 2)** = **floor (15.5)** = **15**. The Revving sound (from the M.V.P. Library) plays on repeat until the machine is returned to its correct position (roll angle less than –94 degrees). Finally, the display is updated to show the reduced score.

After this last addition, the pinball program is complete.

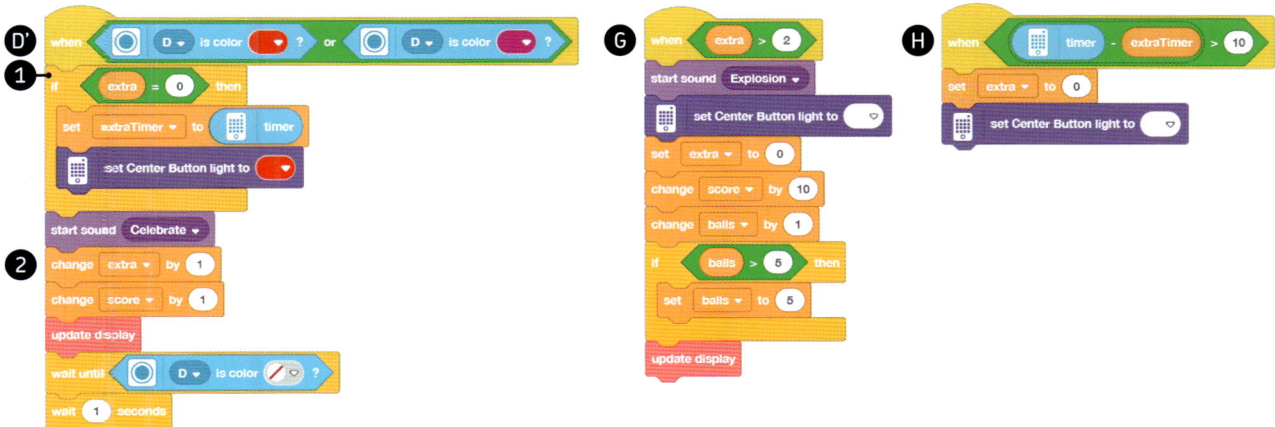

Figure 7-11. These stacks manage the logic to let the player earn extra balls by hitting the ball at the Color Sensor three times in 10 seconds.

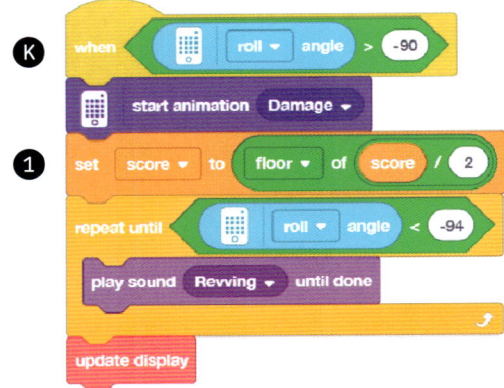

Figure 7-12: This stack detects if Pinball has been tilted up and penalizes the player for trying to cheat.

what you've learned

While building Pinball, you learned how to use the rubber bands as springs so the flippers return to their resting pos - tions. You also practiced using a four-bar linkage to transform the linear motion of the side push buttons into the rotation of the flippers. While programming Pinball, you learned how to control the Hub's display and the Distance Sensor lights with text strings and how to use lists to store a collection of values.

In the next chapter, you'll build a functional guitar that will turn you into a real guitarist, even if you don't know anything about how to play music!

EXERCISE 7-1

1. Currently, Pinball can't display a score greater than 99. Can you find a way to display scores in the hundreds as well? For example, you could change the Hub's center button color to repre-sent different values for a hundreds digit. If you want to do so, remember that the Hub's center button light is changed to red when the player has an opportunity to try to earn extra balls.

2. Do you have other ideas to improve Pinball? What if you make the bumpers react and shoot the ball away when touched by the ball?

8

guitar

There stood a log cabin made of earth and wood
Where lived a country boy named Johnny B. Goode
Who never ever learned to read or write so well
But he could play a guitar just like a-ringin' a bell.
—"Johnny B. Goode," Chuck Berry

The Robot Inventor set includes large elements and frames that allow you to quickly build big, strong models. To prove it, we'll create a working electric guitar. The Guitar is three-quarter size, or 75 percent as long as a real guitar. Instead of having strings that you pluck to make sounds, it'll use the Color Sensor and Distance Sensor to respond to your hand movements, so you can play solos and chords even if you aren't a trained musician.

While building this guitar, you'll learn how to make large yet sturdy structures, like the Guitar's body and neck. While making the programs, you'll see how to use math to transform sensor readings into a different range of values. You'll also learn how to store a lot of numeric data in the items of a list and how to retrieve pieces of that data by using text manipulation blocks.

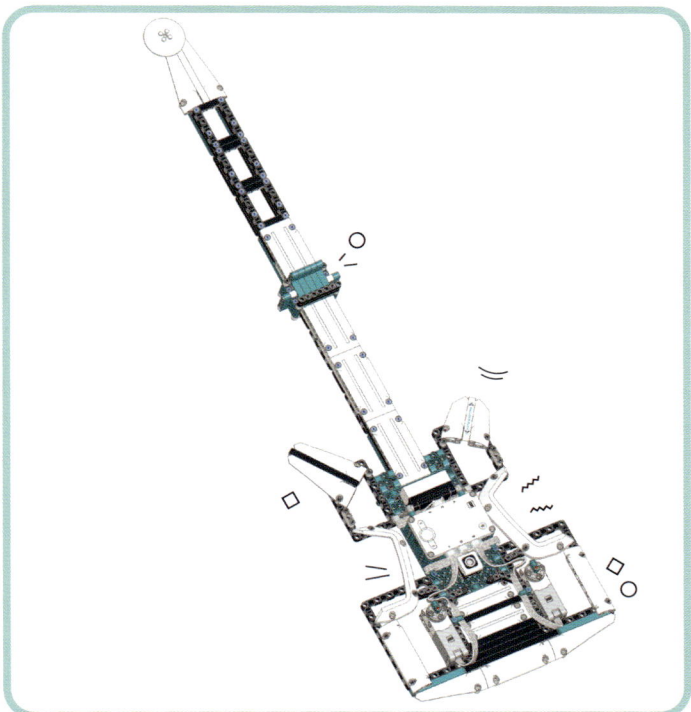

Figure 8-1: You can play solos and chords on this three-quarter-size LEGO electric guitar, even if you've never touched a real guitar!

building the guitar

In this section, you'll find step-by-step instructions to build the Guitar. Along the way, you'll find some notes describing important design choices and interesting techniques. Later in the chapter, we'll add two motors to the Guitar that will function as control knobs so you can quickly adjust settings while you're playing.

1

16x

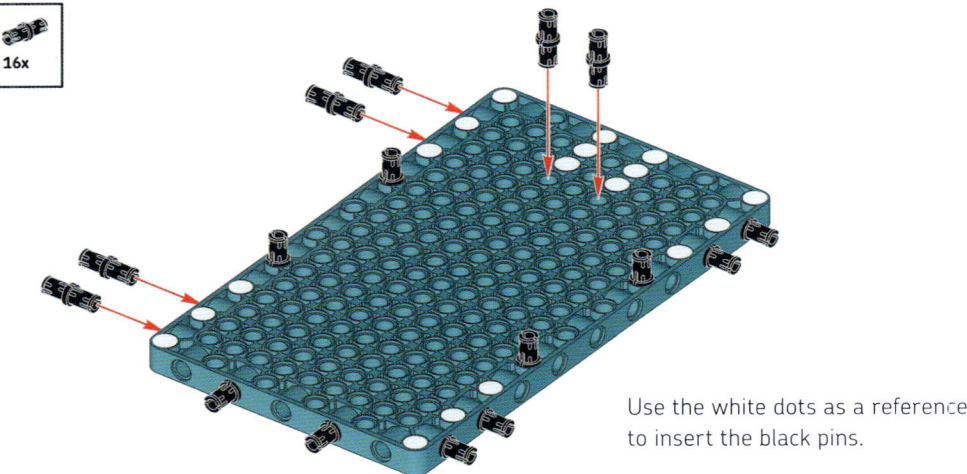

Use the white dots as a reference to insert the black pins.

2

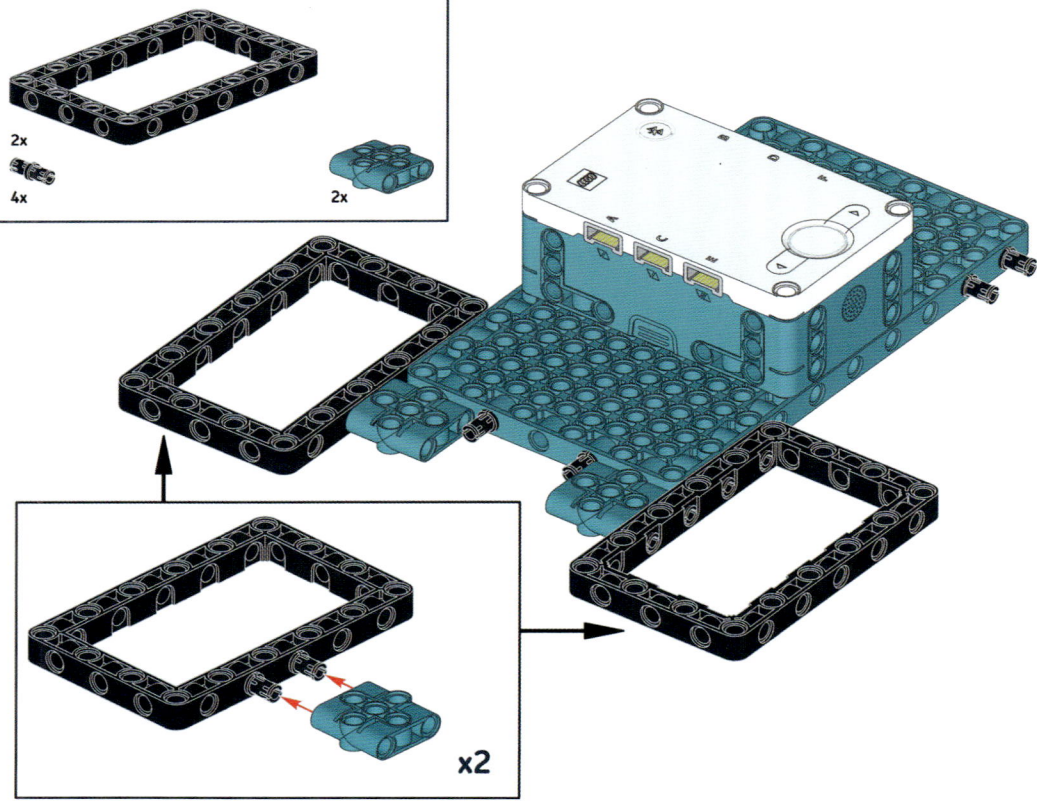

2x
4x
2x

x2

3

4

5

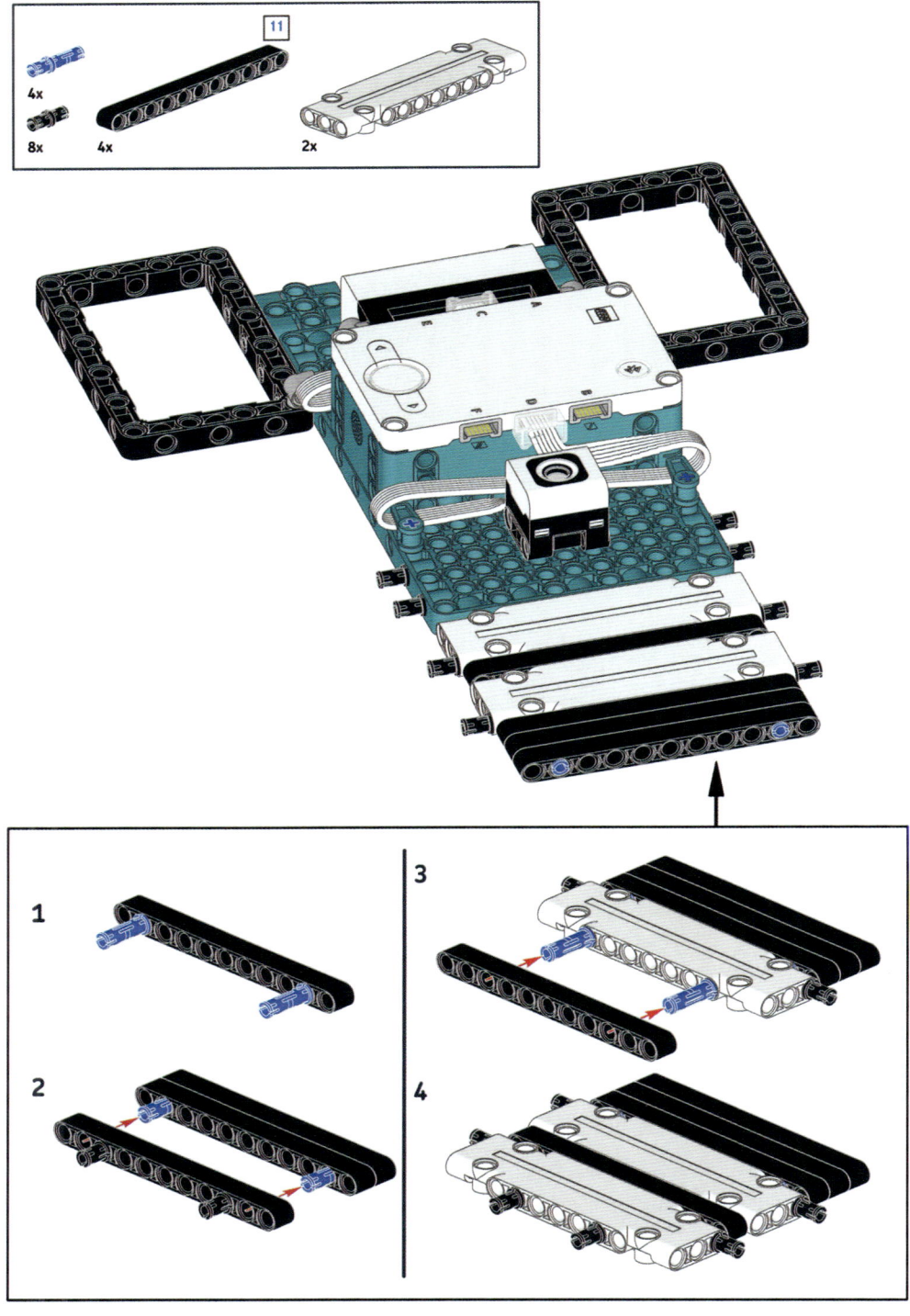

6

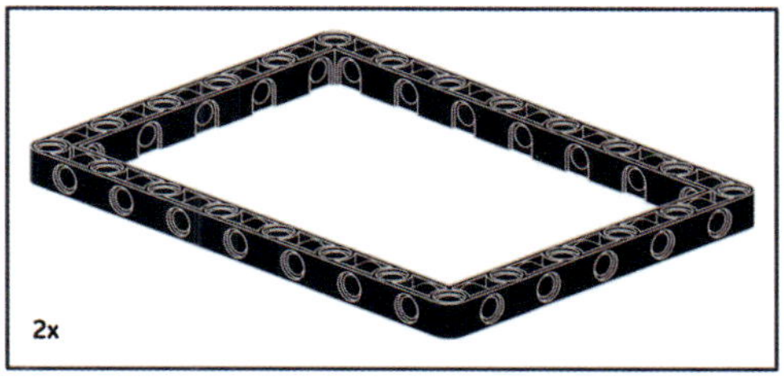

2x

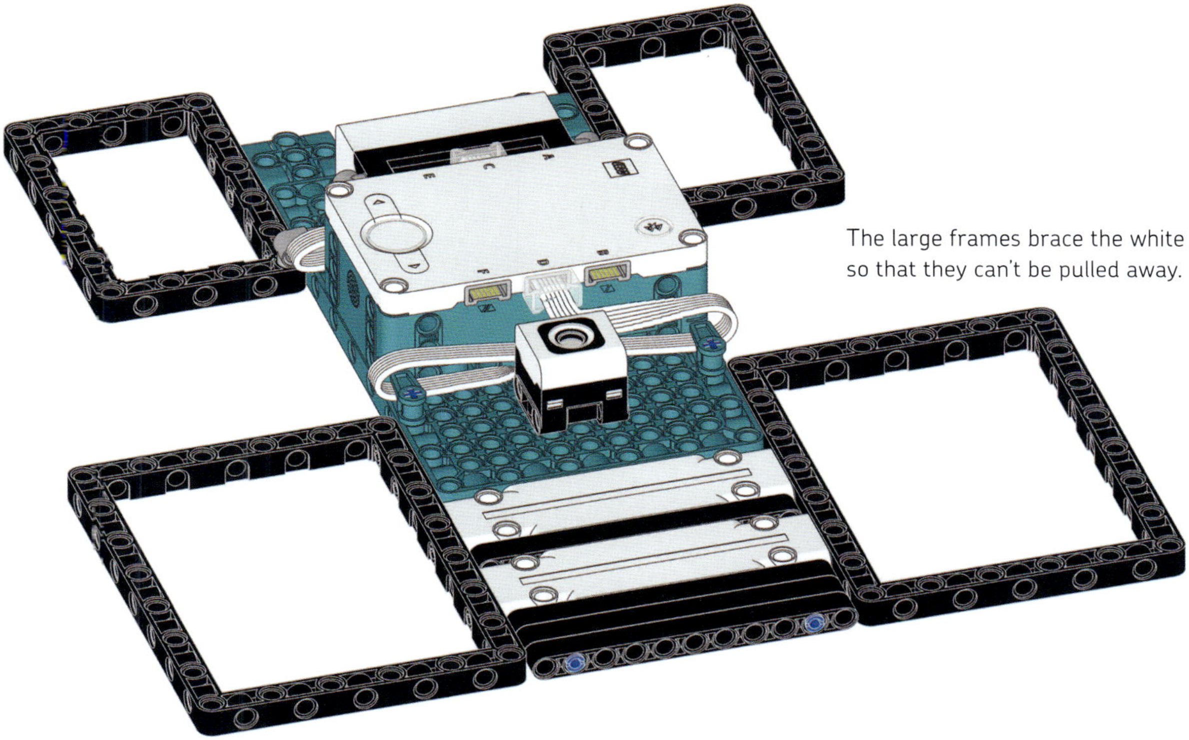

The large frames brace the white panels
so that they can't be pulled away.

7

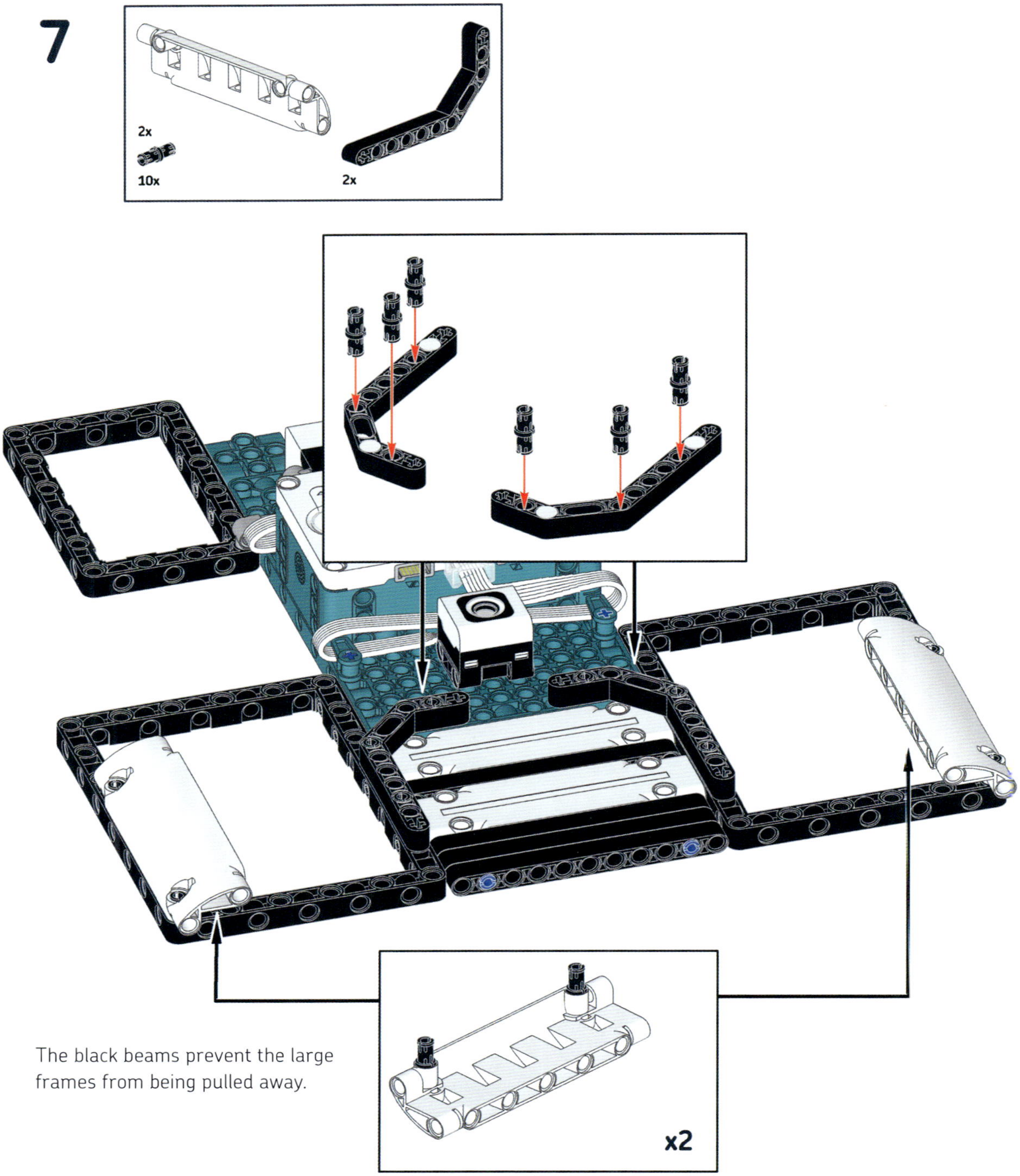

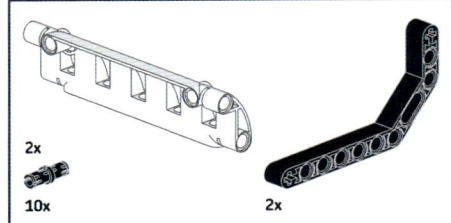

The black beams prevent the large frames from being pulled away.

x2

8

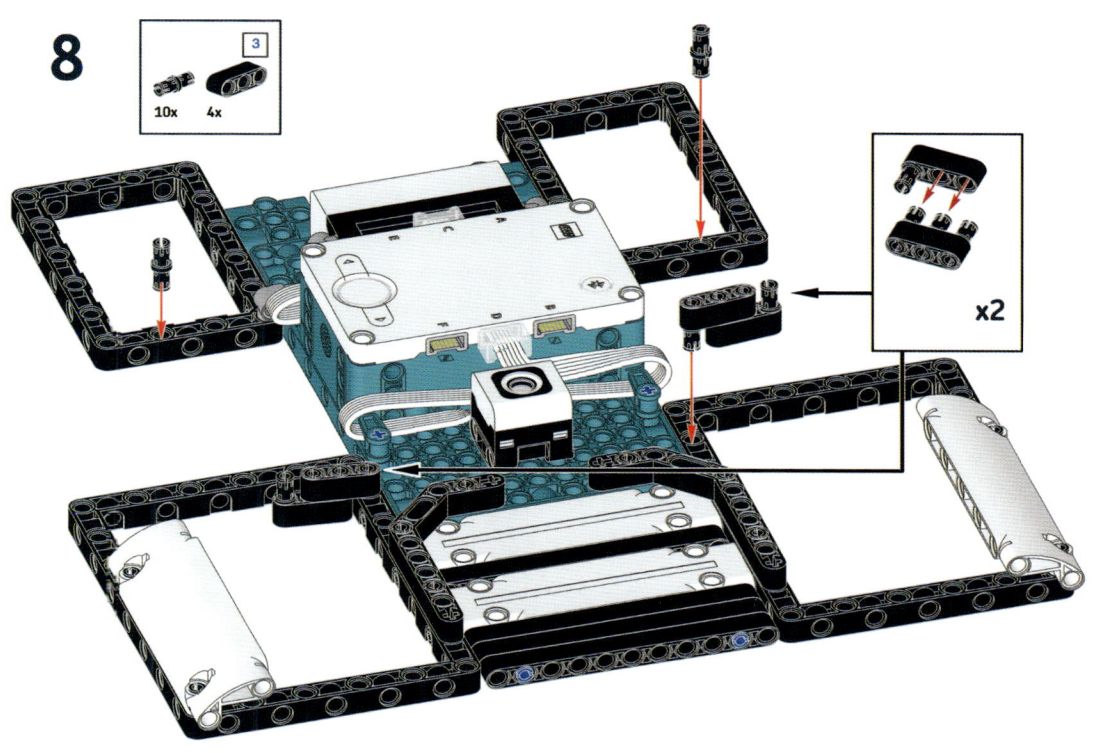

x2

---

bottom subassembly

9

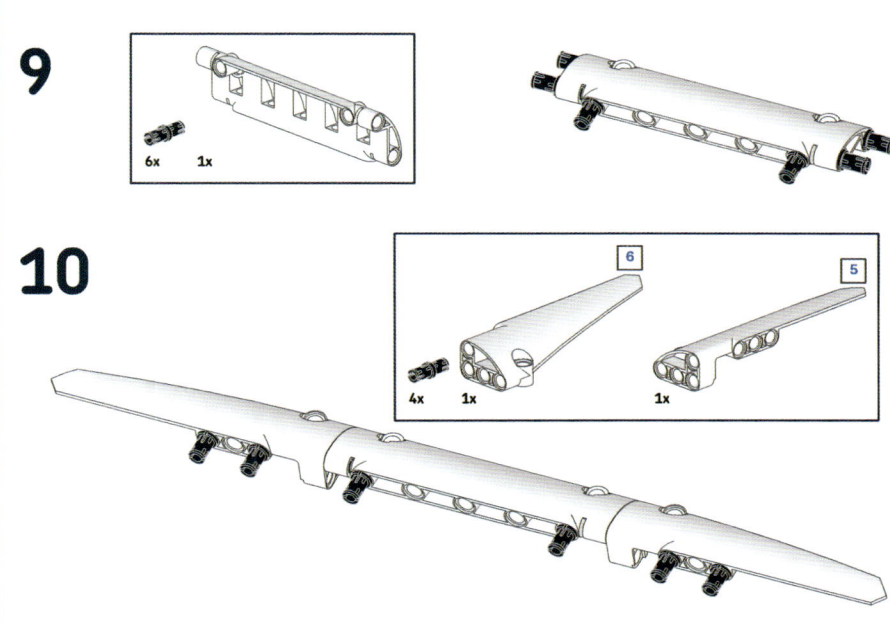

10

11

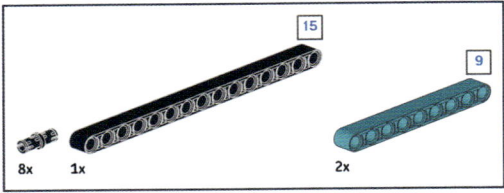

12

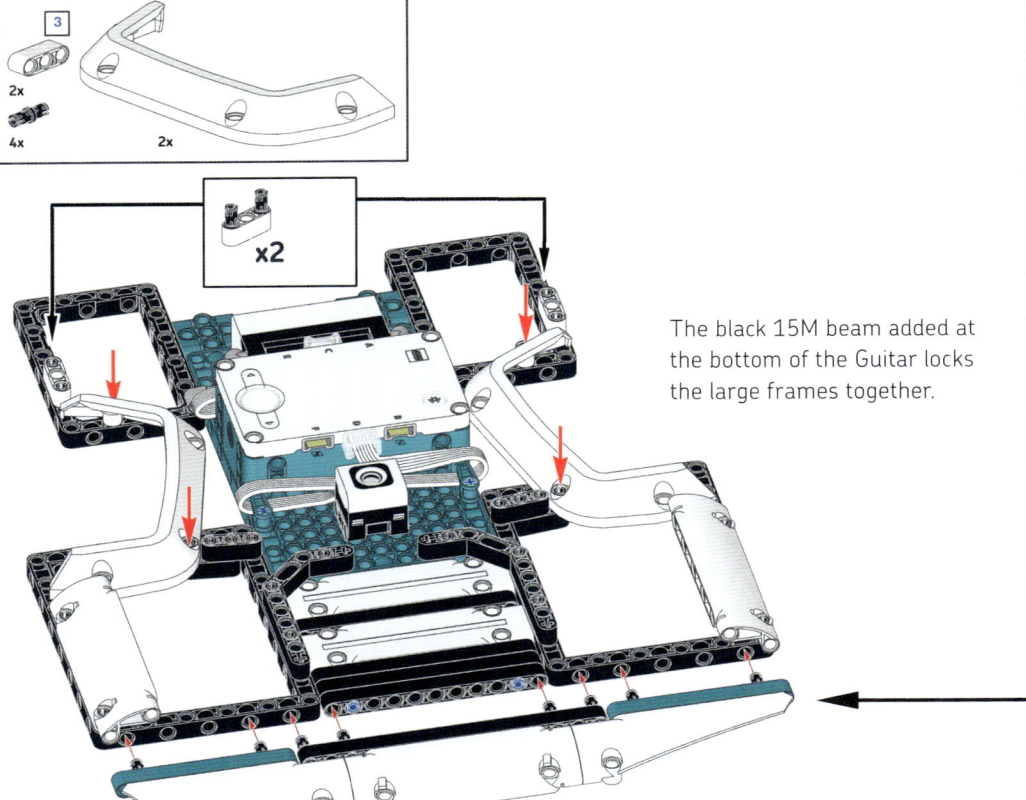

The black 15M beam added at
the bottom of the Guitar locks
the large frames together.

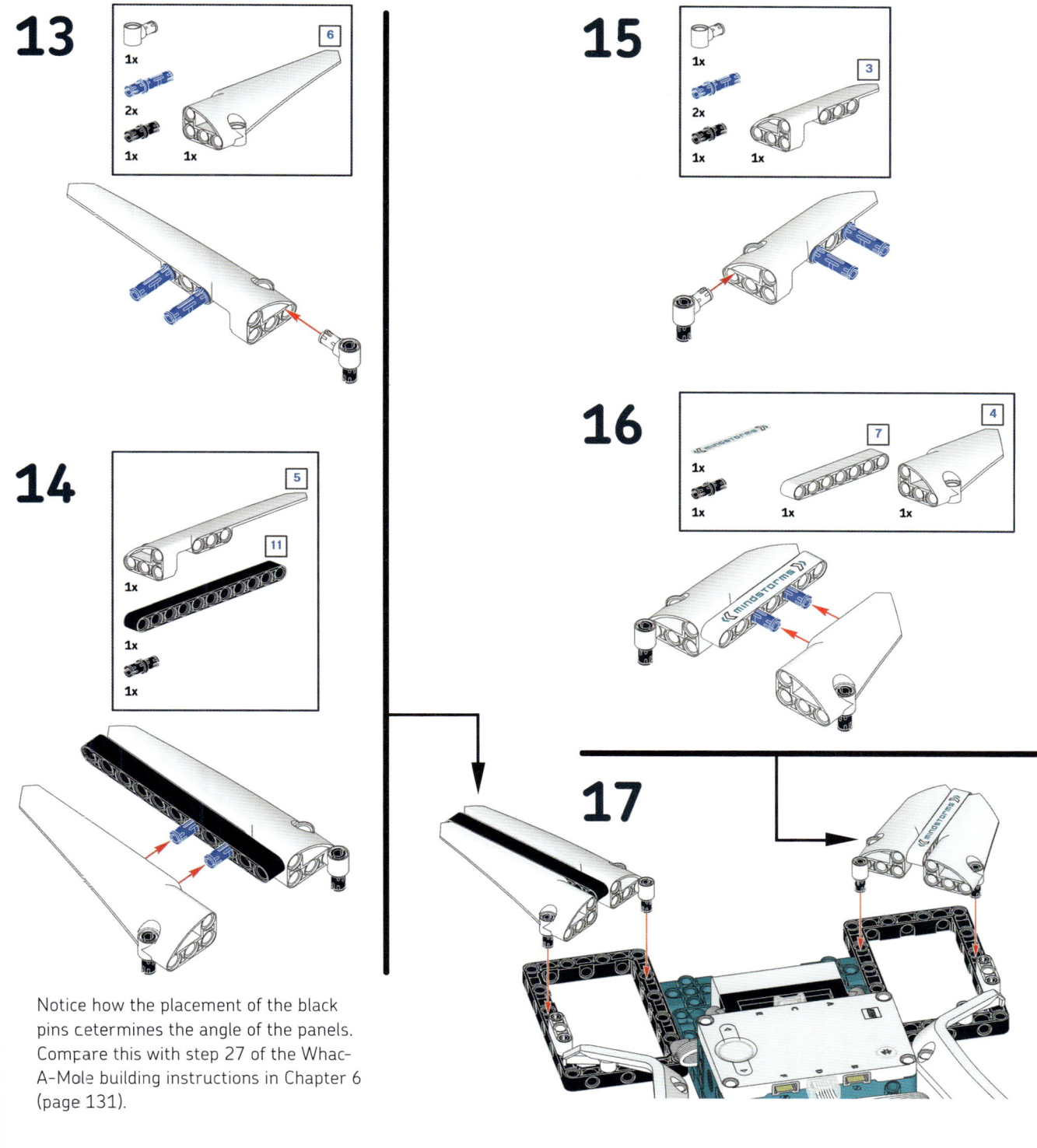

13

1x 2x 1x 1x | 6

14

5 1x 11 1x 1x 1x

15

1x 2x 1x 1x | 3

16

1x 1x 7 1x 4 1x

17

Notice how the placement of the black pins determines the angle of the panels. Compare this with step 27 of the Whac-A-Mole building instructions in Chapter 6 (page 131).

18

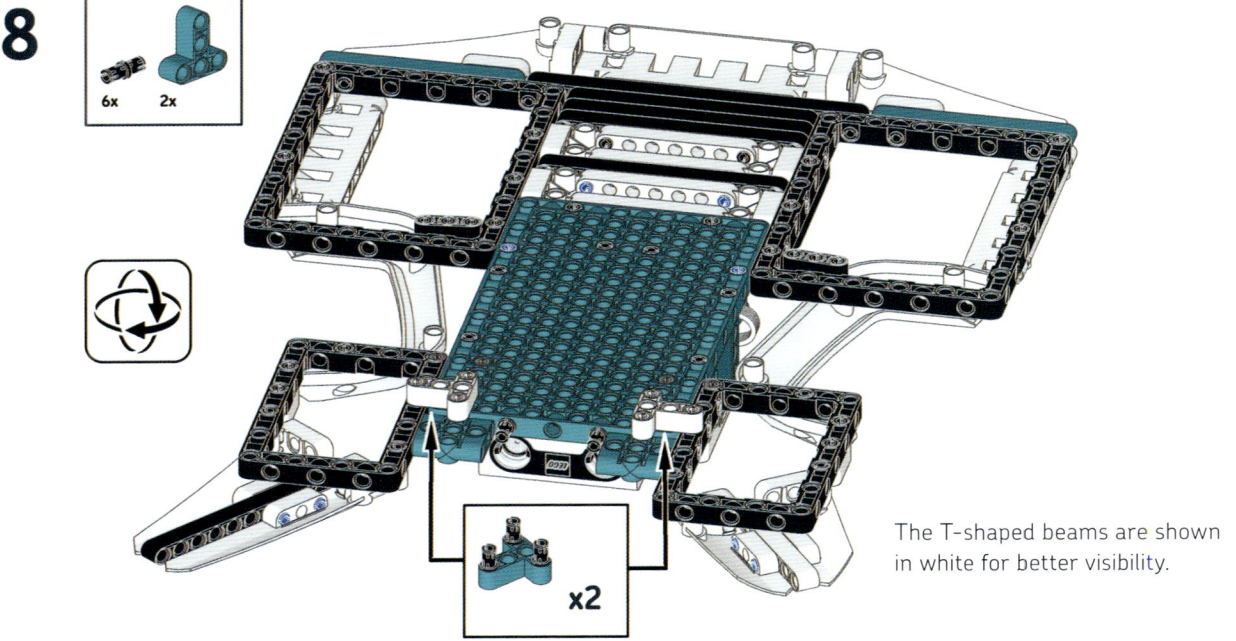

The T-shaped beams are shown in white for better visibility.

---

slider subassembly

19

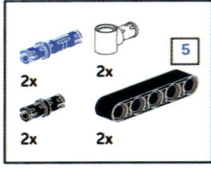

20

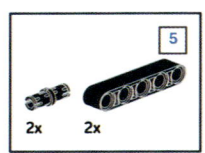

The T-shaped beams are shown in white for better visibility.

21

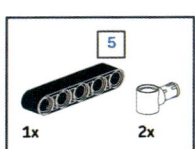

22

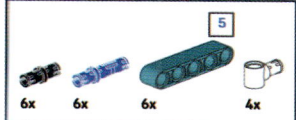

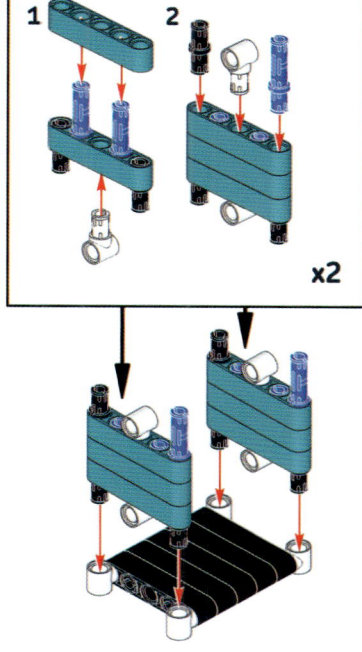

23

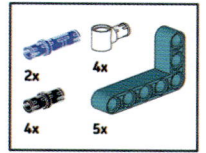

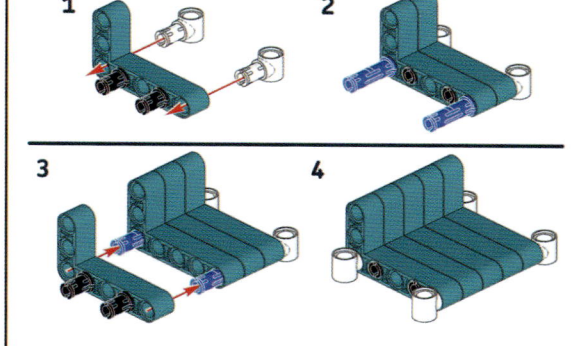

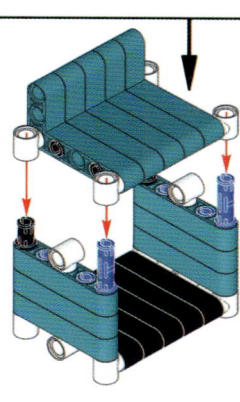

24

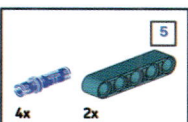

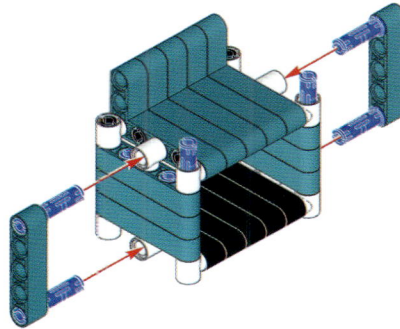

The teal 5M beams prevent the slider from being pulled apart vertically.

25

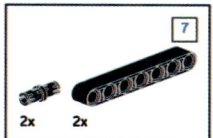

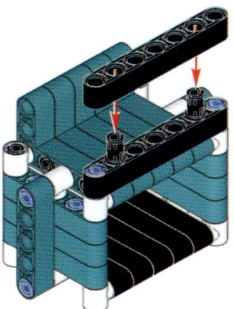

The black 7M beam prevents the slider from being pulled apart horizontally.

By moving this slider up and down the neck of the Guitar, you can control which notes you play.

26

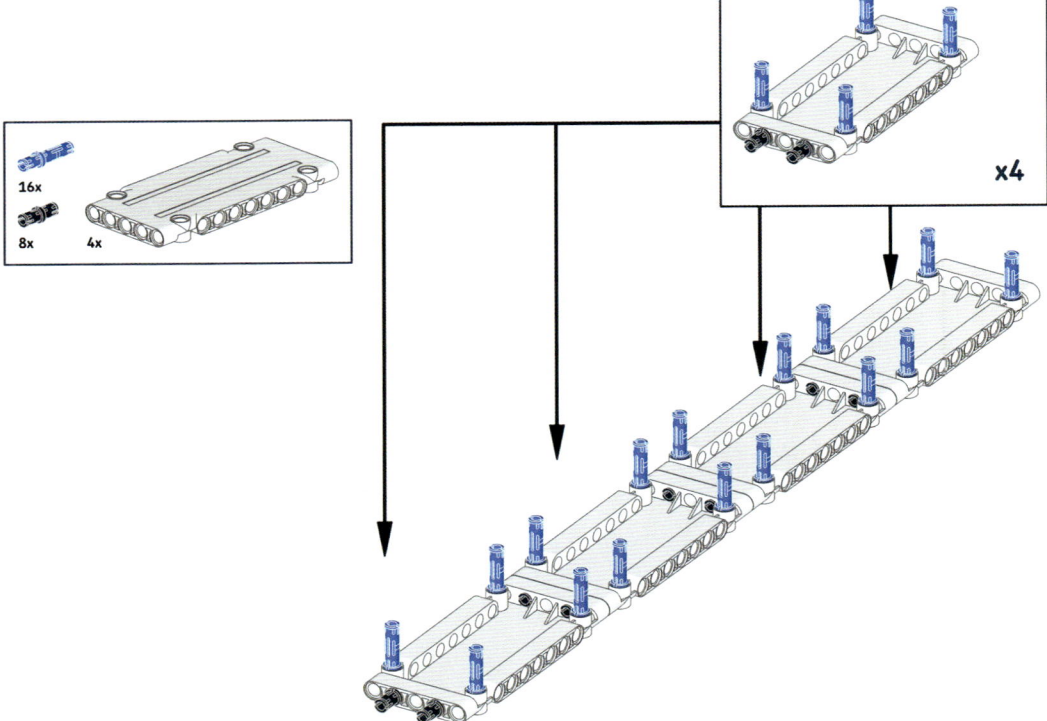

27

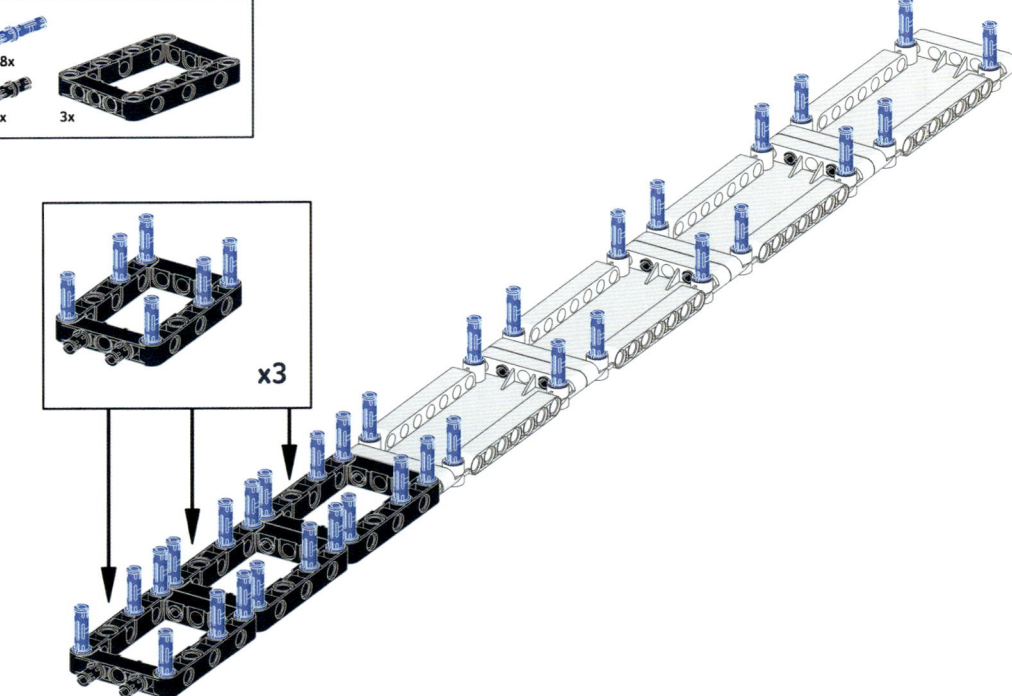

28

13 2x
15 2x

29

13 2x
15 2x
11 2x

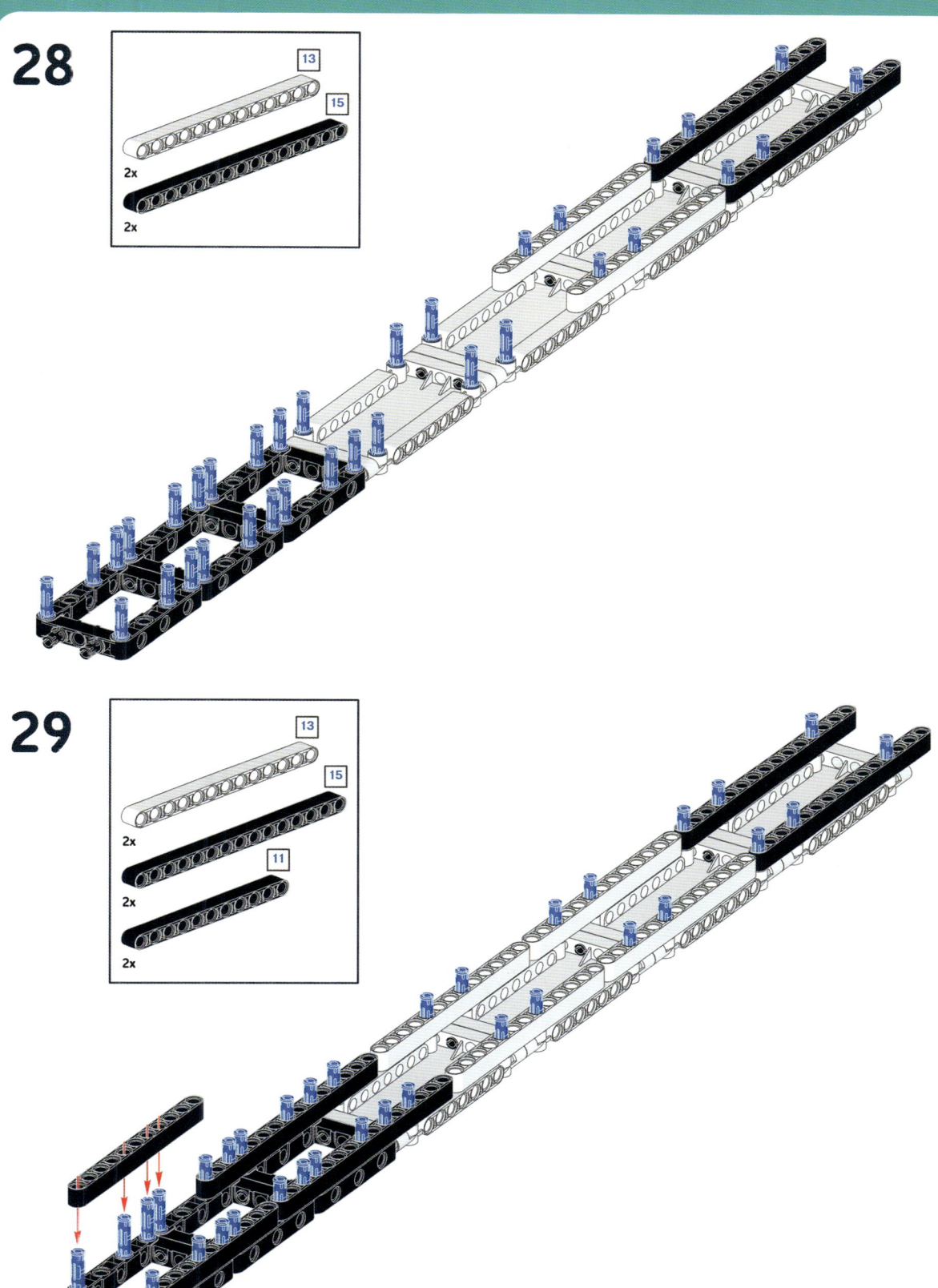

30

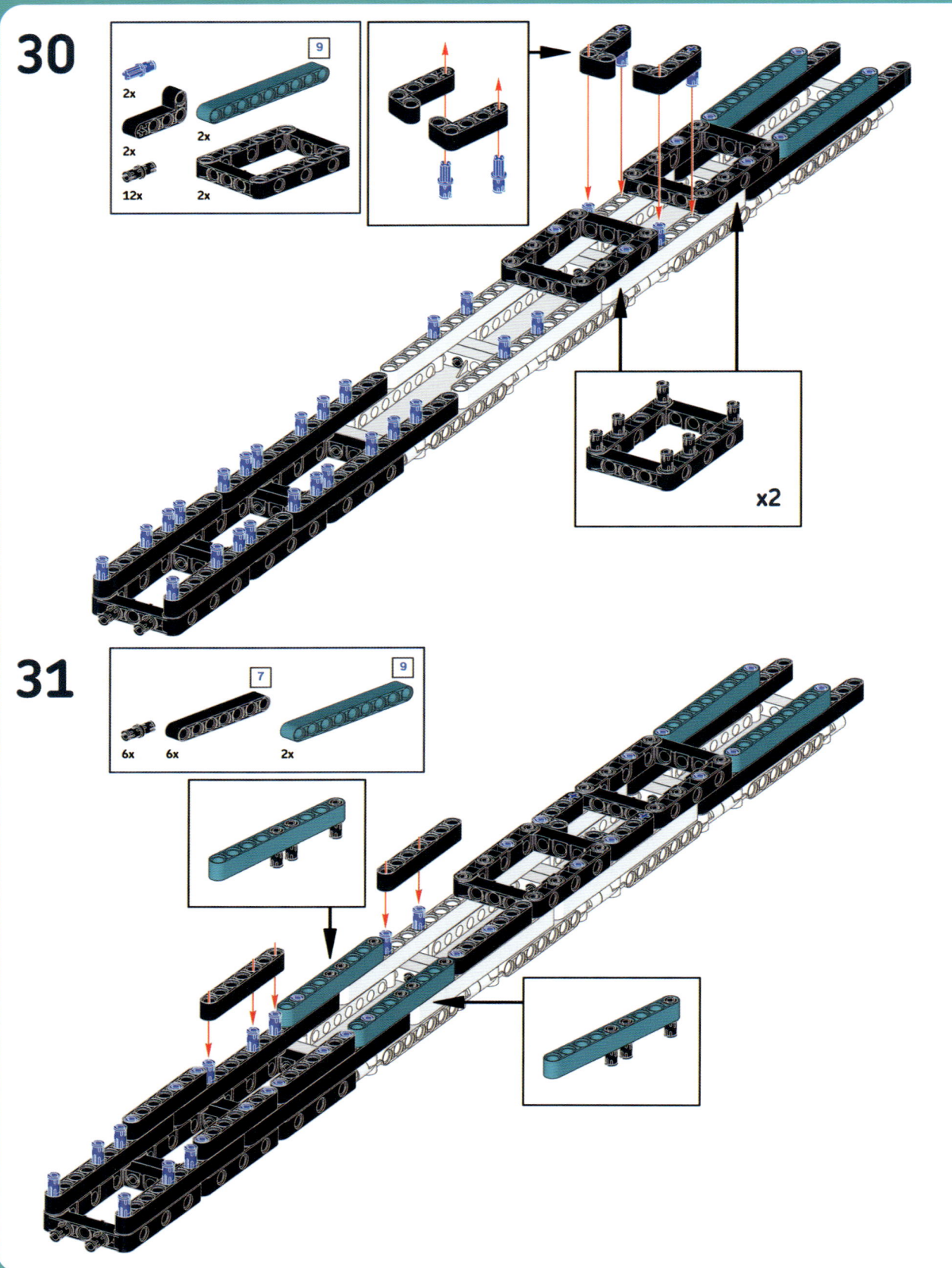

31

headstock subassembly

32

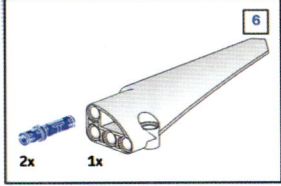

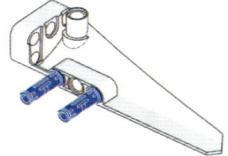

33

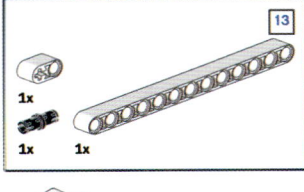

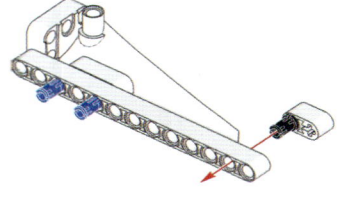

34

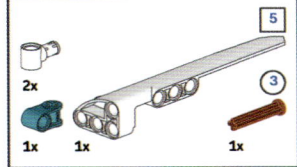

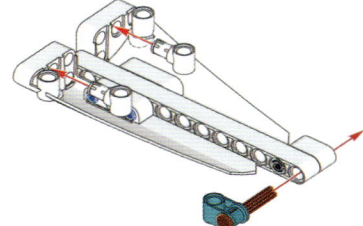

35

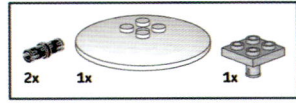

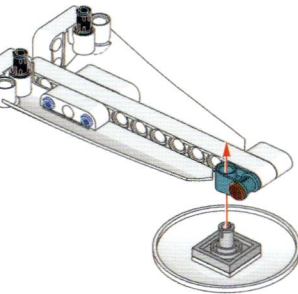

36

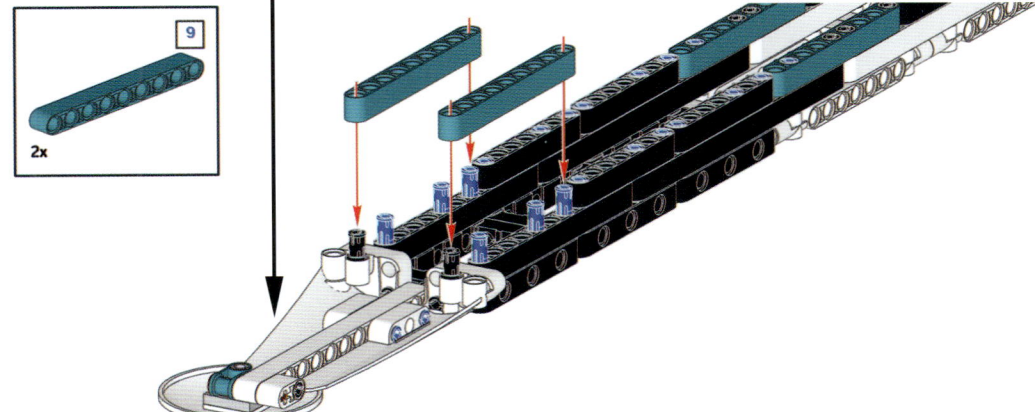

37

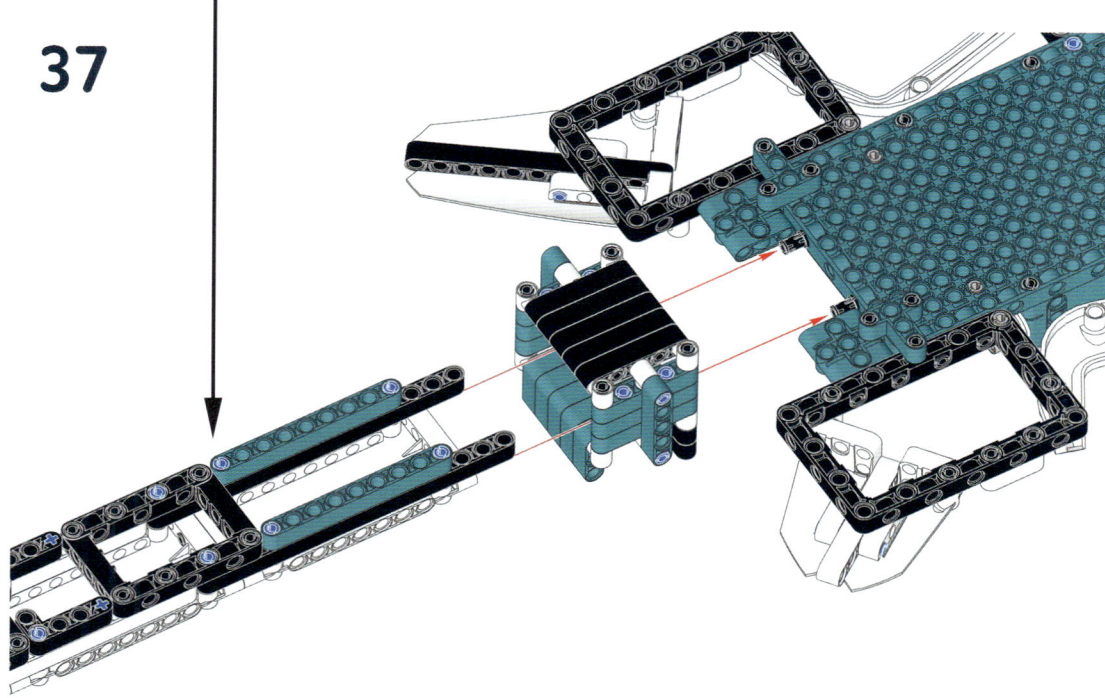

38

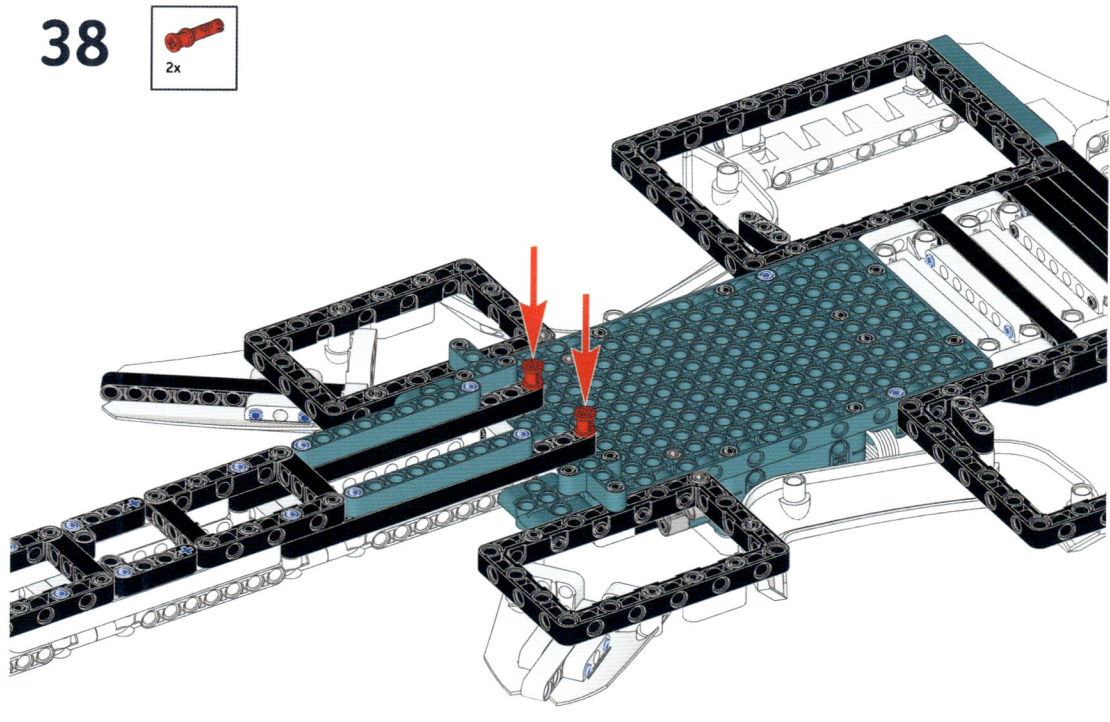

39

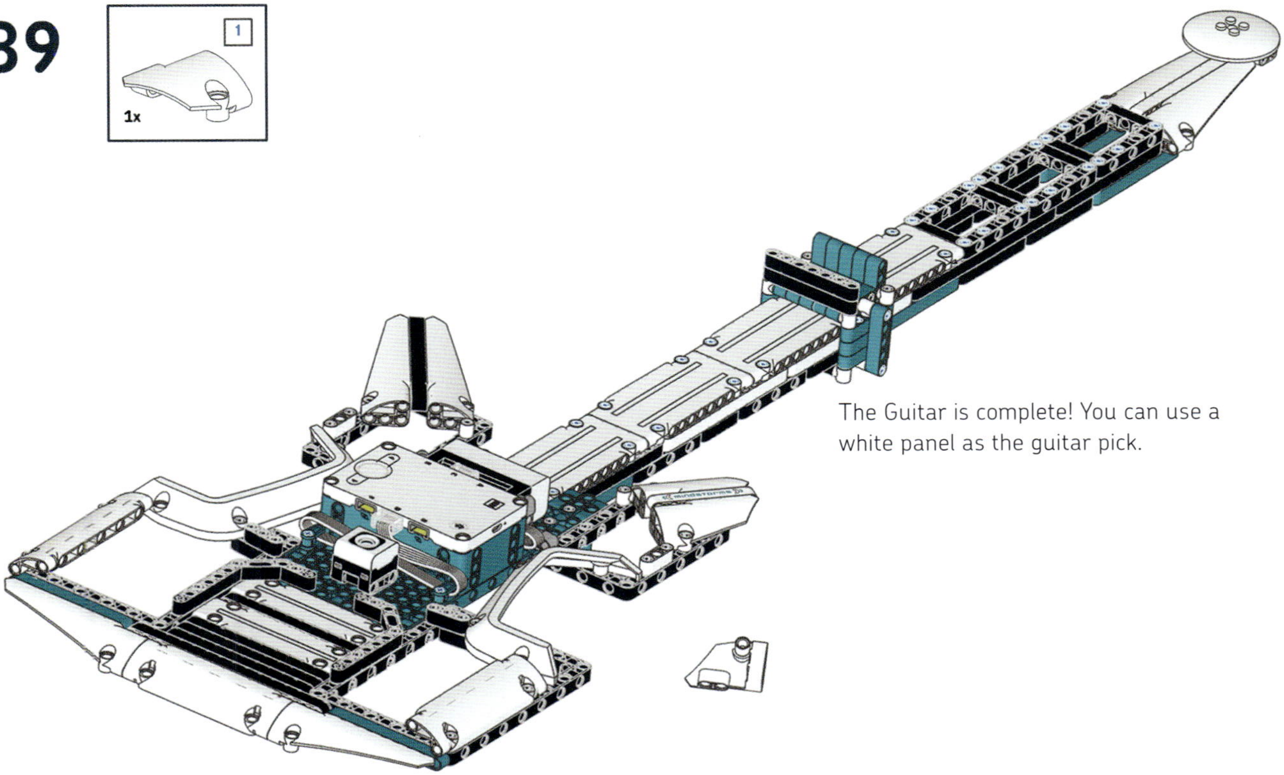

1x

The Guitar is complete! You can use a
white panel as the guitar pick.

playing solos

The first program we'll make allows you to play solos using eight notes from a musical scale. You'll move the slider along the neck to choose which note to play. Just like on a real guitar, the closer your hand (and the slider) is to the base of the neck (where our LEGO Guitar has the Distance Sensor installed), the higher the pitch of the note. Once you've chosen a note, you'll play it by placing the pick (white panel) near the Color Sensor. The distance between the pick and the sensor will determine the volume of the note: the closer the pick to the Color Sensor, the louder the sound.

To allow anyone to play solos that sound good, the Guitar can be set to use three musical scales: major, pentatonic, and blues. You can choose which scale to solo with by using the right button on the Hub. Figure 8-2 shows the musical notation for each scale, but you don't need to know how to read music to try the scales. Each one has its own mood. For example, the blues scale is perfect for playing soulful improvised solos over blues chord progressions. To see what I mean, search YouTube for "blues backing track A minor" and try playing along with one of the recordings you find.

creating the program

Create a new program and name it *guitar_solo*. Then, go to the Variables palette and create a list named `scale` as well as the following variables:

```
distance
mode
note
transpose
```

Now reproduce the custom block definitions shown in Figure 8-3: set up `major` scale (stack A), set up pentatonic scale (stack B), set up blues scale (stack C), and set up (stack D). These blocks work together to fill the `scale` list with a collection of eight notes for you to play.

Use Table 8-1 as reference to set up the items in the lists for the three scales.

Figure 8-2: You can play solos using the notes from these three possible scales, based on the note A. By changing the value of the `transpose` variable, you can play these scales starting from any possible note.

Table 8-1: Numbers for Filling Scale Lists

Major scale	Pentatonic scale	Blues scale
81	76	79
83	79	81
85	81	84
86	84	86
88	86	87
90	88	88
92	91	91
93	93	93

Next, create the custom blocks defined in Figures 8-4 and 8-5. The change scale block (stack E) lets you choose which scale to play using the Hub's right button, and the get note block (stack F) chooses what note to play based on readings from the Distance Sensor.

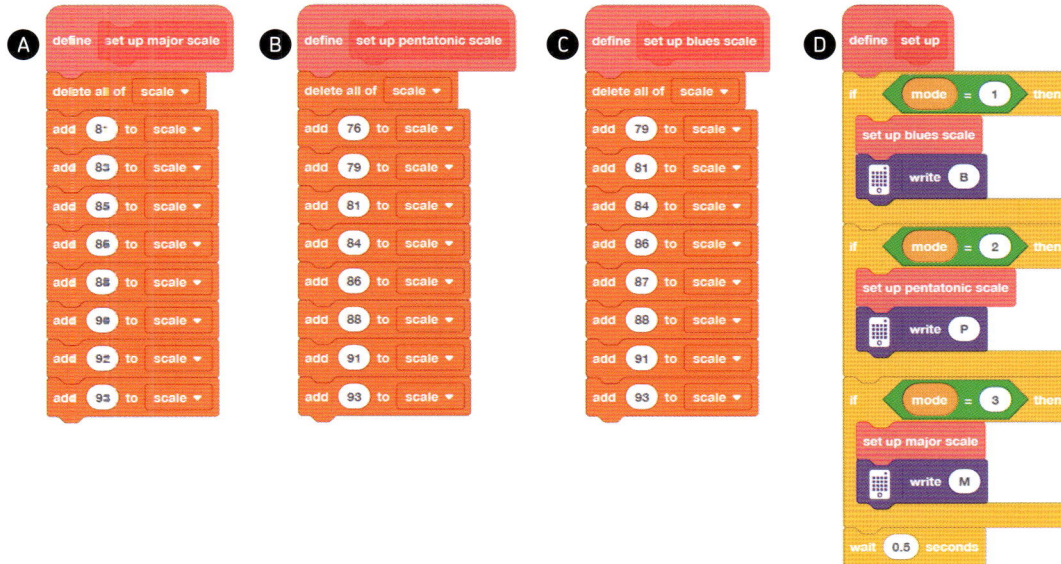

Figure 8-3: The custom blocks to set up the scale to be played

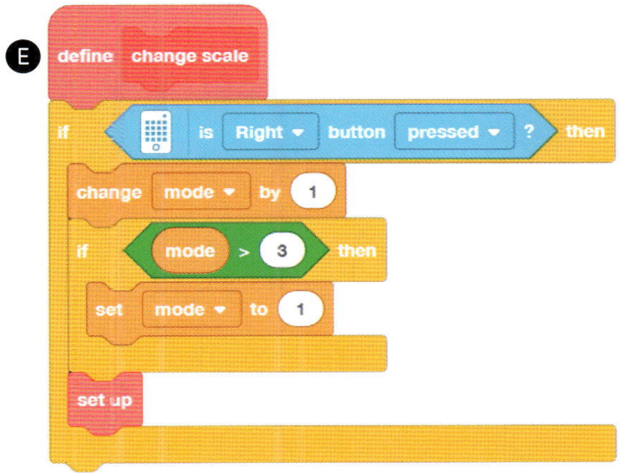

Figure 8-4: The custom block that allows the player to choose the scale

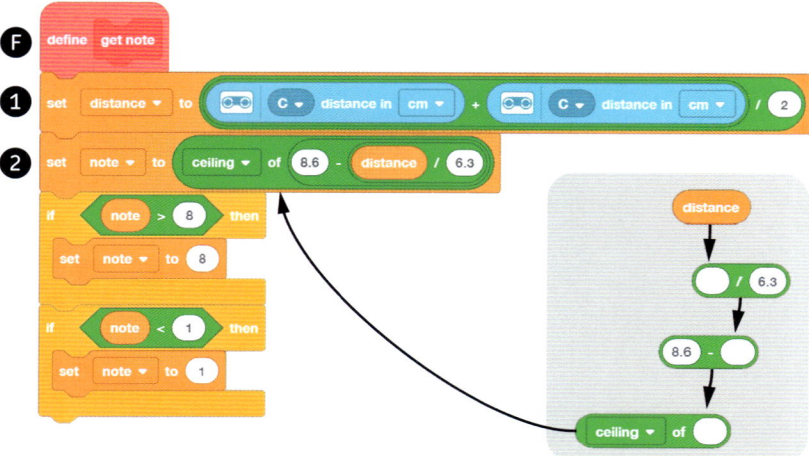

Figure 8-5: The custom block that maps the position of the slider to the note to be played

Finally, create stack G, the main program stack, using Figure 8-6 as a reference. This stack sets up the scales, polls the sensors, and manages the logic for playing notes. When you're done, select program slot 0 and download and execute the program.

playing the guitar

After the text SOLO has been completely shown on the Hub's display, try moving the slider along the neck and watch how the number displayed by the Hub changes. The number corresponds to a note in the scale, from 1 (lowest) to 8 (highest).

Now grab the pick (the white panel) and place it close to the Color Sensor to hear the note you selected with the slider. While keeping the pick near the sensor, try moving the slider up and down and listen as the notes change. Then, keep the slider on one note and slightly move the pick toward or away from the color sensor to change the note's volume. With a little practice, you'll be playing killer solos in no time! To get started, how about trying a famous guitar riff? Play the following notes with the right rhythm and you'll get the opening riff of "Smoke on the Water" by Deep Purple.

understanding the guitar solo program

Let's take a closer look at how the *guitar_solo* program works. As you read, refer back to Figures 8-3 through 8-6.

setting up the scales

Each of the first three custom blocks defined in Figure 8-3, set up major scale (stack A), set up pentatonic scale (stack B), and set up blues scale (stack C), fills the scale list using eight add item to list blocks. Each block adds a number to the list corresponding to a different musical note. Within each stack, we add a different set of numbers to the list, resulting in three different scales. Before adding any notes, we first clear the list of all its items, ensuring that the list will contain only the notes for a single scale.

NOTE To hear which notes these numbers correspond to, pull up a play beep for seconds block, expand the first input, and look for the numbers on the piano keyboard.

The set up custom block (stack D in Figure 8-3) executes one of the three set up scale custom blocks, depending or

Guitar solo, Blues scale

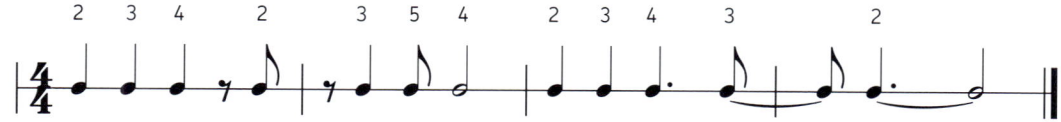

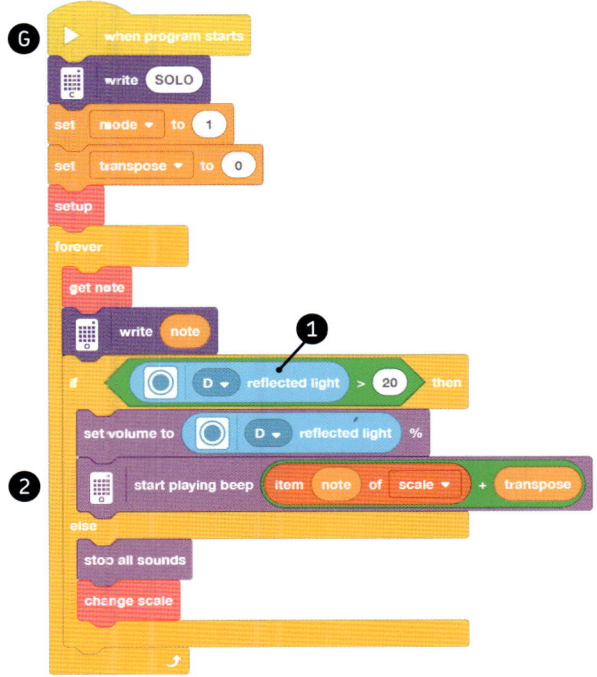

Figure 8-6: The main stack of the guitar solo program

the value of the mode variable. For example, if mode is equal to 1, the set up major scale custom block is executed, loading the notes for a major scale into the scale list, and the letter M (indicating *major*) is written on the Hub's display. The mode variable can't have two different values at the same time, so only one of the if then blocks will be executed each time the set up block is called. The wait for seconds block at the end of the stack allows the player to see the letter M, P, or B for half a second before other blocks in the program change the content of the display.

changing the scale

The change scale custom block (stack E in Figure 8-4) checks whether the player has pressed the right button on the Hub. If so, we add 1 to the value of the mode variable. The valid values for mode are 1, 2, or 3 (see stack D, the set up block). To keep the values in this range, we check whether mode is greater than 3, and if so, we overwrite its value to 1. Then, once mode has been changed, we call the set up custom block to update the content of the scale list with the chosen collection of notes.

choosing a note

Our get note custom block (stack F in Figure 8-5) chooses which note to play based on how far the slider is from the Distance Sensor. Sounds like a simple task, but it's made a little more complicated by *measurement noise*, random variations in the readings from a sensor. In the previous chapter, we dealt with the Color Sensor misreading the red ball as magenta. The Distance Sensor also has trouble taking perfectly accurate measurements. Try watching the Hub Monitor in the top-right corner of the MINDSTORMS App for live readings from the Distance Sensor. The last digit will change frequently, even if the sensor's distance from an object doesn't change.

One good way to reduce, or *filter*, measurement noise is to take multiple readings from the sensor and calculate their average. That's exactly what we do at the start of the get note stack. Instead of using just one reading from the Distance Sensor, we take two readings, add them together, and divide the result by 2. Then we store our average distance reading in the variable distance (❶ in Figure 8-5).

Depending on the position of the slider, the average reading from the Distance Sensor should be somewhere between 4 centimeters and 48 centimeters. We want to convert that value to a number from 1 to 8, corresponding to the index of one of the notes in the scale list. The maximum distance (48 cm) should translate to a note index of 1 (the lowest note in the scale), with shorter distances translating to higher note indexes until a 4 centimeter distance becomes an index of 8 (the highest pitch).

To perform this conversion from Distance Sensor readings to note indexes, we need to use some math. (Did you ever imagine that math would help you create something as fun as an electric guitar?) We make the necessary calculation in block ❷ of Figure 8-5 and store the result into the note variable, which we'll use in the main program stack to select an item from the scale list. If you want to learn about how the calculation works, see the box "Linear Interpolation" on page 204.

After the calculation, we're not quite done. In the event the average Distance Sensor reading falls outside the expected range of 4–48 centimeters, the value of note may end up outside the desired range of 1–8. To prevent this, we finish the get note stack with two if then blocks. First, we check whether note is greater than 8, and if so, we overwrite it with the value 8. Then we check whether note is less than 1 and overwrite it with the value 1 if it is.

LINEAR INTERPOLATION

In the guitar solo program, the get note custom block uses a type of math called *linear interpolation* to convert Distance Sensor readings into the indexes of notes to be played. Linear interpolation uses a set of known values to fill in other values that fall between them. For example, if you know that an empty bin weighs 3.5 ounces (100 grams) and the same bin full of 500 LEGO Technic black pins weighs 6.5 ounces (180 grams), you can estimate by linear interpolation that a weight of 5 ounces (140 grams) would correspond to having 250 pins in that bin. That's because 5 ounces is halfway between 3.5 and 6.5 ounces, and 250 pins is halfway between 0 pins (the empty bin) and 500 pins.

In our case, imagine a graph where the x-axis represents Distance Sensor readings and the y-axis represents note indexes. We already know that a distance of 4 centimeters yields a note index of 8—we'll call that point on the graph (x_1, y_1). We also know that a distance of 48 centimeters yields a note index of 1—we'll call that point (x_2, y_2). Linear interpolation lets us draw a line between those two points—the dashed line in Figure 8-7.

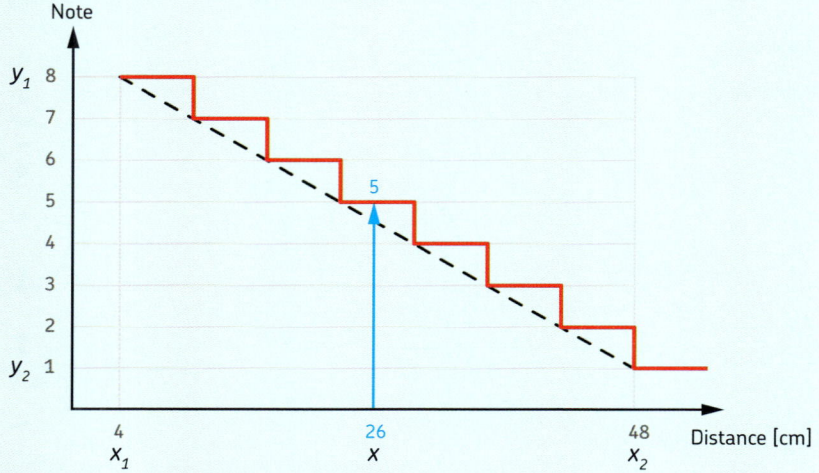

Figure 8-7: Converting distance measurements to notes by using linear interpolation

The equation of that straight line is this:

$$y = y_1 + (x - x_1) \frac{(y_2 - y_1)}{(x_2 - x_1)}$$

Plugging in our known values, we get this formula:

$$\text{note} = 8 + (\textit{distance} - 4) \frac{(1 - 8)}{(48 - 4)} = 8 - \frac{(\textit{distance} - 4)}{6.2857} = 8 - \frac{\textit{distance}}{6.28} + 0.63 \cong 8.6 - \frac{\textit{distance}}{6.3}$$

With this formula, we can take any distance measurement from 4 to 48 and convert it to a corresponding value between 1 and 8. For example, if the sensor reads 26 centimeters, we get this conversion:

$$\text{note} = 8.6 - \frac{26}{6.3} = 8.6 - 4.12 = 4.48$$

As you can see, however, there's a problem: the formula returns decimal values, but we want the note index to be a whole number. We solve this issue in the get note custom block by computing the ceiling of the note value obtained, rounding the value up to the next whole number. In this case, 4.48 becomes 5. The red, stepped line in Figure 8-7 represents taking the ceiling of the note value: the line jumps abruptly from one note value to the next.

Try calculating which notes correspond to 15 centimeters and 40 centimeters. Can you verify your answers by using the graph?

playing notes

The main stack of the program, stack G in Figure 8-6, is where we actually play notes. We begin by writing SOLO on the Hub's display to clearly indicate which program the Guitar is running. Then we set the mode variable to 1 (corresponding to the major scale) and the transpose variable to 0. To *transpose* means to shift a collection of notes up or down. By changing this value, you can play the scales shown in Figure 8-2 in other keys. For example, setting transpose to 1 would make the major scale start on the note B-flat instead of A, while setting it to –2 would make the major scale start on the note G. Later in the stack, the value of the transpose variable is added to the note number retrieved from the scale list to determine the actual note that should be played (❷ in Figure 8-6).

Next we call the set up custom block. As we've discussed, it fills the scale list according to the value of the mode variable. The rest of the blocks are inside a forever loop. At each repetition of the loop, the get note custom block calculates a note index number by reading the slider position, and the value of the note index is written on the Hub's display. Then we use an if then else block to play the note when the Color Sensor detects the pick.

Besides detecting the colors of objects placed in front of it, the Color Sensor has a *reflected light mode*, which lets it return, as a percentage, the amount of light emitted by its circular lights that is reflected back to it. (In the MINDSTORMS App, volume, like reflected light, is represented as a percentage.) The closer an object is to the Color Sensor and the lighter its color, the higher the reflected light percentage will be. We use this feature to control the note's volume. If the percentage of reflected light read by block ❶ is greater than 20, we use the reflected light reading itself to set the volume of the sound that will be played. If the percentage of reflected light is less, no sound is played. This gives players expressive control over their solos: the closer the pick is to the Color Sensor, the louder the note will be.

Finally, the moment you've been waiting for: block ❷ actually starts playing a note! We use the start playing beep block instead of the play beep for seconds block so the sound of the Guitar will be continuous. The note's pitch is determined by the number read from the scale list using the variable note as an index. As explained previously, we add the value of the transpose variable to the pitch to change the key of the scale, if desired.

If the Color Sensor's reflected light reading is less than 20, the pick is too far away, so the Guitar's sound is interrupted by the stop all sounds block. Then the change scale custom block is executed, giving the player the chance to choose a different scale with the Hub's right button. So, if you want to change the scale, you should do it when the pick isn't in front of the color sensor.

playing chords

The next program we'll create for the Guitar allows you to play chords. A *chord* is a set of multiple notes played at the same time. Since the Hub can play only one note at a time, however, we'll program the Guitar to play the chords as arpeggios. An *arpeggio* (from the Italian verb *arpeggiare*, which means "to play a harp") is a kind of "broken" chord, where the notes of the chord are played one after another. Our program plays the notes of the broken chords very quickly one after another as if you were quickly strumming the strings of a real guitar.

For those who can read music, Figure 8-8 shows the chords you can play on the Guitar. Our program will have two playing modes to choose from: one with chords built on a major scale, and the other with chords built on a minor scale. If you can't read music, you can ignore the figure for now. I'll explain how to accompany songs later.

Figure 8-8: The Guitar can play chords built on a major or minor scale, allowing you to accompany many famous songs. By default, the chords will be in the key of C, but you can play in other keys by changing the transpose variable.

creating the program

Let's make the program to let the Guitar play broken chords. Create a new program called *guitar_chords*. Then, go to the Variables palette and create a list named chords, along with the following variables:

```
chordNote
note
noteDuration
transpose
```

Now reproduce the custom blocks defined in Figure 8-9: set up chords major (stack A) and set up chords minor (stack B). These blocks build up lists with the note information for playing chords based on the major or minor scale. Each item of each list is a string including four 2-digit numbers separated by spaces, corresponding to the four notes of a chord. Be careful to reproduce the strings correctly, or the chords you play will sound wrong. Use Table 8-2 as a reference.

Table 8-2: Strings Stored in the chords List

Chords in major mode	Chords in minor mode
60 64 67 72	60 63 67 72
62 65 69 74	62 65 68 71
64 67 71 76	63 67 70 75
65 69 72 77	65 68 72 77
67 71 74 79	67 71 74 79
69 72 76 81	68 72 75 80
71 74 77 80	70 74 77 82
72 76 79 84	72 75 79 84

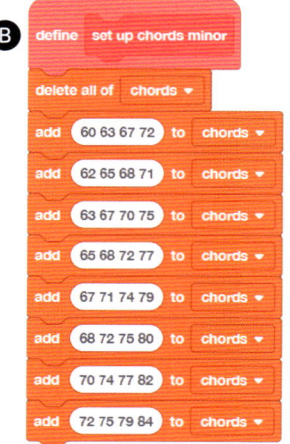

Figure 8-9: The custom blocks to set up chords in a major and a minor scale

Next, create the change mode custom block, defined by stack C in Figure 8-10. This block lets you set whether you'll play chords built on a major or minor scale by using the Hub's left and right buttons. Then, create the get note custom block in Figure 8-5. This is exactly the same block as in the *guitar_solo* program.

NOTE To copy and paste the blocks, click the stack to be copied, press CTRL-C (⌘-C on Mac) to copy it, and press CTRL-V (⌘-V on Mac) to paste it into the new program. Make sure the new program does not already contain custom blocks with the same name as the one you're pasting. Also, when copying a stack containing a call to a custom block, you must copy its definition as well, or the program won't work.

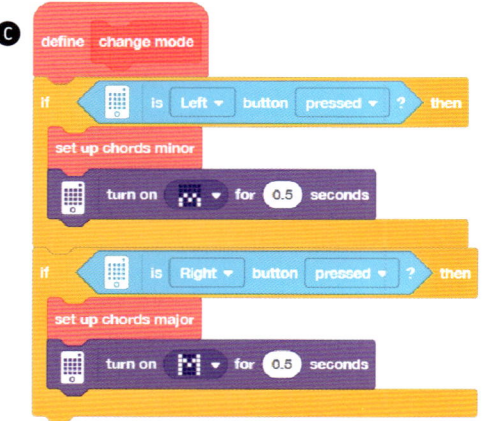

Figure 8-10: The custom block to change between major and minor mode

Finally, create the main stack, stack D in Figure 8-11. The stack is similar to the main stack of the *guitar_solo* program, but it has added blocks to pull the individual note numbers from the strings in the chords list and play the notes as arpeggios.

playing famous songs on the guitar

Most pop songs are built around a few common chord progressions. To see an example, execute the *guitar_chords* program, select the major chord mode by pressing the right Hub button, and try playing these chords in a loop: (1, 5, 6, 4). With these four chords, you'll be able to accompany a truckload of famous songs, including the chorus of "Let It Go" (from *Frozen*), "Let It Be" (Beatles), "Wherever You Will Go" (The Calling), "Can You Feel the Love Tonight?" (Elton John), "She Will Be Loved" (Maroon 5), "With or Without You" (U2), "No Woman, No Cry" (Bob Marley), "Take On Me" (A-ha), "Save Tonight" (Eagle-Eye Cherry), "Torn" (Natalie Imbruglia), and many others. For more ideas, check out *https://en.wikipedia.org/wiki/I-V-vi-IV_progression*. Additional common chord progressions you can play in major mode include (1, 5, 4, 6) and (1, 5, 4, 5).

Other songs rely on chord progressions built on the minor scale. Switch to minor chord mode by pressing the left Hub button. Now play the chord progression (8, 7, 6, 5). (The chord 8

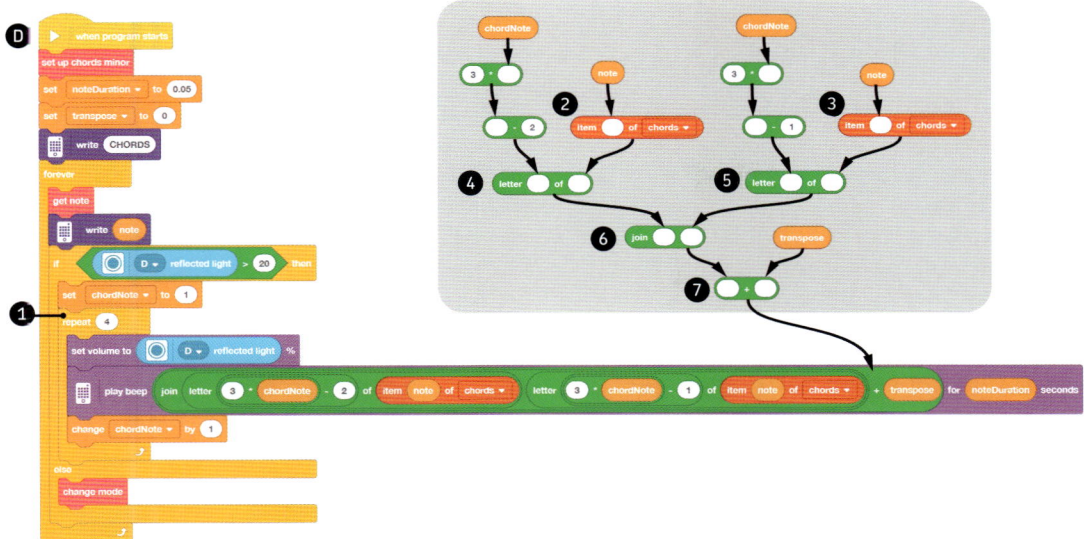

Figure 8-11: The main stack of the program makes the Guitar play broken chords.

is the same as the chord 1, but it's easier to use 8 when the next chord is 7.) With this progression, and the related (8, 7, 6, 7), you'll be able to accompany songs like "Smooth Criminal" (Michael Jackson), "I Will Survive" (Gloria Gaynor), "Moonage Daydream" (David Bowie), "Gethsemane" (A.L. Webber), and the flamenco section of "Innuendo" (Queen).

Search the web for more song ideas. You should find lots of resources showing song lyrics alongside guitar chords. Be aware that chords are often represented with Roman numerals: I for 1, IV for 4, and so on. You should also keep in mind that not all songs are played in the same key, meaning they start from different chords to accommodate different singers' vocal ranges. To change the key of your Guitar, you can adjust the value of the transpose variable, as explained in the box "Transposing to Other Keys" on page 208.

understanding the guitar chords program

Let's walk through the *guitar_chords* program to see how it works. Look back at Figures 8-9, 8-10, and 8-11 as you go.

setting up the chords

The set up chords major and set up chords minor custom blocks (stacks A and B in Figure 8-9) are similar. Both start by clearing the chords list and then add eight items to the list. Each item is a string that includes the numbers of the four notes to be played for a certain chord. For example, the string 60 64 67 72 represents the notes C, E, G, and C again (the second C is an octave higher than the first). Together, these notes make up a C major chord (see Figure 8-8).

choosing major or minor mode

Just as the change scale block in the *guitar_solo* program let you choose among three scales, the change mode custom

block (stack C in Figure 8-10) lets you choose whether to play chords based on a major scale or a minor scale. If the player presses the Hub's left button, the set up chords minor block is executed, filling the chords list with the chords built on a minor scale. A small letter m is shown on the Hub display for half a second to indicate that the minor mode has been selected. Likewise, if the player presses the Hub's right button, the set up chords major block is executed, the chords list is filled accordingly, and a big letter M is shown on the Hub display for half a second to indicate major mode.

NOTE To indicate major or minor mode, I used turn on for seconds blocks to show custom patterns instead of using write blocks, because the small m and big M that the write block displays look too similar. Plus, the turn on for seconds block lets us tell the program how long to display the letter.

playing arpeggios

The main stack of the *guitar_chords* program (stack D in Figure 8-11) handles actually making sounds. First, the set up chords minor custom block fills the chords list with the chords built on the minor scale. Then we set the noteDuration variable to 0.05. This variable controls how long (in seconds) each individual note in the arpeggio will last. Choosing 0.05 means the notes will be played one after another very fast, sounding like strummed guitar strings, but you can try other values as well. We also set transpose to 0 to keep the chords in the key of C. (See the box "Transposing to Other Keys" on page 208 for information about using this variable.) Next, the word CHORDS is written on the Hub's display to remind the player which program is being executed. Then the rest of the stack is run inside a forever loop.

TRANSPOSING TO OTHER KEYS

The chords defined in the chords lists are in the key of C, meaning they're built on the notes of a C scale, major or minor. However, not all songs are in C. To play songs in other keys, you can change the value of the transpose variable, according to Table 8-2. For example, if you want to play along with your favorite song and you find out it's in the key of A, set transpose to -3.

Table 8-3: Transpose Values for C Scale

Key	Transpose value
G	–5
G♯ / A♭	–4
A	–3
A♯ / B♭	–2
B	–1
C	**0**
C♯ / D♭	1
D	2
D♯ / E♭	3
E	4
F	5
F♯ / G♭	6
G	7

Within the loop, the get note custom block sets the note variable according to the slider position along the guitar neck, just like in our solo program (see "Choosing a Note" on page 203), and the note index is written on the Hub's display. Then, as before, we use an if then else block to watch for the guitar pick: if the Color Sensor reads a reflected light intensity greater than 20, the blocks to play arpeggios are executed, once again using the reflected light reading to set the volume. Otherwise, we call the change mode custom block, giving the player a chance to switch between major and minor mode by using the Hub's buttons.

Inside the if space, we need a way to pick out individual notes from the chord so they can be played in an arpeggio, one after the other. To do this, we use the variable chordNote in combination with a repeat loop that repeats four times (❶ in Figure 8-11). We first set chordNote to 1; then we increase its value by 1 with each repetition. In this way, chordNote functions as an index for reading the four notes of the chord individually, allowing us to play just the first note, then the second, then the third, and then the fourth.

Accessing each note requires a little more work, however, since the four note numbers for each chord are stored together in a single text string—for example, 60 64 67 72. We can access part of a string by using the letter of string operator block, but it lets us get only one character of a string at a time by specifying that character's position in the string (1 for the first character, 2 for the second character, and so on). We therefore use two letter of string blocks, one for each digit of a two-digit note number, and then use a join strings block to put the two digits back together.

To calculate the position of the tens digit of a note number within the string, we multiply chordNote by 3, then subtract 2. Similarly, we multiply chordNote by 3 and subtract 1 to calculate the position of the ones digit. This works because each note takes up three characters of the string (including the space after the two-digit number). When we need the second note of the chord, for example, chordNote will equal 2, so $(3 \times \text{chordNote}) - 2 = (3 \times 2) - 2 = 4$. As shown in Figure 8-12, the tens digit of the second note is in position 4 of the string. Likewise, $(3 \times 2) - 1 = 5$, which is the position of the ones digit of the note.

The gray box in Figure 8-11 shows how we put it all together. Using note as an index, we retrieve two copies of a string from the chords list (❷ and ❸). With chordNote, we calculate the position of the tens digit ❹ and ones digit ❺ of the desired note and extract the digits with two letter of string blocks. Then we join the digits back together ❻. The result of

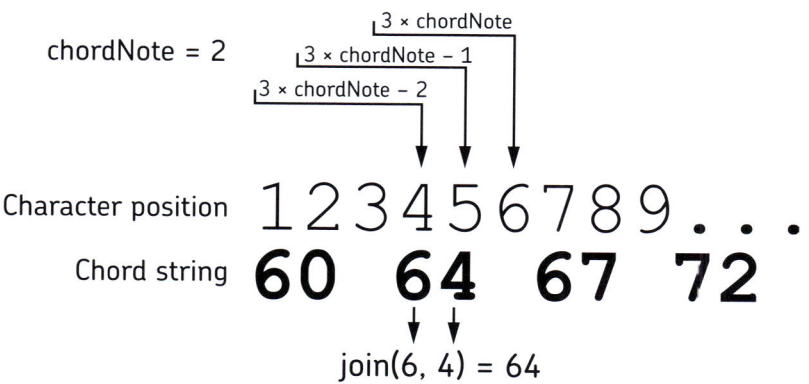

Figure 8-12: Using the chordNote variable to access individual characters of a chord string by position

the `join strings` block is actually a string, not a number, but when we plug the string into the `plus` block to add the value of the `transpose` variable ❼, the string is converted to a number.

We now have a number representing one note from the chord, which we use as the note input of the `play beep for seconds` block, with the duration of the note specified by the `noteDuration` variable. By doing this four times in the `repeat` loop, with the value of `chordNote` changing each time, we're able to play the four notes of the chord in an arpeggio, one after the other. Then the `forever loop` starts again from the beginning.

adding the control knobs

Now we'll add two motors to the Guitar that will function as control knobs. The motors' built-in rotation sensors will read the position of the knobs, allowing you change the transposition (the song key) or the speed of the arpeggios while you're playing.

40

The cable clip should point in the same direction as the marked hole in the shaft.

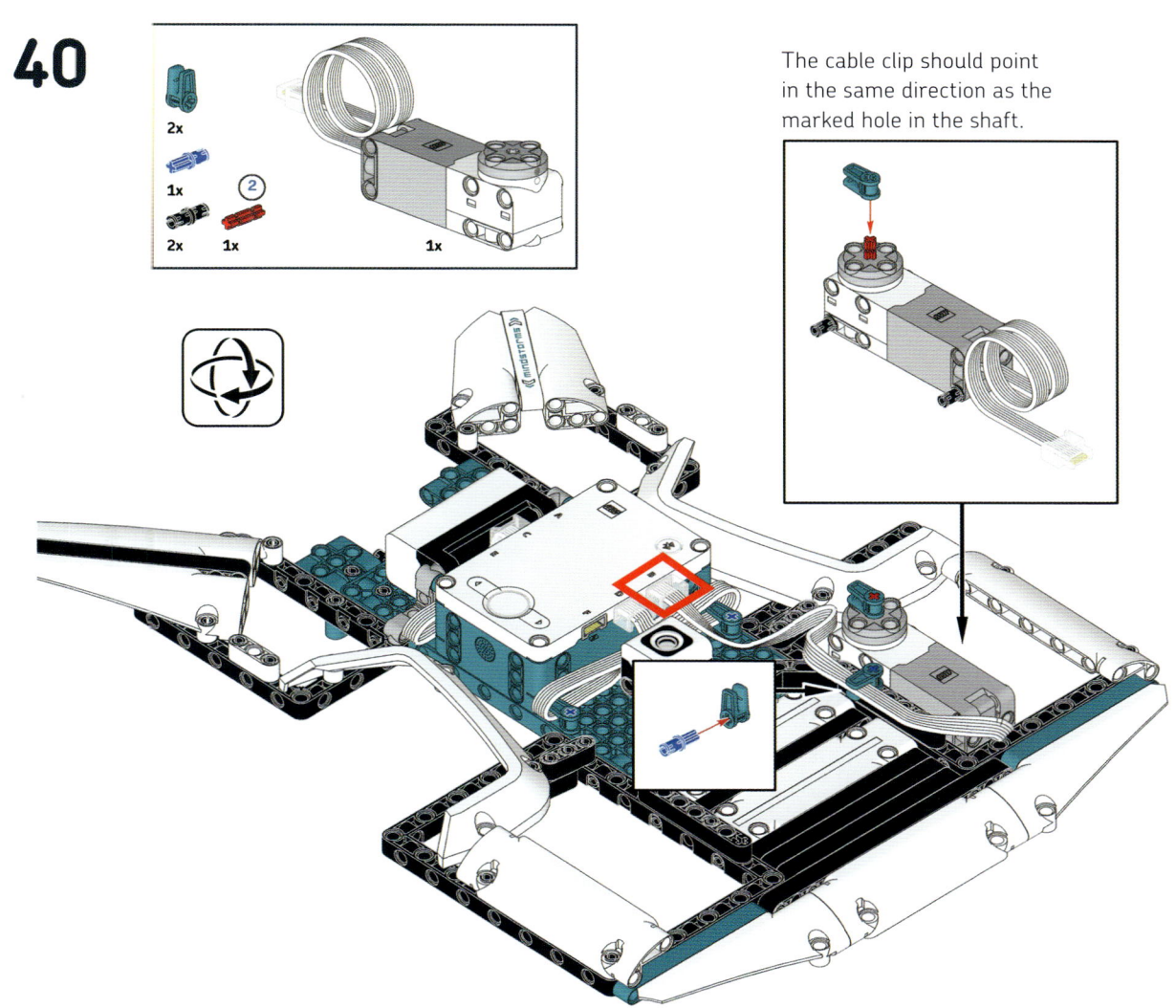

41

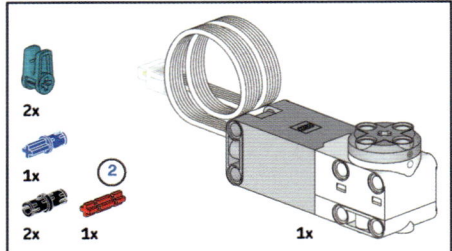

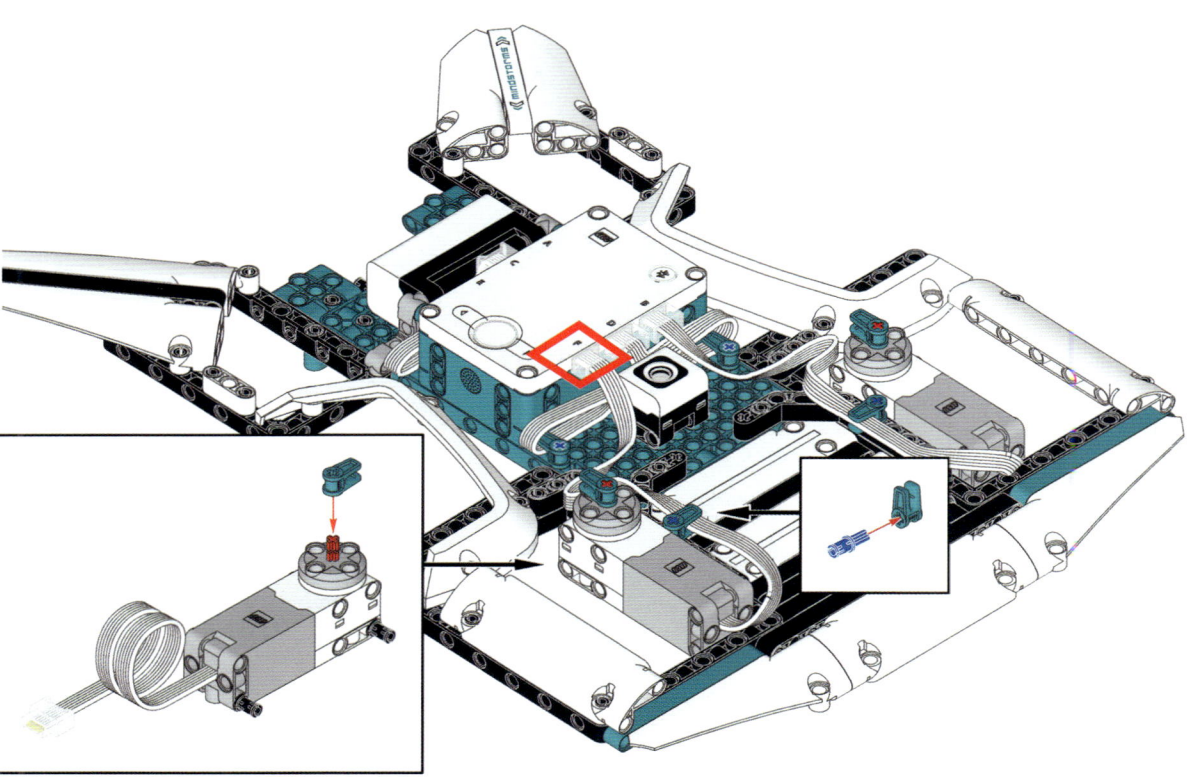

The cable clip should point
in the same direction as the
marked hole in the shaft.

42

6x **2x**

The black beams prevent the
motors from being pulled off.

43

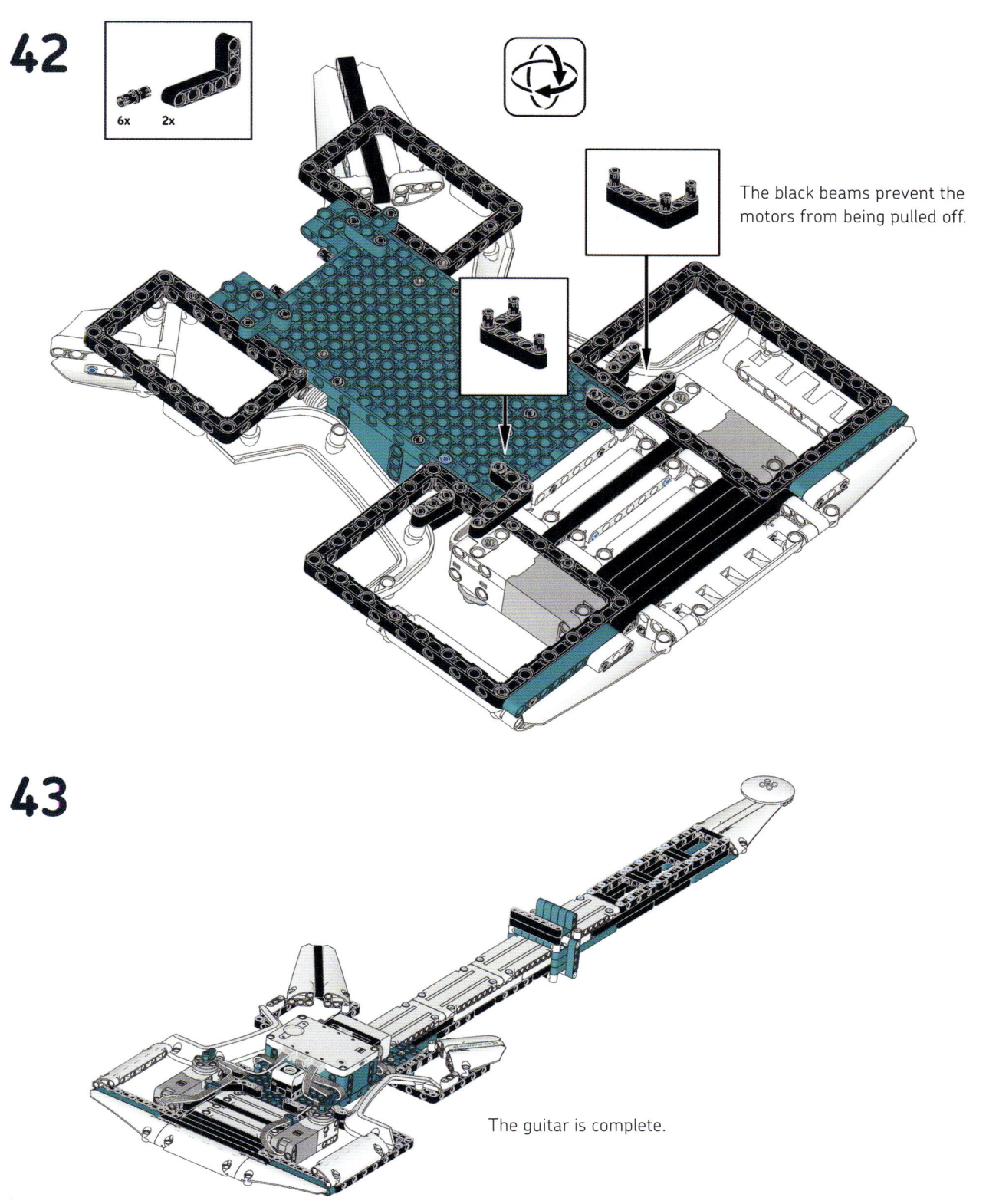

The guitar is complete.

adding transposition control to the solo program

Now that you have the control knobs, we can add stack H shown in Figure 8-13 to the *guitar_solo* program. The new stack lets you control the key of the scale you're playing. Rotating the right knob of the Guitar, you can change the starting note of the scale, with 12 steps across a whole turn of the knob.

Within the stack, we first set the motor attached to the knob to coast when stopped, making it easy to turn the shaft. Then we move the motor to position 0 and set its relative position to 0 as well. Next we enter a `forever loop` where the `transpose` variable is set to a value based on the motor shaft's relative position. We use the relative position since it can be positive or negative, allowing us to transpose up or down, whereas the motor's absolute position is constrained to the range 0 to 359. We divide the position by 30 to create 12 sections along the whole turn of the knob (360 / 30 = 12). Then the result is rounded to get a whole number, which becomes the value of the `transpose` variable.

adding transposition and tempo control to the chords program

With our two knobs, we can control both the key of the chords and the speed (or *tempo*) of the arpeggios in our *guitar_chords*

program by adding stack E, shown in Figure 8-14. Rotating the right knob of the Guitar, you can change the pitch of the chords in 12 steps. Rotating the left knob of the Guitar, you can change the tempo of the arpeggios. The stack is similar to the one in Figure 8-13, though there are some differences.

One difference is that the first few motor blocks control both motors B and F rather than just motor B. The other difference is the addition of block ❶ in the `forever loop`, which sets the value of the `noteDuration` variable according to a calculation based on motor F's position. The calculation

---

`noteDuration = 8 / (120 + (relativePosition / 4))`

---

is designed so that the duration of each note becomes shorter and shorter as the tempo is increased. We start with a base tempo of 120 beats per minute, to which we add the value of the motor shaft's relative position, divided by 4 to make the change more gradual. We divide the result into 8 to get the duration of each note. By default, if you leave the knob untouched, each note will have a duration of 8 / 120, which is about 66 milliseconds.

playing further

Want to get more out of your LEGO MINDSTORMS Guitar? On this book's companion website *https://www.nostarch.com/lego-mindstorms-robot-inventor-activity-book/*, you'll find additional programs. For example, one lets you play a solo melody on top of an arpeggio that the Guitar automatically plays in a loop. Since the Hub can play only one note at a time, the extra program implements the *tremolo* guitar technique, in which you rapidly alternate between multiple notes. When executed fast enough, this technique creates the illusion of a sustained melodic line with a bass accompaniment.

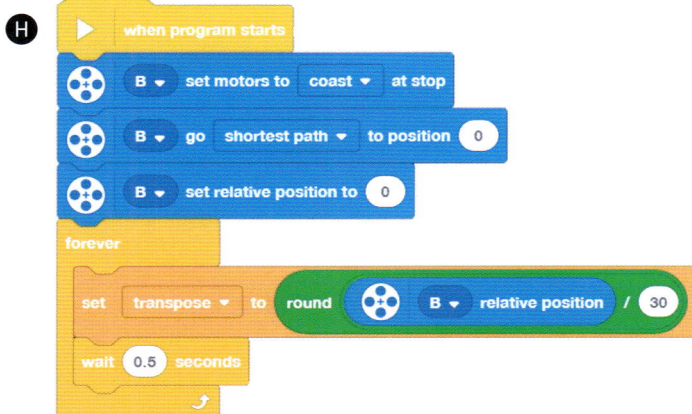

Figure 8-13: Add this stack to the guitar solo program so you can transpose the scale on the fly by rotating the right knob on the Guitar.

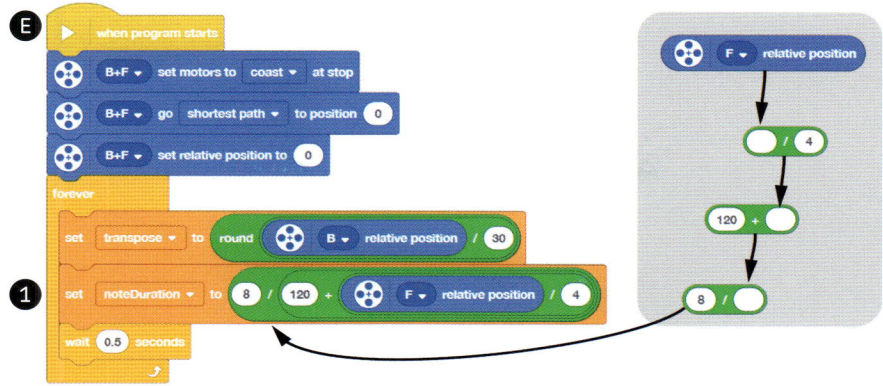

Figure 8-14: Add this stack to the guitar_chords program to be able to control the transposition and speed of the arpeggios with the two knobs.

what you've learned

While building this almost full-size LEGO Guitar, you've learned some techniques to make large and sturdy structures using the frames and baseplate included in the Robot Inventor set. While programming the Guitar, you learned how to use the Color Sensor's reflected light mode and how to play music with the Hub. You also saw how to store groups of numbers as strings in a list and how to read those numbers back out from the string, one digit at a time. I also disclosed the secret of pop songs—that they're built mostly from just a few common sequences of chords—so now you can even compose your own hit! A hint: besides Kraftwerk's song "Die Roboter," there aren't many songs about robots . . .

I hope this project has inspired you to create more musical projects with your LEGO MINDSTORMS Robot Inventor set!

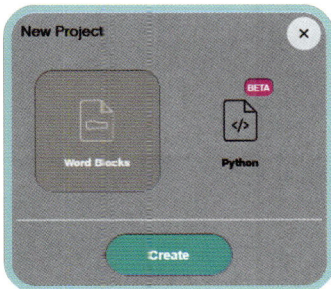

9

where to go from here

You can't teach people everything they need to know.
The best you can do is position them where they can find
what they need to know when they need to know it.
—Seymour Papert (education pioneer and author of the book *Mindstorms*)

I hope you've enjoyed reading this book and that you've learned something useful. But while you've reached the end of the book, your journey into robotics has just begun. There are many ways you can expand and build upon your LEGO MINDSTORMS experience.

expanding the software

Every time you start a new project in the LEGO MINDSTORMS App, the box shown in Figure 9-1 appears. Throughout the book, we've chosen to write our programs using Scratch 3.0 and its word blocks, but you can also choose to write programs with Python, a popular text-based programming language. This feature was in beta at the time of writing this book, meaning it was still being developed, but that doesn't mean you can't try out Python programming for yourself.

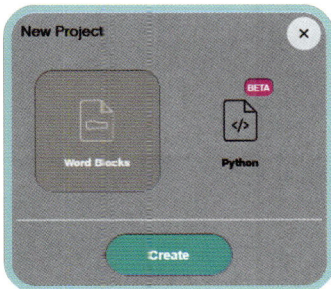

Figure 9-1: When you create a new program, this box lets you select which programming language to use: Scratch 3 word blocks or Python.

To give you a taste of what MINDSTORMS programming with Python is like, here's how our Baseball Batter program (see Chapter 2) might look in the Python language:

```python
from mindstorms import Motor, ColorSensor
import hub # for low-level API

# Initialize the motor.
motor = Motor('F')
motor.set_default_speed(100)
motor.run_to_position(0, direction='shortest path')

# Initialize the Color Sensor.
color = ColorSensor('D')

while True:
    if color.get_color() == 'red':
        motor.run_to_position(40)
        hub.sound.play('extra_files/Hit') # low-level API
        hub.display.show(hub.Image.CHESSBOARD)
        motor.run_to_position(0)
        hub.display.clear() # low-level API
```

Looking at this code, you might notice many similarities to our original Scratch 3.0 program that used word blocks. For example, the command motor.run_to_position works just like the motor go to position block. We use the parentheses after the command to tell the motor where to go, much as we can adjust the inputs on the word block.

For a simple program like the Baseball Batter, it may seem silly to use Python rather than word blocks. As your robots and programs get more complex, however, you might find that you can do more—more easily and with much less visual clutter—with Python.

NOTE Some functionality of the word blocks environment, such as hat blocks that react to events, remote control, and the Sound and Animation Editors, is missing in the Python environment. However, expert users can learn more about the Python Application Programming Interface (API) at *https://lego.github.io/MINDSTORMS-Robot-Inventor-hub-API/*.

A group of experts (including David Lechner and Laurens Valk) created the Pybricks project, a different version of Python than that supported by the MINDSTORMS App. Pybricks provides another way to write Python code for your LEGO MINDSTORMS robots. You can learn more about the Pybricks project at *https://pybricks.com/*.

expanding the hardware

The six ports of the Hub aren't just for plugging in the motors and sensors in the Robot Inventor set. They're compatible with all LEGO Power Functions 2.0 (LPF2) cable connectors. That means you can connect all kinds of other devices to the Hub, including LEGO Powered Up motors and lights, LEGO Technic CONTROL+ motors, BOOST sensors and motors, and the WeDo 2.0 motor and sensors. This greatly increases the flexibility of what you can build.

The Hub's Bluetooth chip also supports Bluetooth Low Energy (BLE), allowing the MINDSTORMS Hub to be connected to other MINDSTORMS Hubs and to other BLE-enabled LEGO hubs, including the WeDo 2.0 Hub, the LEGO BOOST Move Hub, the LEGO Technic CONTROL+ Hub, the LEGO City Hub, the LEGO DUPLO Train Hub, and even the LEGO Super Mario figure. By connecting multiple hubs with BLE, you can control more than six motors or read more than six sensors from the same program, or perhaps make a distributed robotic system with units that can communicate with each other wirelessly. The sky's the limit!

NOTE Connecting to other hubs and using their motors and sensors isn't possible using the Scratch 3.0 or Python environment within the LEGO MINDSTORMS App. You'll have to use a Python library called hub2hub developed by Nard Strijbosch. You can find out more about it at *https://github.com/NStrijbosch/hub2hub/* and read the documentation at *https://hubmodule.readthedocs.io/en/latest/hub2hub/*.

be social

To get support, find inspiration, and meet other people with a passion like yours, you can join the LEGO MINDSTORMS Robot Inventor Facebook community (which I co-founded), as well as the LEGO SPIKE Prime Facebook community. The LEGO SPIKE Prime set is meant for teachers, educators, and students, but it includes the same Hub, motors, and sensors (plus a Force Sensor).

> *https://www.facebook.com/groups/ mindstormsrobotinventor/*
>
> *https://www.facebook.com/groups/SPIKEcommunity/*

I'd love to see how you're enjoying the projects in this book, so please, whenever you post about the book on a social network, tag your content with #LMRIActivityBook.

what's next?

Just because you've tried all the projects in this book doesn't mean you should put it back on the shelf. Instead, you can keep using it as a reference when you need to review how a certain programming block works, how to apply a particular building technique to your creations, or just to check which parts are included in the LEGO MINDSTORMS set (see the inner cover). You can also consult the book's appendix for a handy block reference.

While making the projects included in this book, you learned about the tools the LEGO MINDSTORMS Robot Inventor set has to offer to a robot inventor like you. Now it's your turn: grab your set, mix its rich assortment of parts with other LEGO Technic parts that you have in your collection, and make something! Build something big, something useful, something weird, something fun! Create something that's never been seen before.

And as you work, don't be afraid to make mistakes. Tinker a lot, then tinker some more. Tinkering is the key to inventing new things! Never stop playing! Ciao!

word blocks quick reference guide

In this appendix, you'll find a summary of all the Scratch word blocks used in this book.

The blocks are listed in the same order as they appear in the LEGO MINDSTORMS App, grouped in sections corresponding to each palette. For each block, you'll see a picture and a brief description. For more information, refer to the index at the end of the book, looking for the block name under the *blocks* entry. There you'll find the page where each block is used so you can review a meaningful example of how or when to use it.

motors

Use these blocks to control one or more motors. Choose which motor to control by selecting the port it's connected to. You can also select multiple ports.

Table A-1: Motors Blocks

name	picture	description
motor go to position	A ▾ go shortest path ▾ to position 0 A ▾ go clockwise ▾ to position 0 A ▾ go counterclockwise ▾ to position 0	Turns a motor to a specified angle from 0 to 359 degrees. The motor can turn clockwise or counterclockwise, or it can take the shortest path.
start motor	A ▾ start motor ↻ A ▾ start motor ↺	Starts a motor running clockwise or counterclockwise.
stop motor	A ▾ stop motor	Turns off a motor.
set motor speed	A ▾ set speed to 75 %	Sets the speed of a motor in a range from –100% to 100%. Negative speeds run the motor in reverse. Without this block, the default motor speed is 75%.
motor position	A ▾ position	Reports a motor's current position in degrees from 0 to 359.
motor speed	A ▾ speed	Reports a motor's actual speed, from –100 to 100.

movement

Use these blocks to control two motors in sync to make wheeled robots like Charlie, Tricky, or Blast drive around and turn. For blocks with a steering input, positive values make the robot steer to the right, negative values to the left, and 0 in a straight line. The higher the steering value, the sharper the curve (100 makes the robot spin in place). You should always place the set movement motors block at the beginning of your program if you intend to use Movement blocks.

Table A-2: Movement Blocks

name	picture	description
move for duration		Makes a robot move forward or backward, or spin left or right, for a set number of centimeters, inches, seconds, degrees, or motor rotations.
move with steering for duration		Moves a robot for a set number of centimeters, inches, seconds, degrees, or motor rotations while steering in a given direction.
start moving with steering		Starts moving a robot while steering in a given direction.
stop moving		Turns off the motors.
set movement speed		Sets the speed of a robot's drive motors in a range from −100% to 100%. Negative values run the motors in reverse. Without this block, the default motor speed is 75%.
set movement motors		Defines which two motors will be synchronized and controlled by other Movement blocks. Always place this block at the beginning of your program if you intend to use Movement blocks.

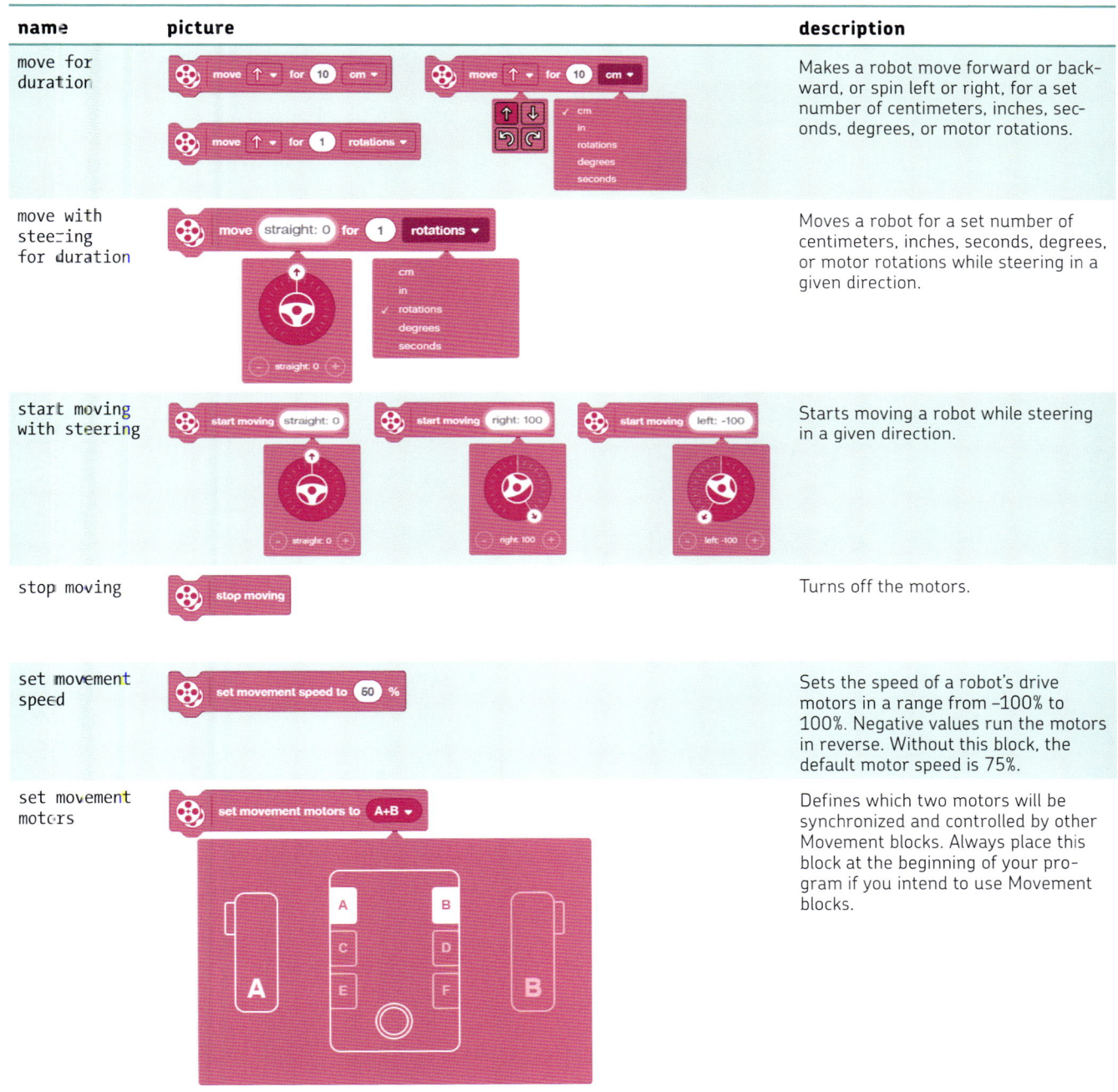

light

Use these blocks to control the Hub's 5x5 LED matrix display, the Hub's center button light, and the lights around the "eyes" of the Distance Sensor.

Table A-3: Light Blocks

name	picture	description
start animation	start animation Celebrate ▾	Plays the specified animation on the Hub display. The program continues executing the blocks after this one while the animation is executed.
play animation until done	play animation Celebrate ▾ until done	Plays the specified animation on the Hub display, pausing the program until the animation is over.
turn on light matrix for seconds	turn on ▦ ▾ for 2 seconds	Lights up a custom pattern on the Hub display for the specified number of seconds. Then the display turns off, and the program continues.
turn on light matrix	turn on ▦ ▾	Lights up a custom pattern on the Hub display. Pattern remains lit until the program ends or another Light block changes the pattern.
write on 5x5 matrix	write Hello	Writes the specified text on the Hub display, scrolling from right to left. If you specify a single letter, it will display without scrolling.
turn off pixels	turn off pixels	Turns off all the lights on the Hub display.
set orientation to (upright)	set orientation to upright ▾	Sets the orientation of the Hub so patterns will be displayed the right way up. Choose upright, left, right, or upside down. The default orientation is upright.
set center button light	set Center Button light to ●	Sets the Hub's center button light color.
light up distance sensor	A ▾ light up ○ ○ ▾	Turns the lights around the Distance Sensor's "eyes" on and off. You can control each of the four lights independently.

sound

Use these blocks to play beeps and sounds from the Hub or from the device you're programming your robots with.

Table A-4: Sound Blocks

name	picture	description
play sound until done	play sound Hit ▾ until done	Plays the selected sound, pausing the program until the sound is over.
start sound	start sound Hit ▾	Plays the selected sound. The program stack continues while the sound plays.
play beep for seconds	play beep 60 for 0.2 seconds	Plays a note from the Hub's speaker for a set amount of time. Select the pitch of the note with the piano keyboard that pops up when you click on the block's first input.
start playing beep	start playing beep 60	Plays a note from the Hub's speaker until some other block in the program stops it.
stop all sounds	stop all sounds	Stops any sounds that are playing.
set volume to (100%)	set volume to 100 %	Sets the volume of the Hub's speaker. The default volume is 100%.

events

Use these blocks to start a stack of blocks when a certain event occurs.

Table A-5: Events Blocks

name	picture	description
when program starts	when program starts	Runs the attached blocks when the program starts.
when color is	A ▾ when color is	Runs the attached blocks when the Color Sensor detects a certain color. Can choose (0) black, (1) magenta, (3) blue, (4) teal, (5) green, (7) yellow, (9) red, (10) white, or (−1) no color.
when distance is	A ▾ when closer than ▾ 8 cm ▾ A ▾ when closer than ▾ 8 cm ▾ ✓ closer than / farther than / exactly at % / ✓ cm / in	Runs the attached blocks when the Distance Sensor detects that an object is closer than, farther than, or equal to a certain distance. Specify the distance in centimeters, in inches, or as a percent (100% is 2 meters, or about 78 inches).
when	when	Runs the attached blocks when the specified condition is true.
broadcast message	broadcast message1 ▾	Broadcasts a message within the program, causing any stacks attached to `when I receive message` blocks set to the same message to run. The program continues after the message has been sent.
broadcast message and wait	broadcast message1 ▾ and wait	Broadcasts a message within the program, causing any stacks attached to `when I receive message` blocks set to the same message to run. Waits until all `when I receive message` stacks have finished before continuing to the next block in the stack.
when I receive message	when I receive message1 ▾	Triggers the attached blocks when the specified message is received from a `broadcast message` or `broadcast message and wait` block.

control

Use these blocks to change the way the program flows. For example, you may want the program to wait, to take different actions depending on conditions, or to loop.

Table A-6: Control Blocks

name	picture	description
wait for seconds	wait 1 seconds	Pauses the program for a set number of seconds.
wait until	wait until	Pauses the program until a certain logic condition becomes true.
repeat loop	repeat 10	Repeats the blocks inside this block a set number of times.
forever loop	forever	Repeats the blocks inside this block forever. Stop the loop with the **stop all** block or by shutting off the program.
repeat until	repeat until	Repeats the blocks inside this block until the specified logic condition becomes true.
if then else	if then else	Executes the blocks inside the first space if the given logic condition is true. Otherwise, executes the blocks in the second space.
if then	if then	Executes the blocks inside this block if the given logic condition is true. Otherwise, the blocks are skipped.
stop other stacks	stop other stacks	Stops all stacks except the one this block belongs to.
stop	stop all · stop this stack · stop and exit program ·	Stops all stacks that are currently running, or just the stack this block belongs to, or exits the program.

sensors

Use these blocks to read the Color Sensor, the Distance Sensor, and the Hub's accelerometer and gyroscope.

Table A-7: Sensors Blocks

name	picture	description
is color?	A ▾ is color ⬤ ?	Returns **true** when the Color Sensor detects a set color. Can choose (0) black, (1) violet, (3) blue, (4) teal, (5) green, (7) yellow, (9) red, (10) white, or (–1) no color.
reflected light	D ▾ reflected light	Reports the amount of light reflected back to the Color Sensor. Values range from 0 (indicating a dark object or nothing) to 100 (when a white object is close to the sensor).
is distance?	A ▾ is closer than ▾ 10 cm ▾ ? A ▾ is closer than ▾ 10 cm ▾ ? ✓ closer than / farther than / exactly at — % / ✓ cm / in	Returns **true** when the Distance Sensor detects an object closer than, farther than, or exactly at a distance. Specify the distance in centimeters, in inches, or as a percentage of the maximum range (2 meters or 78 inches).
is hub (shaken)?	is shaken ▾ ? is tapped ▾ ? is falling ▾ ?	Returns **true** when the Hub is shaken, is tapped, or falls.
is hub button?	is Left ▾ button pressed ▾ ?	Returns **true** if the Hub's left or right button is either pressed or released.
hub angle	pitch ▾ angle	Reports the Hub's pitch, roll, or yaw angle.
timer	timer	Reports the time since the program started in seconds, down to the millisecond.
reset timer	reset timer	Restarts the timer at 0 seconds.

operators

Use these blocks to perform mathematical operations on numbers, compare numbers, combine logic values, or work with text strings.

Table A-8: Operators Blocks

name	picture	description
pick random number	`pick random 1 to 10`	Returns a random number from the given range, including the lower and upper limits.
plus	`( ) + ( )`	Adds two numbers and returns the result.
minus	`( ) - ( )`	Subtracts one number from another and returns the result.
multiply	`( ) * ( )`	Multiplies two numbers and returns the result.
divide	`( ) / ( )`	Divides one number by another and returns the result. Dividing a number by 0 returns the special value **Infinity**.
less than	`( ) < ( )`	Returns **true** if the first value is less than the second value; otherwise, **false**.
equal	`( ) = ( )`	Returns **true** if the first value equals the second value; otherwise, **false**.
greater than	`( ) > ( )`	Returns **true** if the first value is greater than the second value; otherwise, **false**.
and	`( ) and ( )`	Combines two logic conditions, returning **true** only when both inputs are true.
or	`( ) or ( )`	Combines two logic conditions, returning **true** when at least one input is true.
not	`not ( )`	Inverts the input condition, returning **true** if the input is false, and **false** if the input is true.
is between	`is 0 in between -10 and 10 ?`	Returns **true** if the first number is between the other two numbers, including the lower and upper limits.
join strings	`join pine apple`	Combines two text values and returns the result. For example, given the inputs **pine** and **apple**, it returns **pineapple**.
letter of string	`letter 1 of apple`	Returns the character at a particular position of the given text string. For example, **letter 1 of apple** returns **a**.
mod	`( ) mod ( )`	Returns the remainder of the division of the first value by the second value. For example, **10 mod 3 = 1**, since 1 is the remainder of 10 divided by 3.
round	`round ( )`	Rounds the specified number to the nearest whole number. Decimals below 0.5 are rounded down, and decimals of 0.5 or above are rounded up.

(continued)

Table A-8: Operators Blocks *(continued)*

name	picture	description
math functions	abs ▾ of ( )	Applies one of several math functions to a number and returns the result. The available functions are these: • abs (absolute value) • floor (round down) • ceiling (round up) • sqrt (square root) • sin (sine) • cos (cosine) • tan (tangent) • asin (arcsin) • acos (arccos) • atan (arctangent) • ln (natural logarithm) • log (base 10 logarithm) • e^ (power of Euler's number) • 10^ (power of 10)

remote control

These blocks appear only if you add widgets to the virtual remote-control interface in the App. Use these blocks to react to input from the remote-control widgets. The hat blocks won't retrigger if the widget remains in the state that triggered the event.

Table A-9: Remote-Control Blocks

name	picture	description
when joystick is	when joystick J1 ▼ is up ▼	Triggers the attached blocks when the specified joystick widget is in the given state (up, down, left, right, moved, released).
joystick axis	joystick J1 ▼ x-axis ▼	Reports a joystick widget's horizontal or vertical position. The x-axis ranges from –100 for the leftmost position to 100 for the rightmost position. The y-axis ranges from –100 for the bottom position to 100 for the top position. The position is 0 when the joystick is not in use.
when button is	when button B1 ▼ is pressed ▼	Triggers the attached blocks when the specified button is pressed or released.
is button?	is button B1 ▼ pressed ▼ ?	Returns **true** if the specified button widget is currently pressed or released.

variables

Use these blocks to write and read variables and to manage data in lists.

Table A-10: Variables Blocks

name	picture	description
variable	myVar	Reports the value of a variable. Every variable you create gets one of these blocks with its name on it.
set variable	set myVar to 0	Sets the value of a given variable to any number or text string.
change variable	change myVar by 1	Changes the value of a given variable by the specified amount.
list	myList	Reports the items contained in a list as a text string, with the values separated by blank spaces. For example, for the list [1, 2, 3], this block returns the string "1 2 3".
add item to list	add thing to myList	Adds an item to the end of a list.
delete all items in list	delete all of myList	Deletes all of a list's items.
item at index in list	item 1 of myList	Returns the value stored at a particular position in a list.

more movement (extension)

These blocks extend the Movement blocks palette, giving you more options for controlling two synchronized motors. For blocks with a steering input, positive values make the robot steer to the right, negative values to the left, and 0 in a straight line. The higher the steering value, the sharper the curve (100 makes the robot spin in place). As with the Movement blocks, you should always place the `set movement motors` block at the beginning of your program if you intend to use the blocks in this extended palette.

Table A-11: More Movement Blocks

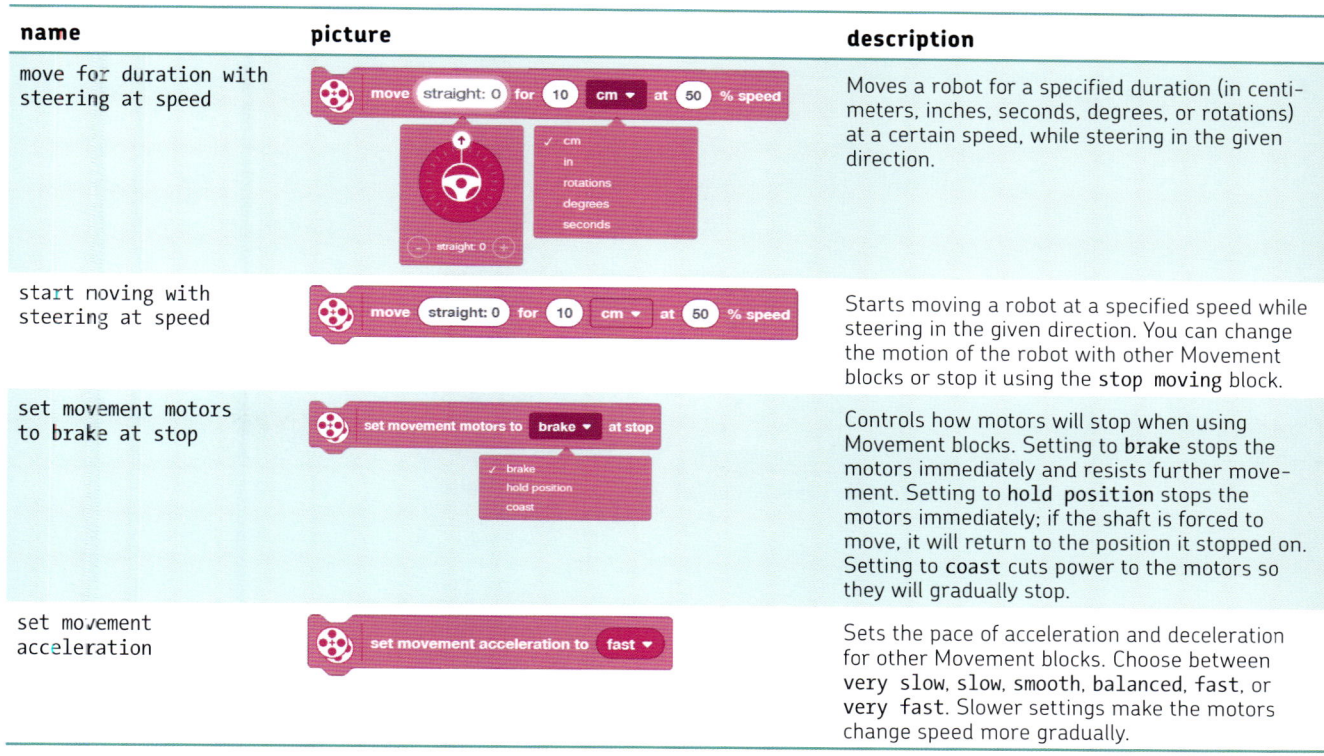

name	picture	description
`move for duration with steering at speed`	move straight: 0 for 10 cm at 50 % speed / straight: 0 / cm, in, rotations, degrees, seconds	Moves a robot for a specified duration (in centimeters, inches, seconds, degrees, or rotations) at a certain speed, while steering in the given direction.
`start moving with steering at speed`	move straight: 0 for 10 cm at 50 % speed	Starts moving a robot at a specified speed while steering in the given direction. You can change the motion of the robot with other Movement blocks or stop it using the `stop moving` block.
`set movement motors to brake at stop`	set movement motors to brake at stop / brake, hold position, coast	Controls how motors will stop when using Movement blocks. Setting to `brake` stops the motors immediately and resists further movement. Setting to `hold position` stops the motors immediately; if the shaft is forced to move, it will return to the position it stopped on. Setting to `coast` cuts power to the motors so they will gradually stop.
`set movement acceleration`	set movement acceleration to fast	Sets the pace of acceleration and deceleration for other Movement blocks. Choose between `very slow`, `slow`, `smooth`, `balanced`, `fast`, or `very fast`. Slower settings make the motors change speed more gradually.

more motors (extension)

These blocks extend the Motors blocks palette, giving you more options for controlling one or more motors.

Table A-12: More Motors Blocks

name	picture	description
run for duration with steering at speed	A ▾ run for 1 rotations ▾ at 75 % speed	Runs a motor for a set number of rotations, seconds, or degrees at a given speed.
start motor at power	A ▾ start motor at 75 % speed	Starts running a motor at a specified percentage of power. Unlike the start motor block, this block doesn't change the power applied to the motor in order to maintain a constant speed.
go to relative position at speed	A ▾ go to relative position 0 at 100 % speed	Moves a motor to a relative position at a certain speed. Use with the set relative motor position block. The relative position has no range limit, so values of 360 or higher will make the motor turn more than one full rotation.
set relative motor position	A ▾ set relative position to 0	Sets the position of a motor relative to a given value. The motor does not move.
relative motor position	A ▾ relative position	Returns how many degrees a motor has turned since the start of the program or in relation to the position set by the set relative motor position block.
set motors to brake at stop	A ▾ set motors to brake ▾ at stop	Controls how a motor will stop when using other Motor blocks. Setting to brake stops the motor immediately and resists further movement. Setting to hold position stops the motor immediately; if the shaft is forced to move, it will return to the position it stopped on. Setting to coast cuts power to the motor so it stops over time.

index